AF725693

Ethics of Life:
Contemporary Iberian Debates

HISPANIC ISSUES • VOLUME 42

Ethics of Life: Contemporary Iberian Debates

Katarzyna Beilin and William Viestenz

EDITORS

Vanderbilt University Press
NASHVILLE, TENNESSEE
2016

© 2016 Vanderbilt University Press
All rights reserved
First Edition 2016

This book is printed on acid-free paper.

The editors gratefully acknowledge assistance from the College of Liberal Arts and the Department of Spanish and Portuguese Studies at the University of Minnesota; and the Department of Spanish and Portuguese at the University of Iowa

The complete list of volumes in the Hispanic Issues series begins on page 341.

Library of Congress Cataloging-in-Publication Data on file

LC control number 2015042855
LC classification number GE42 .E8445 2015
Dewey class number 179/.10946—dc23

Full record available at *lccn.loc.gov/2015042855*

ISBN 978-0-8265-2091-3 (hardcover)
ISBN 978-0-8265-2092-0 (paperback)
ISBN 978-0-8265-2093-7 (ebook)

HISPANIC ISSUES

Nicholas Spadaccini, *Editor-in-Chief*

Luis Martín-Estudillo, *Managing Editor*

Ana Forcinito, *Associate Managing Editor*

Nelsy Echávez-Solano, Megan Corbin,
and William Viestenz, *Associate Editors*

Cortney Benjamin, Scott Ehrenburg,
Heather Mawhiney, and Pablo Rodríguez Balbontín
Assistant Editors

**Advisory Board/Editorial Board*
Rolena Adorno (Yale University)
Román de la Campa (Unversity of Pennsylvania)
David Castillo (University at Buffalo)
Jaime Concha (University of California, San Diego)
Tom Conley (Harvard University)
William Egginton (Johns Hopkins University)
Brad Epps (University of Cambridge)
David W. Foster (Arizona State University)
Edward Friedman (Vanderbilt University)
Wlad Godzich (University of California, Santa Cruz)
Antonio Gómez L-Quiñones (Dartmouth College)
Hans Ulrich Gumbrecht (Stanford University)
*Carol A. Klee (University of Minnesota)
Germán Labrador Méndez (Princeton University)
Eukene Lacarra Lanz (Universidad del País Vasco)
Tom Lewis (University of Iowa)
Jorge Lozano (Universidad Complutense de Madrid)
Raúl Marrero-Fente (University of Minnesota)
Kelly McDonough (University of Texas at Austin)
Walter D. Mignolo (Duke University)
*Louise Mirrer (The New-York Historical Society)
Mabel Moraña (Washington University in St. Louis)
Alberto Moreiras (Texas A & M University)
Bradley Nelson (Concordia University, Montreal)
Michael Nerlich (Université Blaise Pascal)
*Francisco Ocampo (University of Minnesota)
Antonio Ramos-Gascón (University of Minnesota)
Jenaro Talens (Universitat de València)
Miguel Tamen (Universidade de Lisboa)
Teresa Vilarós (Texas A & M University)
Iris M. Zavala (Universitat Pompeu Fabra, Barcelona)
Santos Zunzunegui (Universidad del País Vasco)

Contents

Introduction
Ethics of Life: Contemporary Iberian Debates
Katarzyna Beilin and William Viestenz ix

PART I
Genealogies of Ecological and Animal Rights Movements in Modern and Contemporary Iberia

1 The Environment in Literature and the Arts in Spain
Carmen Flys-Junquera and Tonia Raquejo 3

2 *Nunca Máis*:
Ecological Collectivism and the *Prestige* Disaster
John H. Trevathan 35

PART II
Ecological Crisis and the Neoliberal Appropriation of Public Space

3 Tourism and "Quality of Life" at the End of Franco's Dictatorship
Eugenia Afinoguénova 59

4 Die and Laugh in the Anthropocene: Disquieting Realism and Dark Humor in *Biutiful* and *Nocilla experience*
Katarzyna Olga Beilin 89

5 Cultivating the Square: Trash, Recycling, and the Cultural Ecology of Post-Crisis Madrid
Matthew Feinberg and Susan Larson 113

6 Degrowth and Ecological Economics in Twenty-First-Century Spain: Toward a Posthumanist Economy
Luis I. Prádanos 143

PART III
Iberian Bio-Power: Life as a Political Matter

7 Reproductive Rights in Spain:
From "Abortion Tourism" to "Reproductive Destination"
Pablo de Lora 163

8 *Mar adentro* and the Question of Freedom
Paul Begin 183

9 Still Different?
Biotechnology, Politics, and Culture in Spain
Sainath Suryanarayanan and Katarzyna Olga Beilin 203

PART IV
Reassembling the Archive through the Concept of Life

10 Iberian Cultural Studies beyond the Human: Exploring the Life History of Marcos Rodríguez Pantoja in Spanish Anthropology and Popular Film
Daniel Ares López 231

11 The Bull Also Rises: The Political Redemption of the Beast in *La pell de brau* by Salvador Espriu
William Viestenz 257

12 Animals in Contemporary Spanish Newspapers
John Beusterien 279

13 Accounting for Violence, Counting the Dead: The Civil War and Spain's Political Present
Sebastiaan Faber 295

Afterword
Spain: Taking the Alternative?
Martín López-Vega and Luis Martín-Estudillo 319

Contributors 331

Index 337

Introduction

Ethics of Life: Contemporary Iberian Debates

Katarzyna Olga Beilin and William Viestenz

Among all the factors which have emerged with increasing visibility in the twenty-first century that contribute to changes in the definition of "life" and open new questions about its ethical treatment, the following are the most important for this volume: the raising of collective consciousness regarding ecological crises and especially climate change; the emerging biotechnologies of plant, animal, and human life enhancement developed in part as an answer to this crisis; and the consideration of nonhuman species as deserving of rights, which is possibly a resistance to the growing industrialization of agricultural practices. As these are debated in the context of present and future policy development, the historical archive is also revised as bioeconomy reconsiders the worth and the purpose of life. In the neoliberal economic framework, life becomes not only a means but also a material of production and is integrated into the market exchange processes driven by new biotechnologies. Today's multifaceted crisis, happening simultaneously in ecological, political, and economic contexts, intensifies the alliance between science and economy, which claim to be able to maintain and enhance present "quality of life" through "innovations" focused on integrating all life, including intellectual life, into the market exchange. The need for financial profit motivates and manipulates the mission of research at the university and the notion of common good. In this framework, the humanities are expected to form public opinion in favor of scientific innovations of life. In the best tradition of cultural studies, however, this volume proposes to question rather than praise the relation between life and these new hegemonic discourses.

This volume focuses on the transformations in the understanding of the ethics of life resulting from the debates in the contexts of ecological crisis, biotechnological innovations and animal rights movement activism on the Iberian Peninsula. Methodologically, the articles in the volume will present interdisciplinary perspectives from the humanities, law, social sciences, and sciences in relation to existing debates and research, as well as the widely understood world of art: film, novels, and poetry. We propose to think about these changes in a way that connects them, as forms of resistance, to dominating discourses of economy and nation. We believe that changes in culture occur through a buildup of connections between diverse frames of new social movements, whether they are resisting or attempting to transform the status quo. But, we also want to honestly present disagreements and contradictions between philosophers, movements, and perspectives in order to provide food for thought about these so-timely matters to our prospective readers. We also hope that this volume may open new perspectives for Iberian cultural studies, which are immersed in a ferment of debates just as the essays of this volume are being conceived and written.

This Volume and the Field

Federico García Lorca (1932) famously argued that the Castillian and Andalusian "duende" that feeds off of Spain's intimate relationship with death and is best represented in bullfighting, is superior to the "Galician angel" and "Catalan muse" while Spanish art as such is superior to the traditions of Germany, Italy, and others. José Bergamín, a friend of Lorca's and one of the greatest propagators of bullfighting, in his exile in Mexico founded an editorial house, Séneca, and a journal, *España peregrina,* becoming an intermediary between the world of Spanish literature and American Hispanism. Another key figure providing foundations for the discourses of Hispanism was Américo Castro. In his influential lecture at Princeton University in 1940, Castro did not mention bullfighting, but similarly to Lorca and Bergamín, suggested that Spanish culture, due to its focus on questions of life and death, can offer a solution to the crisis of Western modernity excessively dominated by material questions. This idea of a violence-ridden spirituality has been repeated by various Hispanists during the twentieth century and still can be heard today. In the first chapter of *El espejo enterrado* (1997) (Buried Mirror), entitled "La virgen y el toro" (The Virgin and the Bull), Carlos Fuentes claims that it is in the bullfighting ring where Spaniards find their cultural self and he opposes Spanish attitudes towards death to the hypocritical concealment of death in other Western cultures (an idea which appears also in Hemingway's *Death in the Afternoon,* as well as in writings by Bergamín and many others). In an article published in *Lenguaje y textos* as recently as in 2002 in Granada, Spain, a U.S.-trained young Hispanist, María Hernández, compares Rafael Alberti's and Pedro Almodóvar's *Matadors,* claiming that their portrayal of a dissolution in death that liquidates all differences can be thought of as a solution to "una moral en crisis" (a moral in crisis). This recycling of the discourse of Spanish superiority that solves crises through irrational violence like that in bullfighting seems somewhat uncritical and its argument strikes us as incomplete, unilateral, unconscious of its naïve anthropocentrism. Given the centrality of an animal for this construction of Spanish superiority, it is surprising that Hispanism has not incorporated critical approaches from Animal Studies as they have become increasingly popular among scholars of English, Italian, and other literary and cultural studies. Possibly, a defensive impulse not to question the anthropocentric vision of the field due to its alliance with bullfighting culture has made Animal Studies approaches unwelcome among Hispanists although some of the greatest Spanish intellectuals and writers have been sensitive to the question of human/animal relations.

Anti-bullfighting writers, such as Jovellanos, Larra, Noel, Martín-Santos, and Goytisolo note that a bullfighting worldview is discursively connected to wars, social hierarchies, political torture, machismo, and generalized indifference toward suffering. Those writers refuse to justify bullfighting violence based on the species status of its object and connectedly also turn against other forms of state violence. Analyzing their works, numerous scholars in our field, such as, for example, Michael Ugarte, Jo Labanyi, and various others, have criticized discourses which contribute to the state's symbolic power and have implicitly criticized and transformed Hispanism's paradigm. However, it has been only recently that Joan Ramon Resina (1996) has openly critiqued Hispanism's complicity with the centralist state and insufficient attention to the diversity of cultural paradigms on the Iberian Peninsula. As a result, a rich debate, often framed as a new paradigm referred to as "Iberian Studies," has emerged regarding how to account for the diversity in academic research, and how to talk about cultures without being subservient to nationalisms.[1] Some critics (Resina, Santana, Epps) stress the importance of "the local" as a point of departure in knowledge production, where both the locality of the studied object and of the studying subject should be accounted for. Others would like to liberate culture from its ethnic connections and its geographical boundaries (Moreiras, Gabilondo, Mignolo). This second group of critics intends to elaborate a discourse that could not be co-opted by any state or other ethnic structure of power, but rather provides points of reference other than the nation, and privileges hybridity, mobility, and transformation simultaneously in various cultural contexts.

The discourses found in this volume privilege hybridity and transformation because they focus on human and nonhuman life conceived as interconnected and interdependent, transcending borders, linguistic limits, and otherwise delimited spaces, but in some cases the local specificity acquires a great significance. As is evident in John Trevathan's and William Viestenz's essays in this volume, a focus on the entanglements between all forms of life and the material world in fact solicits the creation of new spatial collectivities, composed of a plethora of actants, whose borders are misaligned with the official demarcations of the nation state. Thus the resulting ethics calls for respect for all forms of life rather than fetishizing death, for adequate representation of live organisms' realities and needs and for the search for new political solutions, which would assure their peaceful coexistence, minimizing all forms of violence and destruction. The focus on life leads us also toward a trans- or postdisciplinary paradigm because it connects life as the humanities understand it to life as researched by biological sciences and regulated by law, which is reflected by the diverse fields of our contributors. The critics whose essays are featured in our volume are focused on the material realities and complexities of life in its organic, rather than symbolic, context. These essays propose alternative ways of constructing discourses that also intend to present resistance to the hegemonic rhetoric of a neoliberal economy of life.

Life

Life has been thought of in terms of physical processes, researched by biology, and also in terms of the so-called "meaning of life," which has been mostly a human domain analyzed by religion and the humanities. It is hard to understand how the processes and their meanings could be considered separately, how they could be the subject matter of different disciplines. In this way, the human being has also been split in half: into an animal body, tested in labs, but with a superior soul, analyzed by soul gurus. This division persisted in our field even after the religious worldview was largely displaced by a cultural turn. The cultural, until recently, has been invariably understood as purely human, consisting of built environments and social relations and opposed to the natural.

The "non-human turn" that, as a number of conference sights announce, has taken place in the twenty-first century is based on an emergence of numerous theories that reconceptualize the culture/nature divide and pretend to elaborate non-anthropocentric frameworks. Among them are Bruno Latour's actor-network theory (ANT), animal studies, environmental studies, critical science studies, new materialisms, vitalisms, object-oriented philosophy, posthumanism, as well as new media theories. Bruno Latour, Donna Haraway, Timothy Morton, Cary Wolfe, and various other contemporary thinkers reformulate the relationship between humans, animals, and other forms of life, and Jane Bennett questions even the habitual distinction between live and nonlive matter. Latour's actor-network theory (ANT) describes reality as series of networks of relations between humans, animals, plants, and objects that are all viewed as possessing agency (although not intentionality) and where processes occur as a result of the accumulation of interactions. In his widely quoted *We Have Never Been Modern* (1993), Latour criticizes modern social sciences for building false divides between nature and society, object and subject, as well as for various other artificial conceptual divides built across disciplines. He postulates rethinking disciplinary discourses and building new concepts to avoid those misleading separations. "Natureculture" is one of those concepts emerging from Latour's work that allows for seeing humans and nonhumans linked together functionally and materially as they have always been in life. "Natureculture" acquires new meanings and political applications in Donna Haraway's work.

In the feminist framework of Haraway, the concept of "naturecultures" challenges human treatment of animals in scientific labs, serves to criticize science's attitude towards life in general, and to develop a political vision. In Haraway's politics a "situated knowledge" (*Simians 183–202*), conscious of its limitations and communicable through alliances, is preferable to an objective point of view from nowhere, which is responsible for hierarchies and the subjugation of life. For Haraway, the vulnerability of the knowing subject is a criterion of the validity of her knowledge. She proposes a transformative criticism that would refuse to disregard suffering, both human and animal, and that in political terms would amount to substituting the capacity to control with the capacity to produce change, to nurture and empower others. For her, the production of innovating knowledge can be compared to "A Game of Cat's Cradle" (1994), where discourses are restructured by teams of thinkers, taking better and better shapes like the threads tangled around the fingers of a group of girls who pass the game to each other in a courtyard. Haraway's belief that "knowledge is better from below" (*Simians* 190) openly connects to the intellectual agenda of the movement against cruelty for animals which aims at the deconstruction of discourses justifying harm and destruction of life from above.

Like Latour and Haraway, Morton criticizes the concept of "Nature" as distinct from culture and suggests substituting the vision of separate domains with the concept of "a mesh," which expresses the idea of ecological interdependence. The interdependence begins on the conceptual level, where anything that exists acquires its identity as different from something else, as in Derrida's *Of Grammatology* (1967), and also etymologically, because everything derives from other entities preceding it. Interdependence is also a result of the fact that all that exists consists of the same physical particles on some basic level. Similar to Latour, Morton compares the structure of life to that of language as imagined by Derrida, and he shows that an infinite and illogical multiplication of differences characterizes both language and the system of life forms. Because they contain otherness, Morton calls all life forms in the mesh "strange strangers," a name which is ethically and existentially consequential. "Strange strangers" cause curiosity, bring respect, and demand hospitality. In spite of its strangeness, however, every nonhuman life form is perceptibly familiar to humans because we have descended from it. This mixture of familiarity and difference that all life presents us creates, in Morton's view, the disquieting sensation of "unheimlich."

Iberian Debates and the Crisis

As a result of the progressive destruction of the environment, apocalyptic scenarios announcing the end of our civilization appear in film and fiction with growing frequency. It is an object of current debates whether the ecological and economical crises that we are facing are just another opportunity for humans to overcome their difficulties, or rather an announcement of the end of human progress and a limit to human freedom. Furthermore, if the crisis is considered as an opportunity for change, there is no agreement on what kinds of transformations need to occur. According to ecologists, if the detrimental change of climactic conditions on Earth is the result of human production and consumption patterns, these patterns need to be transformed to ensure that our planet remains habitable. According to allied science and economy, only the most advanced biotechnologies can save life.

While ecologists argue that the limited character of resources on Earth entails the need to decrease consumption and revise the economic model based on infinite growth, new technologies of life attempt to overcome the limits of nature, extracting a surplus of energy out of living organisms through their genetic modification. In the neoliberal framework, where the environmental problem is often reduced to the insufficiency of resources, biotechnology and the new discoveries related to DNA are thought to be able to infinitely stretch the capacity of the biosphere to feed a growing humanity. According to Sheehan and Tegart, this means a new stage of capitalism where what is exploited is not human labor, but rather the generative and regenerative capacity of live organisms. Genetically modified organisms (GMOs) are integrated into the cycle of production and commercialization of the market, revamping it through various crafty strategies. For example, genetically modified salmon grow larger and can multiply all year round, thus remedying the progressive destruction of fisheries. Genetically modified corn or rice, to the contrary, becomes infertile after one cycle of growth, which forces a close relationship between agriculture and chemical corporations, blurring the distinction between agricultural and industrial production. Because GM seeds require a number of chemical enhancements, farmers become steady customers of the biotech industry as they purchase grain and other chemical substances annually from the patent holding concerns. As today's capitalism is focused on bio-value (Pavone, Waldby, Novas, Cooper), it progressively privatizes resources, including public spaces, which, until recently, were considered common goods. Public parks and plazas are commercialized as part of a process of "revitalization" of urban spaces so that the city may become more profitable. At the same time, the huge billboards that take over the privatized space present passersby with objects of desire that only the economy of consumption can help obtain.

Jorge Riechmann, poet, philosopher, and a professor at the Universidad Autónoma de Madrid, is one of the most persistent critics of the style of life promoted by the neoliberal economy as well as of the uncritical enthusiasm for future betterment offered by biotechnology. He describes the fantasies of genetic engineering and other forms of enhancement of life on Earth as a sort of nostalgia to escape the human condition. He would prefer to invest all efforts and resources in sustaining life in its present form, cautiously limiting the adventurousness of scientific experiments by arguing that both humans and computer programs make unavoidable mistakes which can bring terrible yet unforeseen consequences. He believes that we should similarly abstain from cosmic conquests and concentrate rather on achieving uniform conditions of life among all peoples of the globe. As the title of his recent book *Socialismo sólo puede llegar en bicicleta* (2012) announces, we need deep political transformations and self-restraint. That means that questions of justice are inherently connected to environmental issues.

While 90 percent of European agricultural GMOs are produced in Spain, which, unlike other members of the EU, let Monsanto GM corn in during the late 1990s, there is a generalized distrust towards GM food in Spain. A debate by leading Spanish newspapers informed society about the controversies surrounding GMOs although some pro-GMO intellectuals and experts energetically intervened. In July 2006, *La Vanguardia* published an article entitled "Los transgénicos saltan del plato" (GMOs out of the plate), which informed readers that after all transgenic products began to be marked as such, according to EU norms, it turned out that the consumers rejected them. As a result, more than four hundred food producers and distributors, such as El Corte Inglés, Panrico, Bimbo, Nestlé España, and others, publicly announced that they would renounce using GMOs. As Riechmann, Pavone, Garrido, Amorín, López Arnal, and others argue, the GMOs are not only a health hazard, but they also constitute a political problem. Biotechnology, which genetically modifies nonhuman life, serves the interests of the multinationals, which results in insufficient precaution, leads to privatization of life through patents and turns living organisms into resources feeding the global market.

In Spain, as in the rest of Europe, medical biotechnology has more social approval than agricultural genetic modification. It may be due to the fact that, beginning with the Transition, in more Iberian cultural domains transcendence has been placed in this life rather than beyond. Thus in Almodóvar's *All About my Mother* (1999) and *The Flower of My Secret* (1995), organ donation turns into a metaphor for the ethical treatment of others, for thinking of society as a family, and for a general concern for the future. Spain has the greatest number of voluntary organ donations in the world, but on a different note, it has also witnessed an intense debate on euthanasia (analyzed in Paul Begin's essay) reflected by Alejandro Amenábar's *Mar adentro* (2004), and another long-lasting debate regarding abortion (see Pablo de Lora's article), prompting legal challenges at the Constitutional Court, social division and political controversy with each legislative step taken toward its liberalization in 1985 and 2010, and restriction in 2013. In spite of the obvious common denominator, euthanasia, abortion, organ transplants, and life enhancement through genetic engineering need to be discussed separately as each brings its own set of challenges and dilemmas whose meaning has recently changed in crisis.

The crisis transforms the meanings of life. It is when the discursive manipulation of life intensifies, taking to extreme its habitual strategies. As Afinoguénova shows in her article, while the tourist industry in the sixties and seventies took over the Mediterranean coast, the destruction of ecosystems was justified as an improvement of "quality of life" for Spaniards in a manner very similar to the way that today's bioeconomy justifies genetic modifications. The "pact of silence" during the Transition meant to protect life in the future. The reluctance to open common graves and the invisibility of slaughterhouses are motivated by similar emotions and needs: the maintenance of the satisfying lifestyle. The strategy of hiding harmful truth for the sake of comfort and financial gain tends to backfire, however. It is more than ever during crises when gains and comforts are lost, that the truth emerges and resistance is formed, calling for a change. The consciousness of the coincidence of economic, political and environmental crisis around 2008 prompted emergence of new movements and visions. Resistance has called for a defense of life's undisturbed existence (as in deep ecology), for preservation of ecosystems, environmental justice, or, in some cases, for the rights of nonhuman life. These issues are analyzed in the essays of this volume: the struggle for the common use of the public spaces in cities as in Feinberg and Larson's essay; the animal rights movement's objection to killing animals and the intellectual revision of animality as in the essays by Ares, Beusterien, and Viestenz; resistance in times of environmental degradation as in the essays by Flys-Junquera and Raquejo Grado, Trevathan, and Beilin; and the degrowth movement discussed by Prádanos, as well as critical science and technology studies in De Lora's and Suryanarayanan's works. The social need for the reconstruction of history from the perspective of life, as in Faber's essay, may be read as a resistance against the neoliberal transformation of life into a source of energy.

Human-animal relations were viewed as a model of injustice by various famous Spanish writers. Federico García Lorca was not only a poet of bullfighting, but, surprisingly, he was also among the first who condemned the modern city economy based on massive slaughter of animals. An apocalyptic vision of a polis whose financial success is nourished by the blood of millions of animals and the poor that moves the grids of the machines appears in *Poeta en Nueva York,* written between 1929 and 1930 and published for the first time in 1940, after the poet's death. While for the Wall Street offices, numbers refer to economic growth, for the poet it is the meaning of a massacre. The screams of killed animals pierce the heavens open, but they are not heard in the modern city, which keeps all its killing hidden and its victims silenced in slaughterhouses whose walls are not transparent, but soundproof. A Lorca-like sensitivity has been a mark of today's critique of animal consumption. Paul McCartney famously stated that "if slaughterhouses had glass walls, everybody would be vegetarian," and afterward PETA and also various Iberian organizations against cruelty toward animals, such as Ecuanimal, Igualdad Animal, AnimaNaturalis, and others, placed on the Internet videos representing the horrific treatment of animals in farmhouses and of their slaughter, which had a great impact on many of us.

While the debates on bullfighting and, to lesser extent, on limited human rights for great apes have taken front pages of daily newspapers and have been featured in radio and television debates, there has not been public debate on slaughterhouses. Activists are aware that, even though vegetarianism has slowly become popular, meat-eating is culturally so important in Spain that the chances of success in questioning animal slaughter are slim. But this issue has always been present implicitly while bulls and apes were discussed. During the debates in the Catalonian Parliament, Salvador Boix, legally representing a bullfighter, José Tomás, mentioned the possibility that a bullfighting ban could lead to further demands made by the animal rights movement: "¿Cerrarán luego las granjas de cerdos o pollos o prohibirán también la caza?" ("El debate," 2010) (Will they next close the chicken and pig farm factories and forbid hunting?). Most animal rights activists indeed consider the ban on bullfighting as a first step on the way to questioning the practices of factory farming and slaughterhouses. For strategic reasons, however, this could not be stated during the debates on bullfighting. Bullfighting fans argued that slaughterhouses should be questioned before bullfighting in order to block all possible progress on the matter. This, nonetheless, has had a double edge. On one hand, as intended, in comparison with the horrors of the newborn chickens whose beaks are being cut off, the fight of the bulls seems beautiful. On the other hand, however, this argument has turned attention slowly to the hidden evils of today's meat industry.

In Spain, the mistreatment of animals has been institutionally connected to national identity through the National Feast of bullfighting. Because of this, the animal rights movement there catalyzed changes going much further into the grain of culture than in other parts of the world. It synergized with debates on national identity, on the value of traditions, on the relations between Spain and the EU (Casal, De Lora), Catalonia and Castile (Tosko, Viestenz) and even on the War on Terror (Beilin). Between 1980 and 2000, according to Gallup polls, the number of bullfighting fans decreased from 50 percent to only 10 percent (20 percent of those interviewed declared themselves as "somewhat interested") and the polls from 2006 brought very similar results. About one third of small town festivities that featured games involving cruelty to animals became transformed to avoid inflicting pain on living creatures. This makes it evident that not only were important symbolic elements of national identity affected; lived cultures have also changed. Thus Spain presents a very particular set of cultural and political conditions which provided for exceptionally spectacular transformations. Animal rights movements, opposing the mistreatment of animals, challenged the construction of masculinity, which was defined by violence, and humanity, which was opposed to nature. At the same time those who aimed at the abolition of Franco's inheritance for political reasons supported the anti-bullfighting movement, thereby strengthening it. Thus among cultural changes occurring after Franco's death, those regulating human/animal and human/nonhuman relations can prove to be more meaningful than acknowledged by existing scholarship. It is not without significance that the first Spanish NGO founded after the death of Franco was ADDA (Asociación en Defensa de los Derechos de los Animales).

Opponents of the Catalan ban on bullfighting have classified the measure as an anti-Castilian manifestation, empty of real ethical concern because while bullfighting was prohibited, Catalan small town festivities where bulls are tortured—*toros embolados* (in Catalan "bou embolat") and *toros ensogados* ("bou ensogat")—were not banned (Dopico Black, Tosko, Viestenz). These are festivities involving harassment of bulls, popular mainly in small towns of the delta of the Ebro River. Although an analysis of the debates in the Catalan Parliament shows that the ban of bullfighting in Catalonia was not just an anti-Spanish demonstration but rather a reform motivated by a growing sensibility to mistreatment of animals in this part of the Iberian Peninsula, it was clearly due to the electoral concerns of some of the members of the Parliament that "bou embolat" and "bou ensogat" were not banned. The ban on bullfighting could only be given a favorable vote with the support of CiU (Convergència i Unió), a party deriving most of its votes precisely in those small towns in the Ebro delta where "bou embolat" and "bou ensogat" are most popular. This party's members were willing to vote in favor of banning bullfighting only if their festivities were "saved" and without their votes the ban would not have passed. Mosterín called it "vergüenza catalana" (104) (Catalan embarrassment) and most of the other Catalonian anti-bullfighting activists similarly deplored the way in which politics took over ethics even in this ethically inspired law initiative. (However, Viestenz argues that, because the *correbous* are typical of the Ebro delta, and not all of Catalonia, it is an overstep to dub the practice a national pastime, both in the Spanish and Catalan sense of the nation, meaning that Mosterín's broad attribution of "verguenza catalana" is something of a hyperbole). ADDA, devoted to the protection of animals from cruelty, also presented an allegation against the law that maintains the legality of, "bou ensogat" and "bou embolt."

ADDA's journal issue devoted to the celebration of the ban on bullfighting was rather sad, as most articles deplored the limited scope of the bullfighting prohibition. However, these festivities have since in some cases been celebrated in a symbolic way--explosives are placed on a metal frame carried by a man instead of the horns of a bull--or alternatively, they are organized in semi-clandestine conditions, due to the fear of animal rights activists' intervention.

In the Spanish media, bullfighting has been defended as an integral and historically rooted part of Spanish culture and as a form of art (Almudena Grandes, Juan Manuel De Prada, Fernando Savater, Javier Marías). Grandes's argument, for example, was that only selected people can understand bullfighting because it is a miracle. De Prada claims that only Catholics have access to the mystery of the bullfighting celebration because this religion gives them the sensibility to move naturally between life and death. For this reason, De Prada believes that attacks against bullfighting are in fact against the Catholic religion. Savater (2011) argues that animals should not be considered as subjects of ethics because they are devoid of empathy and therefore moral critiques of bullfighting are baseless.

On the other side, Jorge Riechmann, Jesús Mosterín, Paula Casal, Pablo de Lora, Rosa Montero, Pilar Rahola, Rosa Regàs, Elvira Lindo, Manuel Vicent, Oscar Horta, Juan José Millás, Antonio Muñoz Molina, and politicians such as Cristina Narbona aligned against bullfighting. In their respective books and articles they argued that since the times of Ferdinand VII *tauromaquia* has been a ritual of authoritarianism, symbolically connected to the structures of political tyranny, violent masculinity, and an ancient way of understanding the human relation with nature, where the former's survival can only be guaranteed by the destruction of the latter. The symbolic meaning of this traditional ritual is in their view incompatible with that of a democratic transformation that focuses on the elaboration of new models of masculinity, of equitable participation and more sustainable attitudes towards nature. But the understanding of democracy varies.

In an article published in *El País,* Javier Marías compared the debated ban on bullfighting to the prohibition of smoking in public places or the prohibition of gay bars, and he condemned such laws as limiting liberties and being anti-democratic. Similarly, Savater claimed that the banning zeal of Catalonians brings to mind the Inquisition and Franco's censorship on forms of public life, and in *Tauroética* he compared it to a ban on abortion. Both of these intellectuals denounced attempts to regulate citizens' behavior through governmental prohibitions. Ovejero et al. engage with the argument that the state should not intervene in the way of life of its citizens and that, in other words, it should be forbidden to forbid. He reminds readers that every civil law is full of restrictions on individual freedom, which are imposed for the sake of protecting the freedom of others or for the sake of another, greater good. If the life and suffering of an animal is not considered an ethically relevant good; if, in other words, an animal's life does not matter, or matters less than human freedom to poke them, indeed such a prohibition does not make sense. Ovejero et al. argue that in the case of granting ethical importance to animals' lives, it is necessary to protect them by law, limiting human freedom to inflict pain.

Wolff and Savater attempt to show that cruel entertainments, such as bullfighting or hunting, are essential for the maintenance of life enclaves that are not yet taken over for agricultural or industrial purposes. The argument that bullfighting brings revenue was brought to the debates in the Catalonian Parliament by Salvador Boix, the legal representative of the bullfighter José Tomás, who testified that when Tomás appears in Barcelona, everyone earns more: taxi drivers, hotels, restaurants. These arguments suggest that the life of animal species and of their natural environments are only possible as long as they are profitable. It follows that nonhuman life has the right to exist as long as it is transformable into food, entertainment, or other forms of material resources. The visions of these two philosophers and of Boix are obviously reflecting on the dynamic of life under a neoliberal economy. It is surprising, however, that the philosophers find this state of things ethically satisfying.[2] Although the main argument of the anti-bullfighting movement is that ethics are more important than financial gain, it is also true that bullfighting has been heavily subsidized by the state and possibly without these subsidies it would not be profitable. The money used to subsidize bullfighting could be used to maintain ecological reserves in which bulls and other animals could live and die freely. According to Mosterín, the bullfighting industry receives 600 million Euros in subsidies from Public Administration.

The ecological movement in Spain emerges as in other European countries in the 1960s as a reaction to the negative ecological impact of the "dessarrollismo" (developmentalism) of late Francoism. It was a popular, loosely organized social movement connected to the anti-Franco opposition of the political left, pacifism, feminism, and local nationalisms (Ramos Gorostiza). According to Joaquín Fernández, literally hundreds of environmental organizations were founded during the Spanish Transition years. They were mostly focused on local issues, protesting against environmental destruction by dams and developing anti-nuclear ecological activism. Most of those organizations, however, were short-lived and did not manage to exercise an important political impact. This provided the impression that in terms of environmentalism Spain lagged behind other European countries, immersed in the enjoyment of its (post-Francoist) freedom awakening. The environmentalist enthusiasm of those early years of the democracy was important, though, as it prepared the stage for the second wave of Spanish ecological activism that mobilized against nuclear construction in the late 1980s and 1990s and managed to halt a part of the nuclear development plan. One of the greatest wounds and largest environmental disasters that increased ecological awareness in Spain was the spill of 20 million U.S. gallons of oil from the *Prestige* tanker just off the coast of Galicia in 2002, which is commented on at length in Trevathan's essay in this volume. It caused "la marea negra" (a black tide) as the whole sea was covered with a thick layer of black oil, killing hundreds of thousands of fish, birds, and other creatures. The polluted coast and destruction of life produced a sense of loss, mourning, and the consciousness of a threat; it was compared in *Nocilla experience* to the end of the world. The disaster caused despair, but it also mobilized a great number of volunteers who came to clean up the oil from all over the Iberian Peninsula and abroad. Juan López de Uralde called them "la marea blanca" (a white tide) since most volunteers wore white clothes, sharply contrasting with the black oil. Activists gathered around the platform "Nunca Máis," (Never Again), which brought about a number of protests attended by thousands of people, demanded that Galicia be recognized as a catastrophe zone, and argued for the creation of a disaster prevention system to stop further catastrophes from occurring. But it also mobilized an awareness that black tides will keep occurring as long as energy production depends on oil and thus provided an incentive to search for alternative sources of energy.

Under José Luis Rodríguez Zapatero's government, Spain made notable progress in producing renewable energies and significantly reducing its use of gas and carbon, thus decreasing carbon dioxide emissions. Spain ranks third in the world in terms of covering its needs with wind power (after the U.S. and Germany), and was fifth in 2010 in terms of the coverage of its needs with solar energy. As a result of the economic crisis and subsequent change of government in 2011, however, the subsidies for solar panel production were removed, retroactively leading to great losses as well as many lawsuits in which the legality of such changes in rules was questioned. Several hundred photovoltaic plant operators faced bankruptcy. The crisis became an excuse to intensify the economic measures and processes responsible for the collapse of the international economy, the destruction of the environment and climate change. Riechmann pessimistically reflects that it is not possible to stop the destruction of Earth given existing structures of government and property. While the alarming discourses of environmentalists have not brought about sufficient change, sociologists have already identified the phenomenon of "ecofatigue" (José Antonio Corraliza), which refers to the weariness many feel with all these announcements of future catastrophes. The economic crisis, however, may have proven to be an opportunity for change as it prompted the emergence of a number of alternative movements that slowly renewed political consciousness. On May 15, 2011, a protest movement, known as the "Indignados" (Outraged), occupied the central Plaza del Sol of Madrid. People erected rain shelters and stayed debating day and night for several weeks in the public space. Many of them remain active in regularly meeting assemblies even today.

The Indignados' manifestos ask for development of green energies, for closing of nuclear plants, for sustainable means of transportation, and for the fostering of community life and public spaces. In various publications coming from the movement, the general failure of the system has been diagnosed. The May 15 Manifesto of Indignados reads as follows:

> El obsoleto y antinatural modelo económico vigente bloquea la maquinaria social en una espiral que se consume a sí misma, enriqueciendo a unos pocos y sumiendo en la pobreza y la escasez al resto. Hasta el colapso. La voluntad y el fin del sistema es la acumulación del dinero, primándola por encima de la eficacia y del bienestar de la sociedad. Despilfarrando recursos, destruyendo el planeta, generando el desempleo y consumidores infelices. (10)

> (The current economic model, obsolete and unnatural, blocks the social mechanism in a spiral that consumes itself. While few get rich, the rest sink into poverty. It will collapse. The will and the end of this system is an accumulation of money, which is more important than the efficacy and well-being of society. It wastes resources, destroys the planet, causes unemployment and unhappy consumers.)

The Indignados have been one of the first major social movements on the Iberian Peninsula displaying an awareness that the worsening of life experienced by most is not only due to the neoliberal economy's failure to distribute wealth in a fair way, but also to its destruction of the environment, especially in the poorer communities, both nationally and internationally, where it degrades living conditions of the lower classes. There is a widely spread consciousness that injustice is responsible not only for the social crisis, but also for the environmental one, and that consequently justice becomes redefined in the environmental context.

Spain has been home to "degrowth" economic theories (commented on in Prádanos's essay) that argue for the need to stop the infinite economic growth postulated by neoliberal capitalism. In March 2010, Barcelona hosted an International Degrowth Conference with five hundred participants from all over the world. The basic assumption of the movement is that infinite growth is impossible due to the limited amount of resources on Earth. One of the founders of ecological economy, Joan Martínez Alíer, argues that Western world societies have a false image of the economy as based on GDPs which does not take into consideration the damages caused to the environment, including human health, by economic growth (externalities). According to Martínez Alíer, the real economy is that which deals with energy sources and the environment. The Indignados have been inspired by these debates on alternative economy, which are also entering the academic curriculum in Spanish universities.[3] A slow transformation of the worldview necessary for transformation of the world is underway.

Contributions

Our volume would like to contribute not only to attempts to transform our field, but also to the transformation of the worldview, connecting to Iberian reality, and in solidarity with all those movements whose perspectives we embrace. As some of our contributors argue, animal studies, ecocritical perspectives, urban studies, new social movements studies, and science studies supply refreshing new frameworks for questioning the prevailing paradigms. Their efforts transcend the national; indeed, they oftentimes attempt to deconstruct it. They challenge the worldview based on exclusive anthropocentrism which lies at the heart of current politics of life, but also the optimistic discourse of bioeconomy, which promises to extract life from living beings for a common good and provide profit to multinational corporations. Frequently, the authors in the pages that follow situate Iberian discourses on life within critical movements that feature broadened notions of agency, posthumanist ontology, and the formation of networks that eschew the logic of neoliberal, late-stage capitalism. This is the first edited volume in the field to approach the ethics of life as a significant determinant of Spanish cultural and political processes, and to approach it from the critical perspectives of allied alternative discourses of various fields. The contributions are in many cases the result of collaborative, ongoing research projects and products of the teamwork of multiple scholars.

The volume begins with a history and contemporary overview of important animal rights and environmentalist movements in Spain. The essays in this section, "Genealogies of Ecological and Animal Rights Movements in Modern and Contemporary Iberia," analyze cultural movements that have made their appearance on the Iberian stage during the last decade, influencing politics and lived culture, but have had a growing presence since Franco's death. Since the fall of the Franco dictatorship, the social and natural sciences in Spain have shown a significant interest in environmental issues, as Carmen Flys-Junquera and Tonia Raquejo Grado argue in their expansive opening essay, "The Environment in Literature and the Arts in Spain." This awareness, however, has been historically unmatched in literature, the arts, and criticism, a contrast with aesthetic production in Latin America. Flys-Junquera and Raquejo Grado survey a budding corpus of art works and literary initiatives with a strong ecological purpose that have begun to deploy ecocritical modalities of analysis in the last decade in Spain, due in large part to the rising tide of ecological crises, particularly the environmental impact of increased land speculation and sprawl on the coasts and rural areas. Theirs is one of the few attempts to both acknowledge this resurgence and explain its historical conditions of possibility. In short, their essay contains precious information and analysis of current environmental activism through art in Spain.

In "*Nunca Máis*: Ecological Collectivism and the *Prestige* Disaster," John Trevathan challenges the notion that Spain's current crisis is purely financial in nature, resolving instead to critique the nation's political operations through the guise of ecological polemics, devoting particular attention to the *Prestige* oil tanker disaster in 2002 off the coast of Galicia. Trevathan intuitively locates two competing discourses in the wake of the spill: one "official" narrative that dissimulates and obscures the impact of the catastrophe; and a second driven by a diverse ecological "entanglement' of human and nonhuman elements operating under the banner of *Nunca Máis*. Through a careful study of the writer Manuel Rivas's notion of *hartismo*, Trevathan puts forth that through the ecological consciousness attached to the *Nunca Máis* movement Galician culture gains solidarity, complexity, and a degree of autonomy from the processes of global capital flow.

The state's protection of the relations of economic production has led to widespread, nefarious violence on the side of material reality and annexed public spaces undemocratically, frequently as a result of the choices of private interests.[4] As the authors of the section "Ecological Crisis and the Neoliberal Appropriation of Public Space" illustrate, this form of the nation state owes to its coextensive, even symbiotic, development with neoliberal economic ideologies.[5]

In "Tourism and 'Quality of Life' at the End of Franco's Dictatorship," Eugenia Afinoguénova follows the course traveled by the notion of "quality of life" in Francoist and post-Franco Spain, locating the initial surge in the term's popularity in the 1970s against the backdrop of the Western economic crisis. Afinoguénova's essay is the first to undertake such a trajectory, and her work importantly recognizes that the debates over the "quality of life" that emerged in the final years of the Franco dictatorship became central to the role environmentalism came to play in the drafting of the 1978 Constitution. Tracing the evolution of the ideologies governing "quality of life" in Spain over the last three decades helps to unveil an inheritance of neoliberal practices that first took shape under Francoism and continue to exert an influence on the nation today; namely, that environmental protection and tourism are deemed coextensive, a discourse that seemed to become obsolete but, upon further inspection, governs values that parties such as the PSOE have championed, such as sustainability, beauty, and leisure, and most recently this discourse can be also found in bioeconomy.

While Afinoguénova investigates those paradisiacal, cordoned off parts of Iberia preserved for the capital expenditure of tourists, Katarzyna Beilin, Matthew Feinberg, and Susan Larson analyze the possibilities that a new ethics of life poses for urban, lived-in spaces. Beilin, in "Die and Laugh in Anthropocene: Disquieting Realism and Dark Humor in *Biutiful* and *Nocilla experience*," analyzes "moods of concern," marked by a sense of disquiet related to the perceived, although not fully realized, mutations that result in our sensation that something that cannot be seen is happening around and inside of us. In *Biutiful*, one contributor to this mode of disquiet is precisely the deep cancer of the environment caused by the poisonous effects of neoliberal politics portrayed by the film as an omnipresence of rule by money. In *Nocilla experience*, where dark humor prevails, the mood of disquiet emerges as a result of the limited character of scientific knowledge and a persistent pulsation of the unknown, where human and nonhuman life links in surprising ways on a molecular level.

Matthew Feinberg and Susan Larson, in "Cultivating the Square: Trash, Recycling, and the Cultural Ecology of Post-Crisis Madrid," put forth that the participatory urbanism in Madrid arising due to the 2008 economic crisis culturally repurposed public spaces, thereby rebuffing the logic of debt-driven financial capital. Through an analysis of "architectural recycling," the aesthetics of trash and other new ecologies of urban space, Feinberg and Larson show that in both high-profile city plazas and in the unassuming crevices of everyday life, grassroots movements in Madrid are reconfiguring the intersection of digital, material and cultural life and creating alternatives forms of economic distribution. Feinberg and Larson's essay is a timely, needed excursion into a crisis that is far from resolution in present-day Spain and a visionary analysis of an alternative economy and urban structure, which today is only imagined by intellectuals and architects, but is already practiced in the urban gardens of Madrid by new-minded citizens. This is truly "cutting-edge" research since it not only analyzes cultural problems, but also provides a narrative for a future reconstruction of social life and politics.

The new spatial practices developed within the confines of Madrid analyzed by Feinberg and Larson are in tune with theories of degrowth, a relation made explicit in Luis Prádanos's essay, "Degrowth and Ecological Economics in Twenty-First-Century Spain: Toward a Posthumanist Economy." Prádanos synthesizes the work of a host of Spanish writers who have explored socially fair and ecologically sound holistic alternatives to hegemonic neoliberal globalization. In a poignant, and provocative, turn of analysis, Prádanos avows that these multiple discourses, taken together, advocate for a "posthumanist economy" based on degrowth theory that acknowledges the agency and interdependency of all forms of life. By posthumanizing the economy, these authors seem to be in line with the "non-human turn and new materialism" that has recently occurred in the environmental humanities.

The critique of neoliberal economic production in the preceding section is, of course, intimately linked to bio-power. Indeed, Michel Foucault asserts that the evolution of biopolitics in the West partly arises from the neoliberal extension of market rationality to noneconomic domains such as those related to medicine, reproduction, punishment, and more. That said, our third section, "Iberian Bio-power: Life as a Political Matter," groups together a set of essays that investigate the political regulation and subjugation of flesh, but in so doing we note the closeness between this section and that which precedes it.

Pablo de Lora studies the governmental regulation of a woman's right to procreation over the span of nearly fifty years in his piece, "Reproductive Rights in Spain: From 'Abortion Tourism' to 'Reproductive Destination.'" Uniquely drawing from legal court cases, the fiction of Luis Martín-Santos, and academic research, Lora trenchantly details the winding evolution of Spain from being a nation hostile toward practices such as abortion to becoming one of the more liberal countries for medically assisted reproduction and the genetic diagnosis of embryonic abnormalities, with the key exception of surrogate motherhood. De Lora demonstrates that in Spain, as in many other ambits, conflicts tied to the various discourses regarding reproductive rights are often resolved pursuant to legal dictates. This speaks to the fact that these various debates on life are frequently interlinked and overlap, such as the rift between abortion and surrogate motherhood in the cases of fetuses that exhibit abnormalities while in utero.

Debates in reference to individual freedom and end-of-life decisions have been no less fraught with debate, as Paul Begin evinces in his essay "*Mar adentro* and the Question of Freedom." Begin's analysis of Alejandro Amenábar's *Mar adentro* queries the ambivalence between rationalist ideals and changing definitions of human nature embedded within complex structures of experience. Amenábar's work plays out the complicated, and at times paradoxical, nature of freedom and liberty in Western philosophy. Any individual bound to the restrictions of a social contract sacrifices a degree of freedom to a higher power that is entrusted with preserving the best interests of the commonwealth. What is good for the commonwealth, of course, is subject to political influence. In *Mar adentro*, Begin analyzes a situation in which differing conceptions of freedom and rationality come into conflict in relation to assisted suicide, and the kinds of routes open to a political subject whose situation falls beyond the parameters of Enlightenment definitions of liberty. This approach, which ultimately interrogates the film's endorsement of euthanasia, breaks new ground with respect to critical analysis of the canonical work.

From the creation and completion of life, Sainath Suryanarayanan and Katarzyna Beilin investigate the biotechnological enhancement of human and nonhuman lives by recent innovations in transgenic sciences in their chapter, "Still Different? Biotechnology, Politics, and Culture in Spain." Critics assert that Spain has implemented biotechnology policies in an uncritical manner, without proper public debate, which has rendered Spanish society beholden to the interests of powerful multinational corporations and governments, and that this may be viewed as a continuation of a Spanish past marked by authoritarian regimes. In this sense, Jorge Riechmann states that "Spain is still different" (23) from other European countries where such debates have taken place. While certain attitudes towards science in Spain can be, perhaps, interpreted in this way, Suryanarayan and Beilin show that they result from a complex set of debates and historical experiences, where hegemonic and alternative discourses switch places and cultural and political changes transform the meaning of enthusiasms and caution towards science. All in all, these authors convincingly show that new approaches to life allow for an interdisciplinarity that spans the natural sciences and humanities, as their essay straddles the line between science and technology studies, cell biology, and cultural studies.

In the final section of the volume, "Reassembling the Archive through the Concept of Life," Daniel Ares, John Beusterien, William Viestenz, and Sebastiaan Faber explore the pedagogical possibilities of Animal Studies for the university curriculum, interrogate the intellectual synergy between the emergent practice of Iberian studies and the "non-human" turn of the twenty-first century, and explore the importance of the concept of life for historical memory. The contributions assert that the cultural and historiographical field is always produced by interactions with objects that mediate human experience. In this sense, one can never think of a cultural object as a mere reflection of human agency, as it includes the participation of a broad number of nonhuman actors. Because agentive objects partly mediate human experience, locality and its attendant materiality acquire increased significance.

Bracketing plays a fundamental role in the first essay of the section, "Iberian Cultural Studies Beyond the Human: Exploring the Life History of Marcos Rodríguez Pantoja in Spanish Anthropology and Popular Film," by Daniel Ares. Following Latour's questioning of culture and nature, Ares critically revises a feral child story and questions anthropologic and filmic renditions of this life history. He argues that institutional bio-power, particularly as expressed in the natural sciences and in anthropology's disciplinary separation of nature and culture, succumbs to "anthropodenial"; i.e., the inability to extend to nonhuman life notions of justice, cooperation, and moral behavior by considering the natural world to be radically transcendent from human concern, as opposed to the immanence of "constructed" culture. The incapacity or unwillingness to do so may, however, prove to be as anthropocentric as anthropomorphism, which allows human properties to be found in nature.

John Beusterien, in "Animals in Contemporary Spanish Newspapers," affirms that the supposed dichotomy between the global and the local is a facile division; rather, the global tends to be articulated by the careful stitching together of specific local nodes. Beusterien's work creates a network encompassing the animal and the artificial as well as the redemption of ecological landscapes as sites of cultural production. Two important local nodes in the essay are Spain, the author's area of specialization, and West Texas, where he teaches. Inserting the rising importance of Animal Studies in Spain into the North American university curriculum provides, according to Beusterien, innovative ways for students to rediscover a sense of place. Moreover, Beusterien lucidly shows how debates on life can reenergize how scholars teach, and market, the humanities.

In William Viestenz's essay, "The Bull also Rises: The Political Redemption of the Beast in *La pell de brau* by Salvador Espriu," the author postulates that bestowing a mediating function onto objects and nonhuman forms of life helps to reveal that the set of rules put into operation by the discursive formations of Hispanism silences a range of cultural networks. Viestenz's essay frames the notion of the human as a fluid concept that can list toward the status of an animal object or legal subject depending on political circumstances. Through a meditation on Salvador Espriu's political project in *La pell de brau*, Viestenz argues that Iberian studies offers an adroit permutability conducive to interdisciplinary intellectual study while maintaining specialization in the myriad cultural histories of the Iberian peninsula.

Sebastiaan Faber's essay, "Accounting for Violence, Counting the Dead: The Civil War and Spain's Political Present," questions history and memory, departing from the concept of life and asking about its worth. He asserts that almost all accounts of the Spanish Civil War coincide in their assessment of the conflict as a massive loss of life. When it comes to determining the how and why of these deaths, however, the disagreements begin. As Faber persuasively concludes, the *conditions of possibility* of these types of arguments, including the linking of Francoism with Nazism, can be traced to the evolution of the global debate on the ethics of life and its repercussion on international law and the Spanish public sphere. The way in which historical archives assess the number of deaths during the Civil War and throughout Francoism is related to the operation of structural violence still in effect in the present moment that obscures accountability for those losses.

Notes

1. Among various publications, *Hispanic Issues Online* devoted to the debates their Volume 1 (2006) and Volume 2 (2007), and Mabel Moraña edited *Ideologies of Hispanism* (2005). In addition to these Hispanic Issues volumes, Brad Epps and Luis Fernández Cifuentes, *Spain beyond Spain, Modernity Literary History and National Identity* (2005). Joan Ramon Resina's articles have been also published in book form in *Del hispanismo a los estudios ibéricos: una propuesta federativa para el ámbito cultural.*
2. De Lora (2003, 2010), Mosterín (2011), and Vicent (2001) counter the argument of financial gain, stating that it is not right to earn money at the expense of the suffering of sentient creatures. Ovejero et al. counter the argument of species extinction, stating that if we breed species just to torture and kill their members for entertainment purposes it seems better that they become extinct. No individual in particular will suffer if the species becomes extinct because unborn life does not suffer. He claims that alternatively species could be preserved in existing reserves turned into national parks for their conservation.
3. In his blog tratarde.org, during the last few months of 2013, Riechmann announced three free courses on degrowth theory and on ecological economy.
4. This process often operates in a fashion that renders this material destruction invisible and, as Rob Nixon argues, inflicts "slow violence."

5. This thought is in line with Ernest Mendel's analysis of the late stage of capitalism. One of the State's roles in the stage of late capitalism has been to exclude dangers to the modes of production that favor the stability and predominance of the ruling ideology of dominant classes (474–75).

Works Cited

Amorín, Carlos. *Las semillas de la muerte. Basura tóxica y subdesarrollo: el caso de Delta & Pine*. Madrid: Catarata, 2000.

Beilin, Katarzyna Olga. "Culture and Torture; Debates on Bullfighting in the Times of War on Terror." *International Journal of Iberian Studies*. 25.1 (July 2012): 61–72.

Bennett, Jane. *Vibrant Materialism: A Political Ontology of Things*. Durham: Duke University Press, 2010.

Casal, Paula. "Is Multiculturalism Good for Animals?" *Journal of Political Philosophy* 11.1 (March 2003): 1–22.

Castro, Américo. *The Meaning of Spanish Civilization*. Ed. José Rubia Barcia. Berkeley: University of California Press, 1977.

Cerrillo, Antonio. "Los transgénicos saltan del plato: Los fabricantes renuncian a los alimentos modificados genéticamente." *La Vanguardia*. 31 Jul. 2004.

Cooper, Melinda. *Life as Surplus: Biotechnology and Capitalism in the Neoliberal Era*. Seattle, WA: University of Washington Press, 2011.

Corraliza, José Antonio. "Ecofatiga: Hartos de reciclar." *Razón.es*. 11 Nov. 2011. Web. 30 Dec, 2013.

De Lora, Pablo. *Justicia para los animales. La ética más allá de la humanidad*. Madrid: Alianza, 2003

De Prada, Juan Manuel. "En defensa de los toros." *ABC*. 23 Nov. 2009. Web 3 Dec. 2009.

Derrida, Jacques. *Of Grammatology*. Baltimore: John Hopkins University Press, 1998.

Epps, Brad. "Who Works Where: 'The State of Our Disciplines' in a Fragmented Global Frame." *Hispanic Issues on Line*, 15.1, 2006. Web. 29 Sept, 2011.

Epps, Brad and Fernández Cifuentes, Luis. *Spain beyond Spain: Modernity, Literary History, and National Identity*. Lewisburg: Bucknell University Press, 2005.

Dopico Black, Georgina. "The Ban and the Bull: Cultural Studies, Animal Studies and Spain." *Journal of Spanish Cultural Studies* 11.3/4 (2010): 235–49.

Fernández, Joaquín. *Ecologismo español*. Madrid: Alianza Editorial, 1999.

Fernández Mallo, Agustín. *Nocilla experience*. Madrid: Santillana, 2007.

Foucault, Michel. *The Birth of Biopolitics*. New York: Palgrave, 2008.

Fuentes, Carlos. *El espejo enterrado*. Madrid, Taurus, 1997.

Gabilondo, Joseba. "Spanish Nationalist Excess: A Decolonial and Postnational Critique of Iberian Studies." *Prosopopeya: revista de crítica contemporánea*. (Forthcoming)

García Lorca, Federico. "Teoría y juego del duende." 1932. *Biblioteca Virtual Universal*. Web. 12 Oct. 2009.

Garrido, Francisco. Javier. "Biotecnología S.A. Una aproximación sociológica." *Política y Sociedad,* 39. 3 (2002): 641–659.

Goytisolo, Juan. "La fe de erratas." *El País,* November 27, 2004.

_____. *Reivindicación del conde don Julián.* Madrid: Cátedra, 2003.

Grandes, Almudena. "Arte." *El País.* 9 March 2010.

Haraway. Donna. *Simians, Cyborgs, and Women: The Reinvention of Nature.* New York: Routledge, 1991.

_____. "A Game of Cat's Cradle: Science Studies, Feminist Theory, Cultural Studies." *Configurations* 2.1 (Winter 1994): 59–71.

_____. *Modest_Witnes@Second_Milinnium. FemaleMan_Meets_OncoMouse.* New York: Routledge, 1996.

_____. *When Species Meet.* Minneapolis: University of Minnesota Press, 2008.

Hernández, María. "Matador: El deseo transgresivo en Rafael Albertí and Pedro Almodóvar." *Lenguaje y Textos* 18 (2002). Web. 4 March 2011.

Indignados, 15-M. Ed. Fernando Cabal. Madrid: Mandala Ediciones, 2011. Web. 20 Aug. 2011.

Jovellanos, Gaspar Melchor De. "Toros." *Memoria Para El Arreglo De La Policía De Los Espectáculos Y Diversiones Públicas Y Sobre Su Origen En España.* Ed. José Lage. Madrid: Cátedra, 1977.

La flor de mi secreto. Dir. Pedro Almodoóvar. Prod. Deseo, 1995. Film.

Labanyi, Jo. *Myth and History in the Spanish Contemporary Novel.* Cambridge, UK: Cambridge University Press, 2011.

Latour, Bruno. *We Have Never Been Modern.* Cambridge, MA: Harvard University Press, 1993.

_____. "On actor-network theory. A Few Clarifications Plus More than a Few Complications." *Soziale Welt* 47 (1996): 369–81.

_____. *Reassembling the Social: An Introduction to Actor Network Theory.* Cambridge, MA: Harvard University Press, 2005.

_____. "Where Are the Missing Masses? The Sociology of a Few Mundane Artifacts." *Shaping Technology/Building Society: Studies in Sociotechnical Change.* Ed. Wiebe E. Bijker and John Law. Cambridge: MIT Press, 1992. 225–58.

López Arnal, Salvador. "Conocimientos científicos y decisiones políticas." *Aporrea,* 17 Dec, 2011. Web. 23 Dec, 2012.

López de Uralde. *El planeta de los estúpidos: Propuestas para salir del estercolero.* Madrid: Planeta, 2010.

Mendel, Ernest. *Late Capitalism.* London: Verso, 1987.

Mar adentro. Prod. Alejandro Amenábar and Fernando Bovaira, 2004. Film.

Marías, Javier. "Los exterminadores de toros," *El País.* 3 Jan. 2010.

Martín-Santos, Luis. *Tiempo De Silencio.* Ed. Alfonso Rey. Barcelona: Crítica, 2000.

Martínez Alíer, Joan. *Environmentalism of the Poor.* Edward Elgar, Cheltenham, UK. 2003.

Mignolo, Walter. *Local Histories, Global Designs; Coloniality, Subaltern Knowledge and Border Thinking.* Preface. New Jersey, Princeton University Press, 2000.

Moraña, Mabel. *Ideologies of Hispanism.* Nashville: Vanderbilt University Press, 2005.

Mosterín, Jesús. *A favor de los toros.* Pamplona: Laetoli, 2010.

Morton, Timothy. *Ecology without Nature; Rethinking Environmental Aesthetics.* Cambridge, MA: Harvard University Press, 2007.

_____. "Romanticism and Ecology: 'Twinkle, Twinkle Little Star' as an Ambient Poem." In "*Romanticism and Ecology,*" Ed. James McKusic. *Romantic Praxis* 232. 2007.

_____. *The Ecological Thought.* Cambridge, MA: Harvard University Press, 2010.

_____. "Queer ecology." *PMLA* 125.2 (March 2010): 274–82.

_____. *Hyperobjects: Philosophy and Ecology after the End of the World.* Minneapolis: University of Minnesota Press, 2013.

Nixon, Robert. *Slow Violence and the Environmentalism of the Poor.* Boston, MA: Harvard University Press, 2011.

Noel, Eugenio. *Diario íntimo.* Vol. I and II. Madrid: Taurus, 1968. Print

Novas, Carlos. "The Political Economy of Hope: Patients' Organizations, Science and Biovalue." *Biosocieties.* 1.3 (Sept, 2006): 289–305.

Ovejero, Félix, Pablo de Lora, and José Luis Martí. "De toros y argumentos." *El País,* 19 Aug. 2010.

Pavone, Vincenzo. "Ciencia, neoliberalismo y bioeconomía." *Revista Iberoamericana de Ciencia, Economía y Sociedad* 7.20 (April 2012): 145–61.

Ramos Gorostiza, José Luis. "Gestión ambiental y la política de la conservación de la naturaleza en la España de Franco." *Revista de Historia Industrial* 32.3 (2006): 99–138.

Resina, Joan Ramon, *Del hispanismo a los estudios ibéricos: una propuesta federativa para el ámbito cultural.* Madrid: Biblioteca Nueva, 2009.

Riechmann, Jorge. "Hacia una agroética: Consideraciones sobre la ética ecológica y actividad agropecuaria." *Cuidar la T(t)ierra. Políticas agrarias y alimentarias sostenibles para entrar en el siglo XXI.* Barcelona: Icaria, 2003.

_____. *Qué son los alimentos transgénicos?* 2002. Barcelona: RBA, 2011.

_____. *El socialismo sólo puede llegar en bicicleta.* Madrid: Catarata, 2012.

Santana, Mario "Implementing Iberian Studies, Some Paradigmatic and Curricular Challenges." *Iberian Modalities: A Relational Approach to the Study of Culture in the Iberian Peninsula,* Ed. Joan Ramon Resina. Liverpool: Liverpool University Press, 2013: 54–61.

Savater, Fernando. "Las cornadas de Europa," *El País.* 22 Dec. 1991. Web. 12 Nov. 2009.

_____. "Zoología moral," *El País.*10 Sept. 1991. Web. 2 Oct. 2010.

_____. "Rebelión en la granja," *El País.* 16 Mar. 2010. Web. 13 Feb. 2011.

_____. *Tauroética.* Madrid: Turpial, 2010.

Sheehan, Peter, and Greg Tegart. *Working for the Future: Technology and Employment in the Global Knowledge Economy.* Melbourne: Victoria University Press, 1998.

Todo sobre mi madre. Dir. Pedro Almodóvar. Prod. Deseo, 2004. Film.

Tosko, Katherine. *The Bull and the Ban: Exploring the Catalan Ban on Bullfighting.* Bexhill-on-Sea, East Sussex: Independent Publishing Platform (Suerte Publishing), 2012.

Ugarte, Michael. *Trilogy of Treason. An Intertextual Study of Juan Goytisol.* Columbia and London: University of Missouri Press, 1982.

Vicent, Manuel. *Antitauromaquia*. Madrid: Santillana, 2001.

Viestenz, William. "Sins of the Flesh: Bullfighting as a Model of Power." *Iberian Modalities.* Ed. Joan Ramon Resina. Liverpool: Liverpool University Press, 2013: 143–61.

Wolff, Francis. *Cincuenta razones para la defensa de los toros.* Almuzara, 2011.

PART I

Genealogies of Ecological and Animal Rights Movements in Modern and Contemporary Iberia

1

The Environment in Literature and the Arts in Spain

Carmen Flys-Junquera and Tonia Raquejo

According to Joaquín Fernández, Spain has not had a strong environmentalist tradition, partially due to the Franco regime (*Ecologismo español*). In his study, Fernández notes that the Civil War marked a fracture between any previous incipient Spanish activism, and that real consciousness only started in the 1970s, together with democracy and the economic boom known as the "Spanish miracle" (44). An early study by Josep Vicent Marqués, *Ecología y lucha de clases* (1978), illustrates the leftist, Marxist influence in most initial environmental movements, clearly linked to social injustice. Only with the increase of urban speculation as the result of massive construction, which threatened the entire coast and large tracts of rural space, and the uncontrolled growth of cities did the environmental impact become an issue within the greater population. And as such, most debates are closely tied to social and economic problems rather than with the agenda to preserve wilderness which resonates so strongly in the United States. In the last two decades, environmental debates have appeared in political, social, economic, and scientific discourses; but, sadly, the financial and economic crisis has stunted this momentum of environmental awareness and dried up all funding. Despite the urban sprawl, there is still no strong "green" political party with parliamentary representation in Spain, although there are prominent activists, such as Juan López de Uralde, longtime president of Greenpeace Spain and founder of a new environmentalist political party, Equo, whose major test was to have been the European and local elections in 2014 and 2015, respectively. However, this test was weakened due to the rise of new populist parties with which Equo made alliances. These alliances were very successful in both elections, but the actual weight of Equo and environmental issues was diluted.

During this period there has been significant environmental awareness in the sciences and social sciences, but the humanities and arts in Spain were largely disconnected from environmental concerns. In the journal *Ixquic* 2 (2000), critics Jorge Paredes and Benjamin McLean assert that environmental literature in Spanish could only stem from Mesoamerica and was virtually impossible in the Iberian Peninsula, given its strong Judeo-Christian tradition, and the importance of humanism, rationalism, capitalism, and positivism, which made the culture insensitive to the environment (25). Even if their concept of environmental literature as direct denunciation of environmental degradation is restrictive, it is true that there has not been much environmental concern in the humanities. However, things are changing. Research groups with an environmental focus have been created within both literary and art criticism. Various novels and poetic works have recently addressed environmental concerns and there are numerous artists with a clear ecological purpose in their creative work. This essay aims to present an overview of the recent research within the fields of literature and the fine arts.

Literature and Ecocriticism

While ecocriticism has become a significant literary approach in the United States and the United Kingdom, an ecocritical school of thought is merely incipient amongst Spanish Hispanists. It is clear, as Paredes and McLean state, that literature coming from the indigenous peoples of the American continent has cried out in protest over the degradation of the environment and natural landscapes (20), while this has not happened in Spain, with the exception of a few recent cases. Most Spaniards have had a strong sense of place-attachment to their region, but the approach from criticism has usually been either one of nationalism and identity or some variation of the classic *locus amoenus* and the pastoral.

Prior to the 1970s, Spanish literature took up the thematization of nature without, however, being ecologically oriented. Paredes and McLean discuss the

Spanish Renaissance and Baroque literature (9–16) while Pratt and Gordon, in one of the earliest ecocritical articles about Spanish literature, point out that during the Romantic period, Spanish literature tended to focus on the sublime of nature (249). Nevertheless, despite the clear influences of the environment on Spanish culture and literature—which both writers point out—there has been little ecological awareness.

Miguel Delibes's speech upon entering the Royal Academy of Language in 1975 was the major wake-up call. In his speech he interprets his own work based on an environmental approach derived from the report *The Limits to Growth* (1972) commissioned by the Club of Rome, an approach which dominates the little work carried out in the late twentieth century in environmental writing in Spain. This can be considered, as Marrero Henríquez states, the first critical interpretation of literary work based on environmentalist thinking ("Ecocrítica" 208). Delibes's approach, rather than a nostalgic view of rural life, can be interpreted as associating rural life with a healthier and more sustainable lifestyle in contrast to the so-called progress of the cities. Although Delibes does not make a direct denunciation of environmental degradation, his work is informed by clear environmental thinking, opting for an indirect approach, tinged with poetic tones which he considers more effective (Pérez Abad 94). While Delibes is considered the first environmentalist writer (Paredes and McLean 8), other writers have started to deal with these topics. Julio Llamazares shows in his novel *La lluvia amarilla* (1988, based on the flooding of a valley for a reservoir) a clear ecological understanding of life, though he maintains a nostalgic and almost romantic tone, rather than one of direct denunciation. The abandoned village represents the end of the dialogue between nature and humankind because modern city life has turned a blind eye and nature slowly reclaims the village. Another significant contribution is Jesús López Pacheco's poetic volume *Ecólogas y Urbanas* (1996), where the title is a play on ecology and eclogues. The collection is full of witty poems, sayings, absurd skits, and rhymes, which nonetheless have deep social and environmental messages such as his "Epigrama de un ingenio de la corte imperial" (37) (Epigram by a genius of the imperial court), which characterizes our consumer society by the garbage and smoke it produces: Consumer society / has its own culture: / to produce garbage and smoke, / or, if not, smoke and garbage (my translation). Another short poem also illustrates the Spanish concern of linking economic growth and progress to environmental devastation: "Pequeña corriente de conciencia en versos de plástico" (94) (Small current of conscience in plastic verses). This poem plays on the programed obsolescence of our consumer society which produces things that create waste only to produce something new to take care of the waste or to create a demand for more ecological products and processes, in order to make profit out of them.

> Selling products produced
> only to be sold
> produces great pollution
> offering as a solution
> the demand created for recycling
> and alternative products
> natural organic healthy
> reinvesting in the new scheme
> to receive benefits with both hands
> good life for the living
> (above all one must be postmodern)
> only plastic and God are eternal (my translation)

Both poems, which rhyme in Spanish, play with the titles of traditional Spanish poetry, only to highlight the current skepticism of capitalist economic cultures.

Moving into the twenty-first century, a number of writers have directly engaged environmental issues. Foremost among them is the poet and essayist Jorge Riechmann. His book, *Un mundo vulnerable: Ensayos sobre ecología, ética y tecnociencia* (2000), is considered a landmark and his collections of poetry, in particular *Con los ojos abiertos* (2007), are invariably linked to environmental issues, once again with a critical view of consumer society and its effects on culture. For example, in his poem "Tratamiento de residuos" (Waste treatment) he lists the types of waste in which both the human and nonhuman become the waste of progress:

> El principal problema estriba actualmente
> en la eliminación de los residuos
> . . . .
> El agua es un residuo
> de la producción de imprescindible energía
> animada e inanimada. El aire, residuo
> de combustiones en benéficos motores.
> Fauna y flora son ya propiamente residuos

de las industrias cárnica y farmacéutica. El trabajo
humano, residual residuo
de la actividad incansable de las máquinas.
La historia es residuo
de la conquista del espacio y el arte residuo
altamente tóxico
de la televisión por cable.
. . . . (40)

(The main problem currently
is the elimination of waste
. . . .
Water is a residue
of the production
of the necessary energy
animate and inanimate. The air, residue
of combustion in beneficial motors.
Fauna and flora are clearly residues
of the meat and pharmaceutical industries. Human labor,
residual residue
of the untiring work of machines.
History is the residue
of the conquest of space and art is the residue
highly toxic
of television by cable. . . . ; my translation).

In 2001, the well-known poet Juan Cobos Wilkins published a lyrical novel, *El Corazón de la tierra*, later taken to the screen by Antonio Cuadri in 2007. The novel combines the historical narrative of what was perhaps the first demonstration for environmental justice in Spain, at the British owned mines of Río Tinto in 1888, with the fictional encounter between two women, one Spanish and one British, in the 1950s. The character Blanca, witness to the historical events and an old woman at the time, clearly portrays ecofeminist values.[1] Other acclaimed novelists, such as Rafael Chirbes, have also directed their attention to environmental issues. His award-winning novel *Crematorio* (2007), developed into a TV series in 2011, is a clear denunciation of the urban speculation that has ruined the Spanish landscape and economy. Philosopher and writer Marta Tafalla has also published recently a novel with clear environmentalist concerns, *La Biblioteca de Noé* (2007). This futuristic novel develops a plot to free animals being held captive for the fur trade, as it reflects on the importance and fascination of skin, of all species, as the site of the writing of our culture—books, furs, bodies. In the same year, another philosopher, Paula Casal, also published a novel for young adults, *Martina y el mar* (2007), with an explicit environmental and pedagogical spirit as well as an ecofeminist sensibility, much as Laura Gallego's *Donde los árboles cantan* (2011), a fantasy novel for young adults which clearly depicts the agency of nature.

Young novelists are also dealing directly with environmental issues beyond the traditional sense of place. Many of these relate capitalism, growth, and progress with environmental devastation. For example, Alberto Vázquez Figueroa's *El mar en llamas* (2011), examines the relationship between economic growth and fossil fuel dependency. José Luis Sampedro and Olga Lucas, in their novel *Cuarteto para un solista* (2011), also reflect on the lack of sustainability of our lifestyle and the limits to growth.[2] In 2008 Virginia Ferrer published *Recuerda Mundo: Novela ecológica en la tierra de Oliva Sabuco*, where water becomes the central theme among others such as gender, disability, scientific abuse, and the environment. Likewise, there is an incipient trend of Spanish science fiction dealing with environmental issues, such as Rosa Montero's *Lágrimas en la lluvia*, Jose Ardillo's *El salario del gigante*, and Pablo Tusset's *Oxford 7*, all published in 2011.[3] Poetry with environmental concerns has also begun to develop. Other than the previously mentioned Jorge Riechmann, Julia Barella's last collection, *Praderas de posidonia* (2013) has a clear environmental sensibility, as does Sofia Rhei's recent *Bestiario microscopico* (2012), where she highlights the semiosis of nonmammalian organisms, taking her poetry into the realm of the very recent trend of material ecocriticism. There have also been small inroads into theater and the environment. The well-known naturalist writer Joaquín Araújo wrote the script for a theater production *Bosque de Bosques*, a play directed by Oscar Miranda and produced by the theater company Vistapalabra, and funded by the Ministry of the Environment, Marine, and Rural Affairs, which toured 161 small villages throughout Spain during 2010 and 2011, as a cultural measure to aid in the prevention of forest fires. On a whole, it is obvious that environmental issues are significantly increasing in very recent literature in Spain.

With regard to research and criticism in the twenty-first century, there is very little outside the founding of the ecocritical research group GIECO in 2006, backed by the Franklin Institute of the University of Alcalá.[4] This is the first and only formal ecocritical group in Spain and where most ecocritical research is currently being carried out. Its members have published some of the incipient ecocritical work on Spanish literature as well as numerous articles on English and American literatures, which lie outside the scope of this article.[5] The first clearly ecocritical Ph.D. dissertations in Spain were written by members of the group (Montserrat López Múgica, Olga González Calvo, and Irene Sanz Alonso) and there are several more in progress, although none specifically on Spanish literature. The group has introduced ecocriticism at numerous conferences in Spain and hosted the European Association for the Study of Literature, Culture and Environment (EASLCE) conference on "Cultural Landscapes" in 2008, publishing in digital form the conference proceedings, in which there are numerous papers on all European literatures and in three languages: *Paisajes Culturales: Herencia y Conservación*. The first joint publication of the group, other than individual articles or the proceedings, was a special issue on ecocriticism in the journal *Nerter* 15–16 (2010). This journal, devoted to cultural and literary issues for a general audience, is probably the first overview of ecocriticism published in Spain. In the same year, *Ecocríticas: Literatura y medioambiente* was published. This is the first academic ecocriticism reader in Spanish in which there are two articles devoted to Spanish literature: those by Jose Manuel Marrero Henríquez and Julia Barella, who discuss prose and poetry, respectively. These publications are landmarks since they are the first to discuss general aspects of ecocritical movements in Spanish, making this critical school available to Hispanists in Spain, or other Spanish-speaking researchers who do not read English—one of the major handicaps of the school, since well over 90 percent of ecocritical research is published in English. These publications were shortly followed by another, *Literatura y sostenibilidad en la era del antropoceno* (2011). In this book we can find, among others, several articles on Spanish literature: one on poetry by Julia Barella, one on the Generation of '98 by Santos Casado, as well as one on classical Latin literature by Aurelio González Ovies, although these approaches still retain the more classical tendency of the pastoral or sense of place. More recently, an ecocritical book series within the Franklin Collection, CLYMA (culture, literature, and environment), has published its first volume, *La Realidad y Simbología de la Montaña* (2012), and toward the end of 2014 volumes on sense of place and the sea will be coming out. Also significant is the special issue on ecofeminism of the journal *Feminismo/s* 22 (December 2013), with contributions mostly of Spanish ecocritics.

And finally, the research group, together with the European Association for the Study of Literature, Culture, and Environment (EASLCE) has come out with a new academic electronic journal, *Ecozon@: European Journal of Literature, Culture, and Environment* (www.ecozona.eu), which publishes articles in five languages, in the interest of stimulating ecocritical studies in other non-English speaking countries. This journal, which is in its fifth year, has several articles about Spanish literature, both in English and Spanish. The thematic issue for Fall 2013 was precisely on "Mediterranean Ecocriticism," guest edited by Serenella Iovino, and the Spring 2014 special focus topic has been that of translation in the environmental humanities, guest edited by Carmen Valero Garcés. Thus, in the past four years ecocritical production has greatly increased in Spain and a large effort to promote ecocritical studies of Spanish literature is underway. Less than a year ago, a new distribution list, RedCYMA (Red Hispana de Cultura y Medio Ambiente) was launched in order to connect humanistic and cultural environmental studies in Spanish, once again recognizing the important language barrier that ecocriticism among Hispanists is facing.[6] If ecocritical approaches to the study of literature in Spain are still few, despite the recent increase, the academic research on art and nature is even newer. Given that, as mentioned, some overviews of literature and criticism do exist, the major part of this article will focus on the more unknown field of the fine arts.

Land Art and Ecological Art in Spain

Artistic interest in the environment is quite recent in Spain. Actually, it could be said that before the 1980s there were hardly any examples of artists working with nature from an ecological point of view. Nevertheless, there have been exceptions. The alliance between art and nature showed its first signs in the 1960s when painter César Manrique (1919–1992) returned to Lanzarote in 1966 after having lived in New York. In 1968 he built, or rather excavated in the lava, his extraordinary house, which has become a reference for adapting architecture to the peculiarities of climate and landscape. In Manrique's house the natural formation of volcanic bubbles are used as natural rooms. Thus, artificial and natural structures intermingle, creating a sense of ambiguity between the interior and the exterior space which is emphasized by the vegetation that grows in the interior of the rooms and up through the open ceiling. Such a concept of home—in which there is no defined boundary between the interior and exterior space—shows how Manrique understood architecture as part of nature.

In the 1970s some Spanish artists started to experiment with environmental sculpture and ephemeral art following the trends of American and European artists.[7] Nevertheless, these experiences were soon isolated examples since this trend virtually disappeared until the 1990s, when art developed different approaches to nature which were partly inspired by contemporary art scholars as well as encouraged by the local governments.[8] The progressive awareness of the environment had then its origin in the development of public art; however, as the work was placed outside—in the territory—it was still related more to the concept of landscape than ecology.[9] At a later point, publications such as *Ecoarte* (1986) linked art and ecology. *Ecoarte* is a project by Maria Novo that, in the light of environmental education and creativity, proposes the convergence of art and science in order to understand the environment. For Novo, this would provide a necessary advance towards trans-disciplinarity, "which [would] report on the complexity of our environment, on the phenomena of life in its whole extension."

Examples of artistic reclaiming projects in Spain, such as the ones we find in the United States with Robert Morris, Robert Smithson, or Michael Heizer, took some time to develop and there are but a few.[10] Diego Arribas's project for *Ojos Negros* is a paradigmatic example in Spain of reclaiming a mine landscape as an artistic territory to be used by the people of the area. Arribas started in the year 2000, when he organized the event *Arte, Industria, y Territorio* (Art, Industry, and Territory) inside the iron mines of *Ojos Negros* (Sierra Menera, Teruel). The first artistic competition took place that year in order to select a few works of art to be placed, executed, or performed in the mines. Also, an encounter with specialists of different areas (architects, sociologists, and industrial archaeologists) participated in discussing how the mine's enclosure could be transformed and how to carry out these transformations in the whole area. The main objective of the project was not just to reclaim the mines themselves as an entropic aesthetic landscape (a scenery open to artistic interventions), but also to preserve the mine's installations (the industrial constructions) and the miners' district. A mining ecomuseum was proposed in order to keep alive the memories of the mining activity, and also an environmental and archeology center with exhibition halls to display art related to the mines, as well as documents showing the history and development of the mines and any other ethnographical features of the place related to life in that time (Matos 120). By the end of 2005, the old office building had already been remodeled in order to accommodate new necessities: a coffee room, a lodge, a conference hall, and an exhibit hall where a second artistic event took place. Selected artists proposed ephemeral works designed according to a site-specific condition, including both geographical and physical location, but also the local culture and history of the people living there. *Ojos Negros* has been a rare example of success since the mine is now a legally protected place to preserve its historical memory (D. Arribas).[11]

The Río Tinto mine [Fig. 1], to which the previously mentioned novel *El corazón de la tierra* refers, has been also a place of artistic interest, such as that

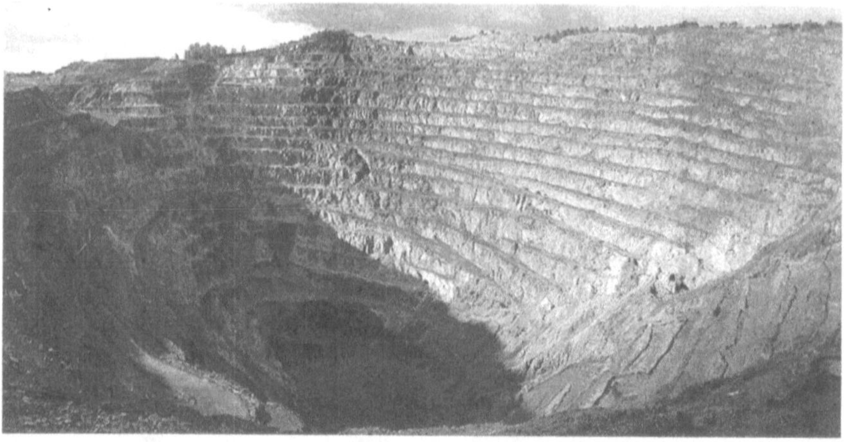

Fig. 1. Corta Atalaya, Río Tinto mines (Huelva). Author's Photography.
Reclaimed Landscape in Rural Territories and Sustainable Ways of Agriculture

of Eva Lootz, one of the earliest artists who was attracted to mining landscapes in Spain. It was in the beginning of the 1980s when she went for the first time to Río Tinto, Huelva, reputed to be the oldest mine in the world. The spectacular man-made crater, which measures several kilometers in diameter, creates a very powerful landscape, enhanced by tinted earth due to the presence of iron, copper, and other minerals. But rather than the visual power of the landscape, Lootz was captivated by what she calls "the theater of matter"; that is to say, the human, cultural and economic implications of the mining activity. For her, the mine is as much a monument as a cathedral or a pyramid, but a "negative monument," because many of the human and cultural implications are hidden.

The progressive emigration of the rural population has been an important problem for Spanish society and local governments. It is frequent to find abandoned towns and rural areas with a very low population density. There have been numerous artistic projects aimed at both consolidating the rural population and restoring abandoned constructions and sites. *Atrapar el Paisaje* [1998, Fig. 2] by Juan Antonio Juarez and Jesús Palmero inaugurated the artistic activity at the *El Apeadero: Centro de operaciones de Land Art*, an art center set up in Bercianos del Real, a town on the Way to St. James. This pioneering project was created by Javier Hernando Carrasco, art critic and professor at the University of León, and Carlos de la Varga, whose art gallery Tráfico de Arte was a reference for contemporary art in León at that time. *El Apeadero* was partially financed by Renfe, the Spanish railway, since one of its main objectives was to remodel the Bercianos train station, out of use by then. By 1999, the architectural

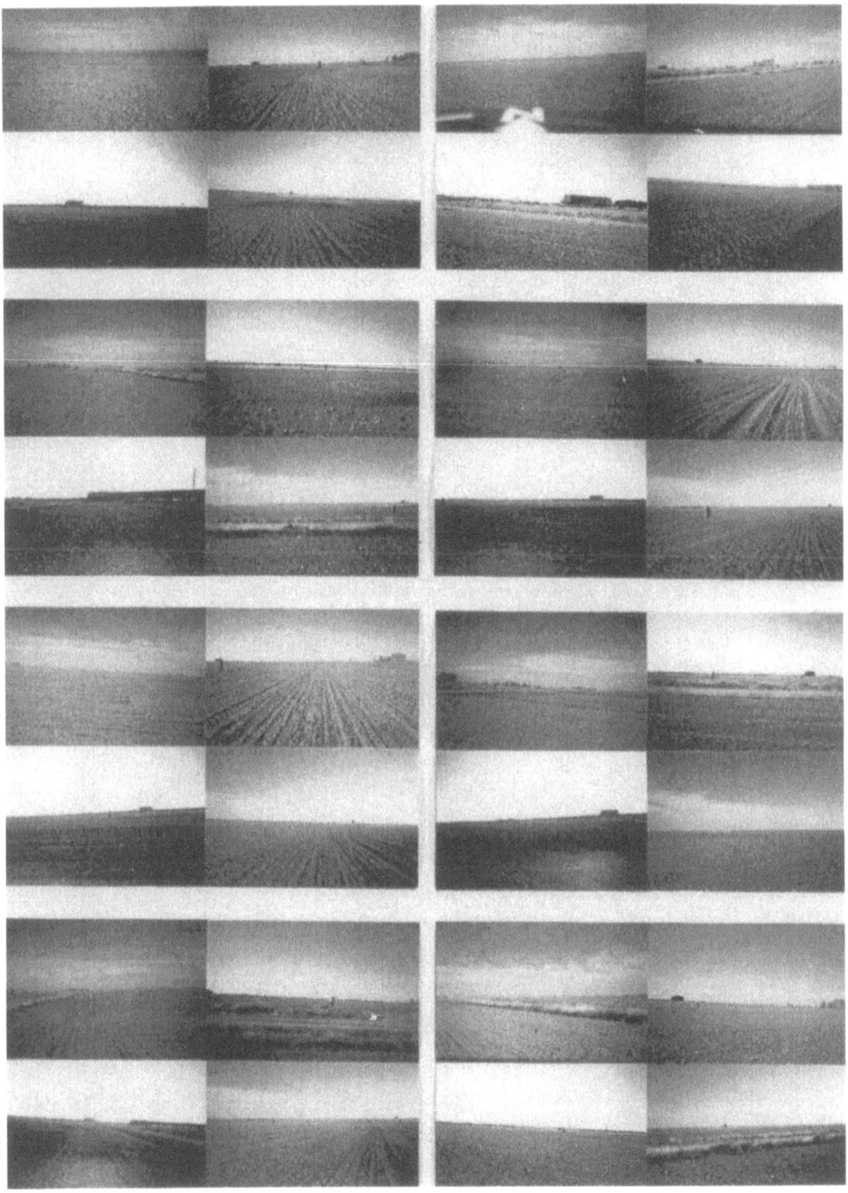

Fig. 2. Juarez and Palmero, *Atrapar el paisaje,* Centro de Operaciones Land Art. Bercianos del Real Camino, León, 1999. Courtesy of the artist.

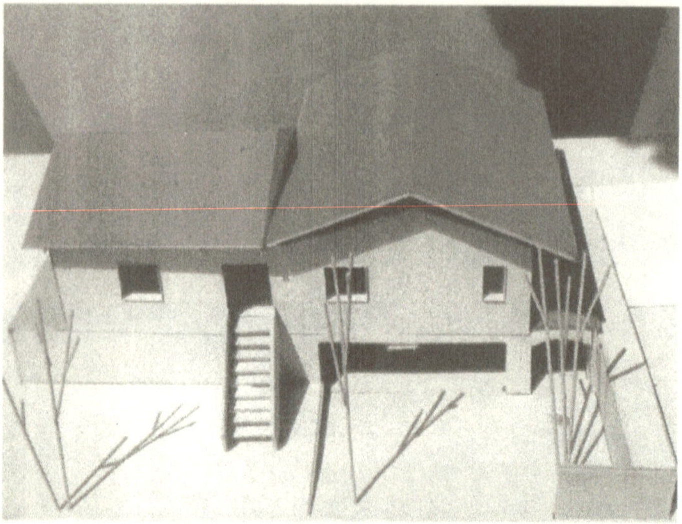

Fig. 3. *Centro de Operaciones Land Art.* Winner desing for remodeling the old railway station at Bercianos del Real. León. María Jimenez, María Iza, Antonio Cantero and Alejandro Tomás, 1999. Published in Román, A., Tascón, J. and de la Varga, C. eds. *Concurso de Ideas. Proyectos Rehabilitación del Apeadero de Bercianos del Real Camino para Centro de Operaciones Land-Art,* León, 1999.

competition was resolved, favoring Maria Jiménez's design [Fig. 3], although the full project was never carried out. Nonetheless, the *Apeadero* project not only remains as an area of work in the open space of Castilian lands, but also has become a model for new artistic generations who have been more and more sensitive to the environment. Moreover, *El Apeadero* has also had an obvious ecological effect on the area since the creation of a spontaneous small lagoon has been detected as a result of the construction of the León-Burgos road. The lagoon has been protected as part of the art center and is now serving as a place of transit for migratory birds (Hernando 66).

The artistic event *Atrapar el paisaje* (Capture the landscape) took place here, along the railroads tracks which run across the grain fields from León to Sahagún. As the title says, this work had the intention of capturing the landscape in images. One hundred people were invited to participate in the event, using both video and still photography. This artistic event was conceived as a collaborative action involving neighbors from the town as well as visitors. In order to distribute people throughout the landscape, a hundred markers were placed all over the area, one for every participant. Atop each marker was a Polaroid camera to be used by each participant, who took four photographs of the landscape at the first sign of a train coming across the land. This action was performed twice in the day: in the mid-morning and in the late afternoon, coinciding with the timetable of trains running from León to Sahagún, and returning from Sahagún to León. Thus, eight hundred snapshots were taken. Later on, they were exhibited on a panel inside a train coach at the León railway station. The resulting composite landscape hinged on the different ways viewers saw, revealing the impossibility of describing nature objectively. Perception is subjective, tinged with the physical and sociocultural position of the viewer, whether it be a farmer, an artist, or a pilgrim walking the pilgrimage. The exhibit questioned the convergence of these realities and whether art was capable of portraying the polysemic nature of place.

We define the artistic happening *Felipe vuelve a casa con las ovejas sonando* (Felipe Comes Back Home with the Sounding Flock) in a similar vein. The event was a sort of "flock concert" proposed and carried out by the shepherd of the town of Bercianos after being inspired by the artistic activity there. The flock concert consisted of three hundred sheep with different cowbells, up to nine distinct tones, all from the herding tradition. The bells were carefully chosen to produce a rich range of pitches and resonances while they were walking along down the hills towards the stable. The musician Nilo Gallego helped Felipe with the rehearsal by controlling the sound in relation to the flock's motion, whether they gathered or dispersed. So on October 23, 1999, Felipe conducted his flock concert in the evening. Five days later, Lucio Infante, the shepherd of Sahagún, repeated the concert with his own flock. With these concerts, the *Apeadero*'s main objective, that of involving the local people in artistic action, was achieved. This is a very important point in relation to the ecological features of this art center because, rather than imposing a pre-established cultural model, alien to the ways of rural life, the project aimed at enacting the wishes and ideas of the community.

The commitment and interaction between artist and local people is also relevant in Asunción Molinos's work, who lives and works between Cairo, Egypt, and her hometown of eighty inhabitants, Guzmán, Burgos. In her hometown, she reconstructed an abandoned traditional shepherd hut and transformed it into a darkroom so that the landscape would project itself onto the hut's wall [Fig. 4]. The inside of this shepherd's hut became an observatory from which to look at the scenery, a public meeting space for people to share. By giving it a different use, the people from Guzmán could experience new uses of old places and could value the surrounding landscape in an aesthetic way, beyond the farmers' view of profitability. More recently, Asunción Molinos's work has fo-

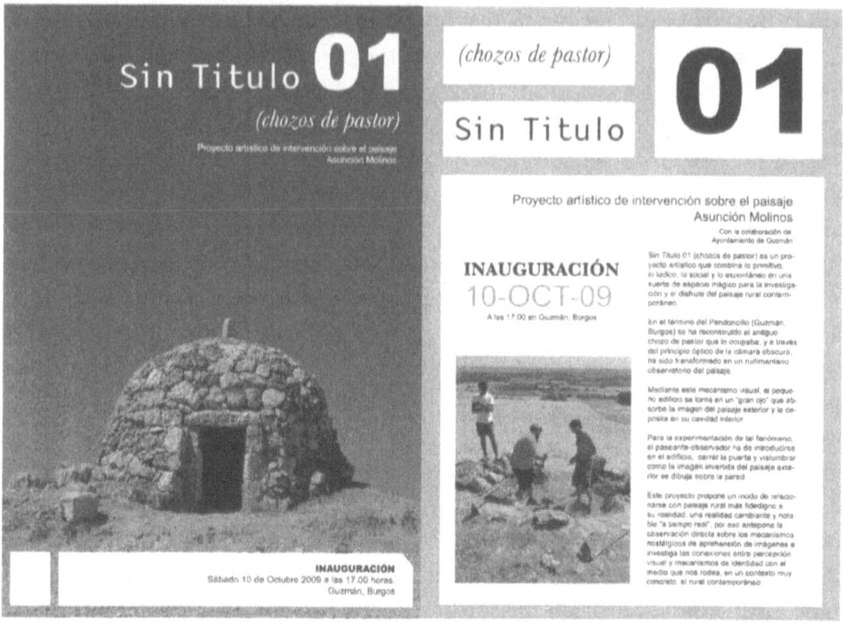

Fig. 4. Asunción Molinos, *Chozos de pastor*, Gúzman, Burgos, 2009. Courtesy of the artist.

cused on food and its agricultural and sociocultural implications, exploring the rural sphere and issues from a transnational approach. According to the artist, her practice is centered on the changes within the rural context, looking at what is being left behind by progress. *Untitled 3 WAM, World Agriculture Museum* (Cairo, 2010) was an installation in which she mixed reality and fiction, recreating the Museum of Agriculture in Cairo. Her installation, which extended over more than six rooms, was meticulously "constructed as a sort of cabinet of curiosities by a team of local artisans, craftspeople, calligraphers, carpenters, and electricians, directed by Molinos" (Pedrosa 306). Her work is very critical of the patented seed banks, and illustrates the unethical world food trade that is responsible for famines. For Molinos, there is enough food for everyone in the world, if only it were properly farmed and distributed.

Another of Molinos's admixture of reality and fiction highlights the inequality of access to food. In *El Matam El Mish-Masry: The Non-Egyptian Restaurant* (2012), she digs in the local area of the restaurant as if she were an archaeologist excavating and looking for ancient Egyptian treasures. This Cairo neighborhood was built on the fertile lands of the Nile Delta; however, the earth found in the uppermost layers is full of wrappers, plastic rubbish and covered by the rubble that has piled up from years of building activity in the area. Aware of the different qualities of food consumed by the Egyptian people, Molinos proposes to cook three menus: one cooked with vegetables that have been irrigated with wastewater and that are the most commonly consumed. Another menu is cooked with first class quality vegetables from the still remaining fertile lands of Nile Delta—most of them being exported and consumed in luxury restaurants. The third menu has a symbolic ingredient, the rubbish collected from the dig. The display underscores that not everyone has access to the same quality of food, some of which is unsafe and consumed by the majority of Egyptians in the area. Local consciousness about the food value is intrinsically related to the regulations of uses of the territory and their political criteria. "The restaurant, as a reading and visualizing machine, will radiate out eventually to a rich past that has been scorched and parched beyond agricultural recognition. [Molinos] will excavate the present for the past in order to produce a future" (Ferguson 104).

Against the patents and monopolies of genetically modified seeds, Spanish artists have also taken action, following the trend of the Seed Savers Exchange (created by Kent Whealy in 1975 in the United States). Fernando García Dory has created a bank of seeds (*Semillas en Red*, 2005) to protect the biodiversity of consumable plants. It is a work in progress based on the free interchange of seeds in danger of extinction among small farmers and artists. As a self-organized structure, this *Semillas en Red* connects people, edible gardens, and computers providing an "ecohacking" system in order to preserve the heritage of the local communities: technology and food are in the hands of the users and consumers. This way of relating two apparently antagonistic worlds, seeds and computers (portrayed as two different moments of cultural development), can be seen also in *Biopollution*, two artistic actions which took place in León (2002) and in Madrid (2003). In both, seeds of Phacellia, a plant with purifying properties, were disseminated next to the private property of a large chemical corporation, Syngenta, devoted to trans-genetic farming; thus, artists acted as "legal bioterrorists" (Matos, *Intervenciones* 213–16), invading the territory as if they were hackers attacking a software program. However, rather than destroying the system, this "attack" would preserve it for the sake of a better future for the human species, since once Phacellia seeds are spread on the genetically modified plantation, they work as a sort of "natural correcting code" that reduces error rate in the developmental human farming.

From 2010 to 2013, García Dory created the *Campo Adentro: Arte, Agricultura, y Medio Rural* (Inland: Art, Agriculture, and Countryside) project to "launch a cultural strategy in support of rural life," as the project's blog states.

Nowadays, *Campo Adentro* is a major project sponsored by the government which "offers the chance to analyze current perceptions and representations of rural life and how these influence the construction of identity . . . . Rural life is that last, silenced but persistent 'otherness,' regarded with apprehension and detachment at some times, and bucolic idealism at others" (Campo Adentro). These projects aim for rural art to emerge and intermingle with urban culture, so that the previous gap between the city and the countryside can become a channel for communication in order to create a more sustainable future.[12] In a reverse mode, rural areas have recently become a solution for opting out of the system and at the moment, there are quite a few artists who are emigrating to the countryside due to the economic crisis in Spain, among them Antonio Ballester, Carmen Cañibano, and Cristobal Martín (Huerta).

Some artists had already been working silently in rural towns or villages since the last decade of the twentieth century, such as Bodo Rau in La Huerta, Salamanca. Rau has organized encounters on culture and ecology, inviting artists such as Helen and Newton Harrison and Herman Prigann (Matos, *Intervenciones* 146). We can also find the work of Iraida Cano in the Arreciado, Extremadura, Sergio Ferrua in Valencia's mountain range, or Lucia Loren in the northeastern mountain range of Madrid. Loren's work at Puebla de la Sierra, Madrid, is particularly interesting inasmuch as she has integrated her work with the natural necessities of the place and the people living there. In her *Bosque Hueco* (2004) (Hollow forest), she stitches together the trunks of trees which have naturally split by sewing up their holes with dried stalks. Her work is so subtle that it imitates natural patterns and animal tracks. The artist understands her work as part of the natural cycle and she argues that it plays with the rhythm of creation-destruction. This connection with the land offers also an aspect of gender, sometimes even with a hint of humor, as Loren is aware of the often oversimplified view that relates motherhood to natural cycles. *Madre sal* (2008) (Mother Salt), Villoslada de Cameros, La Rioja, is an open-air installation of salt sculptures distributed in the land for the cows to lick. Salt is an essential part of the cows' diet and Loren's salt breasts highlight the irony between milking, nursing, and licking [Fig. 5]. Moreover, the title, *Madre sal*, leads to the confusion of the sun (sol) and the salt (sal), thus correlating the patriarchal tradition of the Sun God to breastfeeding.

This interchange between artist and rural life has been, so far, the most representative aspect of ecology in art in Spain. Interesting encounters have taken place in rural areas where the artists meet once a year. In these artistic rural encounters, in Sajazarra or Santa Lucía de Ocón, both in La Rioja, for instance, their works are made in the cultivated lands after the harvest, and they fre-

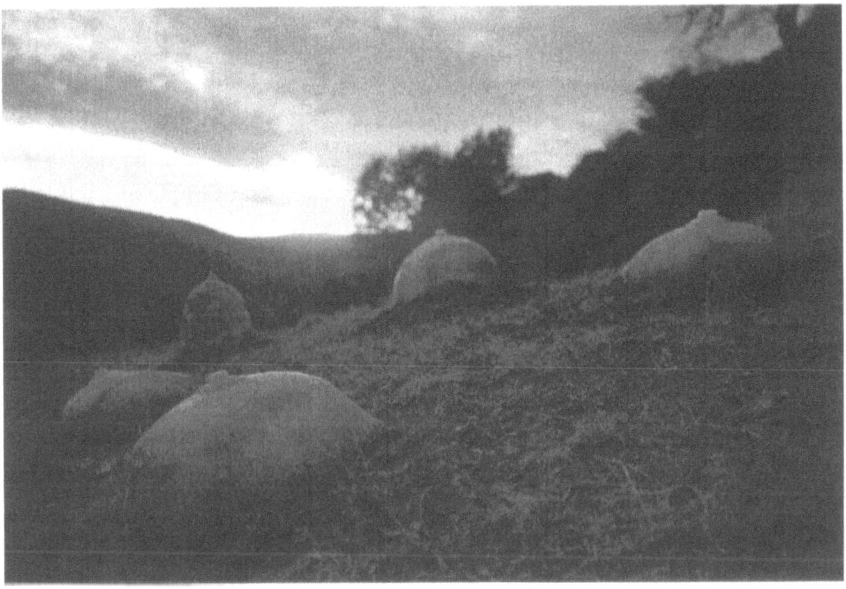

Fig. 5. Lucia Loren, *Madre Sal*, Villoslada de Cameros, La Rioja, 2008. Courtesy of the artist.

quently use the straw and other natural materials left over for artistic purposes (Matos, *Intervenciones* 275–83). Farming has also been used as an artistic creation: the project *Culturhaza* has been founded by farmer-artists such as Protasia Cancho, Agripino Terrón, and Holga Méndez, who have named their activity "Agrolandart" (Cancho 37).

Artivism and Ecology

Photography has been a very effective way to denounce abusive practices in the exploitation of both the human and nonhuman. An almost surreal view of an alien landscape on Earth can be seen in Monserrat Soto's *Invernaderos* (2002). In this photographic series, Soto portrays the greenhouses that extend all over the territory of Ejido, Almería, that grow most of the vegetables for both the domestic Spanish market and exportation. This profitable agribusiness defies the traditional perception of agriculture as a "natural" way of sustainability by producing what is being called "perverse nature" (Miro). The plastic landscape speaks directly of exploitation: the cheap labor of immigrants as well one of the major ecological problems of the area, caused by the huge amounts of pesticides poisoning the aquifers due to their high concentration. Ironically, these sites are next door to the Cabo de Gata Natural Park, so the visitor to Almería experiences the contradiction of contemporary life: on the one hand, the exploited and devastated areas; on the other, the preservation of nature as if it were a vestige from the past that has to be cared for, an example of what is no more, as if this protected area were part of a Museum of Nature. The result is an empty city of plastic with streets and corridors, where human absence provokes a sort of uncanny feeling as in the devastated land of a science fiction film.

Sergio Belinchón is another photographer denouncing harmful practices in the environment. His *Suburbia* (2002) is a photographic series that shows us a devastated, apocalyptic landscape. It is as if Belinchón were a sort of filmmaker, playing with the spectator's imagination, questioning what has happened to the land: a land almost obsessively crisscrossed with highways and construction sites as if the amount of cement were proof of progress (Hernando 82). In Spain, after the reform of land use laws (with Jose María Aznar´s government from 1996–2004), every single piece of land was (and still is) susceptible to being exploited as an urban area, resulting in over-development, one of the major reasons behind the current economic crisis.

Bárbara Fluxá, an artist gathering waste objects from the countryside, oceans, and rivers, criticizes consumerism and campaigns for cleanup. In 2005, she organized a collective cleanup campaign at Esles, Cantabria, during July and August, peak vacation months. The campaign was advertised through posters around the village, asking people to collect any abandoned or discarded object in the nearby area. These objects were later properly classified and taken to the local church, where they were exhibited as a sort of "found treasures," making the village people aware of the waste generated in their own landscape. *Des (h)echos* is a more complex project—and also a work in progress—consisting of several archives from dozens of objects abandoned in the rural environment. As if in an archeological excavation, Fluxá puts into practice a meticulous classifying method: each object gets photographed, without any external manipulation, in the exact place it was found. Also following archaeological techniques, she reconstructs the objects with plaster when broken or incomplete so that they retain their original shape. Finally, they are exhibited in a glass case as if they were part of an ethnographic collection [Fig. 6]: "In the display cabinet, at the museum," writes the artist, "from the cold and distant glass perhaps we can be able to [*sic*] stop time for a while stopping the dizzy pace of our society, to establish and discover new links with our own material culture" (Fluxá). Waste is highly valued by archaeologists and ethnographers since its analysis provides

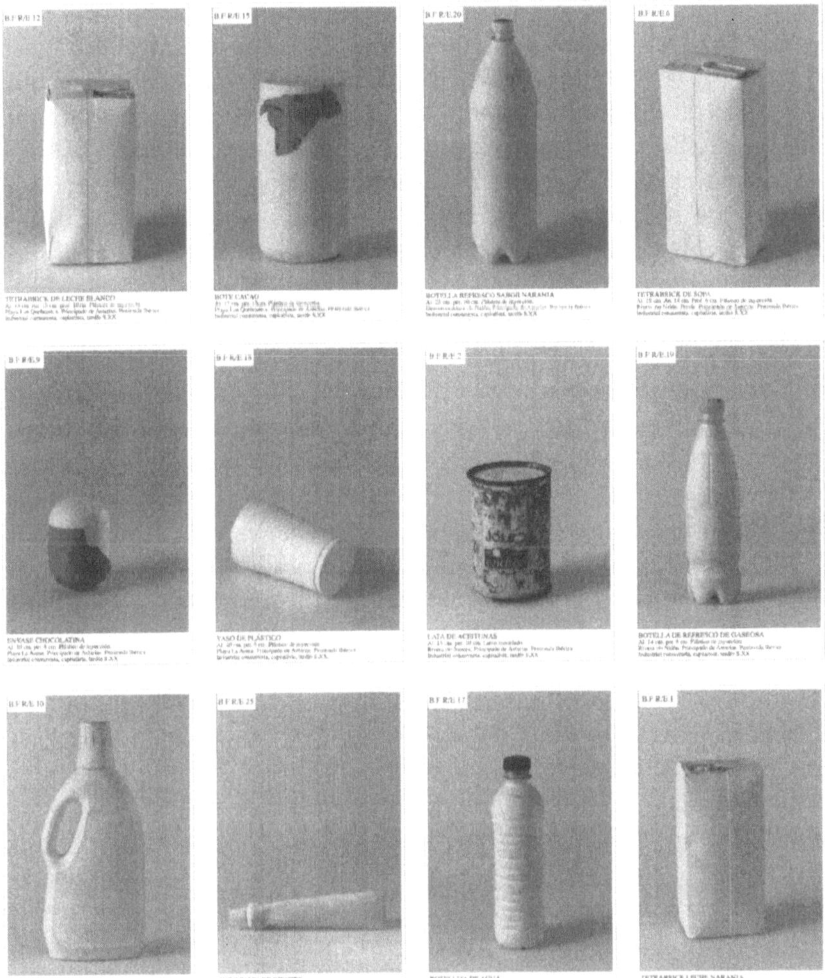

Fig. 6. Barbara Fluxá, *Des (h)echos*, 2005. Courtesy of the artist.

plenty of information about previous ways of living. For Fluxá, the objects she collects are part of our material culture and her exhibit constitutes an effort to understand consumerism and our relationship to the environment.

Similarly concerned, Basurama is a community based in Spain that has been experimenting with different approaches to waste. Mainly formed by architects, Basurama uses waste as a creative resource to produce playful objects or installations. The group conducts workshops, stages public art interventions, creates maps or data visualizations, and works with a wide range of stakeholders, ranging from neighborhood communities and local artists to regional governments. As they describe in their blog, "Through practice, participants become aware of the implications of using waste as a source material: properties, availability, and transportation. The process changes the way designers and communities approach the design project, and provides a better way to understand how to work with what already exists [. . .]." (Basurama). Recently the Basurama project has been extended to R.U.S. (Residuos Urbanos Sólidos; Solid Urban Waste) all over Latin America, operating in cities such as México D.F., San Juan, Puerto Rico, Buenos Aires, Lima, Cordoba, Asunción, and so on (Basura y Espacio). Basurama's approach to ecological artivism is based on collective participation and collaborative work, common strategies nowadays to combat the destruction of the environment. Artivist groups also use the digital web as a platform, so global actions can be taken beyond political frontiers. Art activity has also become global, and may take place in different countries at the same time. This is, for instance, the case of *Guerrilla Gardens*, originally from the United Kingdom but extended now in several countries, among them Spain. Spanish *Guerrilla Gardens* act through a network system on waste grounds, mainly in urban areas, where actors, that is to say, the people who get involved in that specific action, throw earth balls containing seeds, deliberately, albeit illegally, reclaiming nonurbanized gaps in the cities as new sites for living plants, making communities aware of the importance of green spaces.

Luzinterruptus is another anonymous artistic group who began to act in Spanish cities together with neighborhoods to solve specific problems in those areas. This collective started around 2008 with the idea of focusing people's attention on problems that are found in the city and that seem to go unnoticed by authorities and citizens. However, not everything they do has a subversive aim: "Sometimes we simply want to embellish, or to highlight anonymous places or corners that seem special or objects to which we think extraordinary artistic value, although they have been left on the streets for unknown reasons, with artistic intention, by anonymous people" (Luzinterruptus). Citizen collaboration is one of the major features of these artistic groups, as is shown in their collective river created in Caracas by filling plastic bags with water and introducing in each of them toy fish and plants [Fig. 7], thus creating an artificial ecosystem that displays artificially what is natural and needs to be preserved.

This relationship between water and ecological art has centered the attention of other artivist groups. *75% water* is a project by S.D.F. (Sindomiciliofijo)

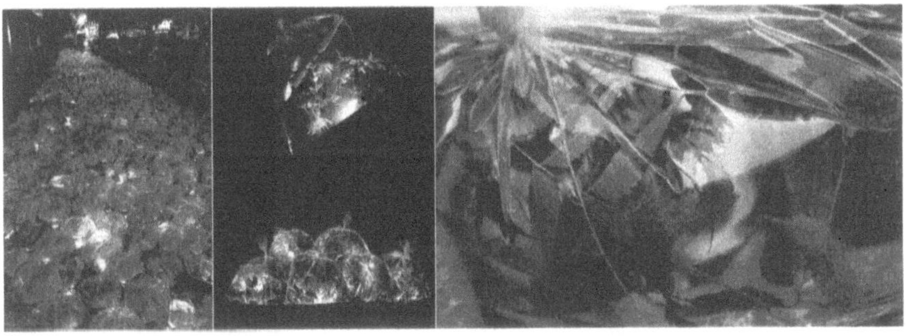

Fig. 7. Luzinterruptus, *Río en Caracas*, Juannary, 2013. Fotografía de Gustavo Sanabria. Courtesy of the artist.

(Without-permanent-home) that hopes to bring attention to the essential importance of water for life on Earth, since three quarters of the planet is water and roughly the same percentage can be found in the human body. For S.D.F., water is one and the same all over the planet since it is globally interconnected, in spite of great distances. Water flows through the planet as veins do throughout the human body; however, we are not always conscious of the basic role that it plays, neither in life nor in how climate change is affecting its distribution and salubriousness. Thus, this art collective has taken to the web as a digital channel to flow *75% Water* over the consciousness of people, requiring their active participation [Fig. 8]. During the last five years and with open public collaboration, S.D.F. has created a sort of atlas with four hundred and fifty displays of water, each coming from different parts of the world: from Australia, the Americas, Asia, Africa, and Europe. Each unit of water is part of its landscape of origin, which is photographed and shown, and exhibited in bottles, each of them placed on a large map that is displayed on the floor of the art gallery. The resulting map has, on the one hand, the intention of reducing cultural, geographical, and cognitive differences of the places of origin, and on the other, to generate a collaborative group of people building strategies to look after rivers and oceans on Earth and expanding awareness of a responsible use of drinking water (S.D.F.). Thus, the exhibit highlights the importance of water, anchored in the very heart of our culture.

Lootz's last work has also dealt with the importance of water. In her project *Dis-Cursos de Agua* (Water Dis-course), started in 2005, she has taken the rivers of the Iberian Peninsula as her artistic material. Specifically, she explores the ways that rivers have built and developed orographic forms. In several lightboxes she shows the water distribution of each river and its tributaries. All of

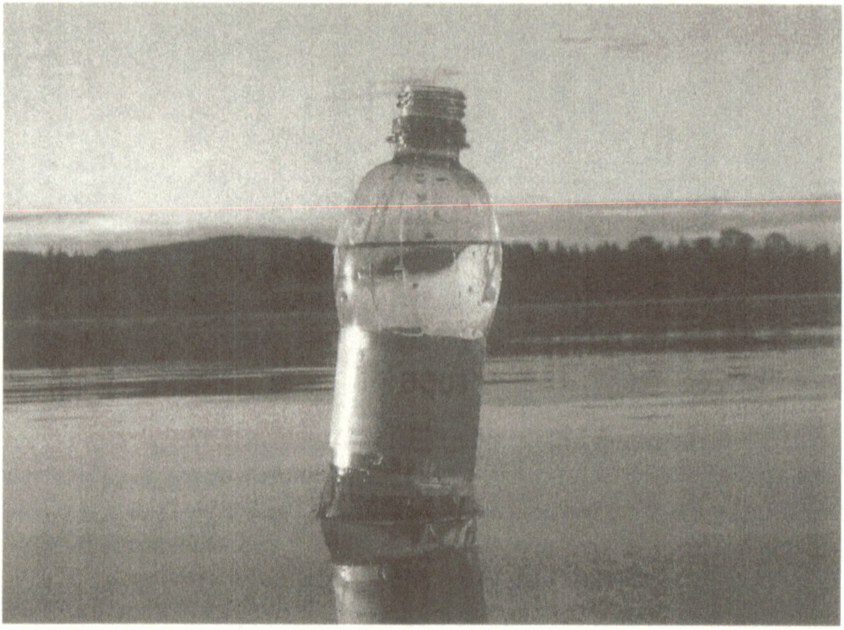

Fig. 8. SDF, *75% water,* water collected from Ivalo river (Finlay and Norway frontier) by Alberto López Pérez. Exhibited at Casa Encendida (Madrid) 2008. Courtesy of the artist.

them form a pattern which has a close similarity to organic structures, "for rivers are living things," Lootz says. The names of the tributaries are described so that the work resembles the digital maps from which the lightboxes were created. But names also have a meaning, and many times they describe features of the places they flow by or the sources they come from. The very course of the water contains a sort of map of its own evolution that Lootz catches in a series of sculptures in marble. "By processing parameters such as changes in the river course, floods, variation in water level, time periods, she created three-dimensional models, 3D archives, which afterwards can be cut in marble by using rapid prototyping tools. Through this she aims to transform temporary sequences in the evolution of riverbeds into sculpture" (Lootz 66). With this project, Lootz calls attention to the water shortages in the world and the political responsibility for caring and distributing that essential element which is in danger of being converted into a commodity. This project emphasizes awareness of that liquid which makes life possible, and particularly in Spain where scarcity of water is one of the more serious threats of climate change.[13]

Conclusion

This essay has highlighted contemporary developments in literature and the fine arts, as well as criticism in Spain, in relation to environmentalist and ecological thought. While literary criticism has become relatively active, much less exists in the field of art criticism; however, the very existence of the research group and grant, "Art and Ecology,"[14] with its close ties to the ecocritical group GIECO, illustrates the growing concern and interest in bringing the humanities into environmental discourse. We have also shown that the main concerns of Spanish artists tend to converge. Peter Barry, in what is the first book of literary theory and criticism to include a chapter on ecocriticism, points out the main difference between United States and United Kingdom ecocriticism: celebratory (going out to nature to seek peace and identity) versus minatory (denouncing environmental threats), respectively (242). Although there are numerous works that celebrate nature, such as Barella's collection of poems that reflect on the importance of the Mediterranean and water for Spanish character, many contemporary artists seem to tend toward a critical attitude, more similar to the British, denouncing capitalist greed and excessive, unsustainable growth, much in line with Marxist trends in the early seventies. Issues such as the abuse of urban development and its subsequent corruption, the loss of rural life, or consumerism and waste management loom large in both literature and artistic events, as has been illustrated. And certainly, one of the more pressing issues is that of water. Excessive growth of intense agriculture and urban development have depleted and contaminated fresh water supplies and every year some region of the country faces severe drought and water shortage. Intense political battles between the countries' regions, where even major political parties break discipline, are fought over control of water rights. In the above-mentioned novel by Rosa Montero, *Lágrimas en la lluvia*, the lack of water becomes a major theme, where people must shower and wash with vapor in order to not use much water, a very expensive commodity in her futuristic Madrid. Eva Lootz's exhibit, like that of many others, focuses on water. The exhibit catalogue brought together art and poetry, evidenced by a poem written by José Parreño for the occasion, "Sutra del Agua" (Sutra of Water):

sangre de planeta
madre de lo verde

que en las alcobas de los troncos
yaces con el sol

y tu hijo es el aire
que respiramos

partera de los huesos
de todos los frutales
músculo transparente
que eleva las maderas
y pule
las montañas
con caricias de siglos

espejo blando que se vuelve vapor
vapor que acaba siendo
el mármol de las nubes
. . . .
tu desapego de las formas
tu diligencia tu discreción tu fuerza

escúchame/cuando
rezo a una gota (Parreño,"Silencio")

(blood of the planet
mother of greenness

you, who lay with the sun
in the alcove of every log
and your son is the air
that we breathe

midwife of the stones
of every fruit tree
transparent muscle
that raises the wood
and polishes
the mountains
with age long caresses

soft mirror that turns to steam
steam that will become

the marble of the clouds
. . . .
your indifference to shapes
your diligence your discretion your strength

listen to me / when I say
my prayer to a drop. (Translation by Ángela Suárez and Guillermo George Hardy)

This poem is one more example of recent Spanish art with environmental concern, one where literature and art merge together. Although it is not one of denunciation, it does highlight the importance of water and the need to respect it. The shape of the poem, as a verbal sculpture, fixed in print, with its lack of defined borders or sentence definition, emulates the continuous ebb and flow of water, its strength and pliability, its essence in all beings. The poem shows the power of art in increasing awareness about nature: a slow but sure rise in both rational and emotional affect.

William Rueckert, in a landmark essay, "Literature and Ecology. An Experiment in Ecocriticism" (1978), suggests looking at poetry (and, we would add, any art) as "stored energy . . . as energy pathways which sustain life." He continues, comparing poems to fossil fuels, but renewable since they come from "ever generative twin matrices, language and imagination." This "stored energy," released upon each reading, consists of an energy pathway to multiple human communities (108). Rueckert continues asserting that "reading, teaching, and critical discourse all release the energy and power stored in poetry so that it may flow through the human community"; thus, these activities constitute inexhaustible energy transfers to the language centers and creative imagination of the readers. Teaching and critical discourse increase and intensify the process, improving the pathway to the community. Rueckert claims that poetry (or any artistic creation) is like green plants which "arrest energy," transform the matter to another order, and help to create a "self-perpetuating and evolving system," one which creates "creativity and community" (110–11). Thus writers and artists, teachers and critics, infuse readers and spectators with creative energy; they stimulate our perception and sensitivity, and reach our creative imagination which is one of the most effective ways to transform our relationship with the environment. This role of the humanities has not always been acknowledged, particularly in Spain, and this article hopes to reflect the timid beginnings and increasing tide of environmental concerns in Spanish literature, art, and research in the humanities.

Notes

The research for this article was funded by a Research project of the Ministry of Science and Innovation (Ref. HAR2011-23678) and by the CLYMA project (Ref. IF 2011-009) of the Franklin Institute of the University of Alcalá.

1. See Flys Junquera, "Blanca Bosco's . . ."
2. For a more detailed analysis of the limits of growth and recent Spanish novels, see Luis Prádanos "Euro-Mediterranean . . . "
3. See Prádanos, "Decrecimiento . . . "
4. i.e., Niall Binns published *Callejón sin Salida* on Latin American poetry and Manuela Palacios has two articles on environmental issues comparing Irish and Galician women poets.
5. See Margarita Carretero on Pereda and Jose Manuel Marrero on Francisco Gónzalez.
6. To subscribe, see http://www.rediris.es/list/info/redcyma.html
7. For instance, Nacho Criados's *Ajustes* (1973) (Adjustments) shows a resemblance to Dennis Oppenheim's works, while Angels Ribé´s *Intersección de agua* (1969) (Water Intersection) evokes Hans Hacke's *Condensation Cube* (1963) and Josefina Miralles's *Mujer árbol* (1973) (Tree Woman) refers to Ana Mendieta's work. The former is concerned with the way the body interacts with nature; the second displays the natural process of matter, such as the performance of water drops on a glass surface. Also in the early 1970s, two academic publications, Simon Marchan's book *Del arte objetual al arte de concepto* (1972) and Victoria Combalia's *La poetica de lo neutro* (1975), contributed to the spread of foreign artist's environmental sculptures in Spain. Coincidentally, in these years the artist Wolf Vostell visited Los Barruecos (Malpartida), an exceptional granitic area in Cáceres that impressed him so much as to name it "a Nature work of art" that now is a Museum (Vostell-Malpartida Museum n.p.).
8. In 1997 José Albelda and José Saborit, teaching at the Fine Arts College of the University of Valencia published *La construcción de la naturaleza*, where ecological art is already discussed and encouraged. In the artistic context the book on *Land Art* (1998) by T. Raquejo helped to divulge the tendencies of art in nature as well as those few Spanish artists who were already concerned about the environment. Touching specifically on ecological aspects, one needs mention in the first place the work of Javier Hernando Carrasco "Visiones de la naturaleza: el arte y la sensibilidad ecológica" and, secondly *Intervenciones artísticas en espacios naturales. España (1970–2006)* (Artistic Interventions in Nature. Spain [1970–2006]), a Ph.D. dissertation by Gregoria Matos Romero that contains valuable information about Spanish environmental art works. Moreover, from September 19 to December 10, 2006, the exhibit *Naturalmente artificial. El arte español y la naturaleza 1968–2006* (Naturally artificial. Spanish Art and Nature), curated by Jose Parreño at the Esteban Vicente Contemporary Art Museum (Segovia) took place. In 2008 an interesting collection

of four volumes—entitled *Construcción y Destrucción del Territorio. Memoria de lugares Españoles*, was edited by Aurora Fernandez Polanco, Cristina Peñamarín and Magdalena Mora and published by the Complutense University—with the collaborations of the Ministry for Environment, Marine and Rural Affairs. In the biological context Joaquin Araujo's book, *La Cultura Ecológica* (Ecological Culture) is also of great interest since creates a bridge between science, culture, economic habits and organic farming. From 1994 The Beulas Foundation (Huesca) played an important role in bringing art closer to nature through landscape under the projects of Javier Maderuelo and Teresa Luesma at the C.D.A.N. (Centro De Arte Naturaleza) in *Arte y Naturaleza* (Art and Nature) first, and *Pensar el paisaje* (Think Landscape) later. Since then many other projects have emerged to interact with nature. See for instance: *Idensitat* created in Calaf in 1999 by Ramón Parramon, (Idensitat Web Site); and the Observatori del Paisatge of Catalunia (Catalonian landscape observatory), directed by Joan Nogué (Observatori Web site).

9. For example, Agustin Ibarrola´s *Bosque de Oma* (Oma's Forest) (1930–) (Fig. 1) in the natural reserve of Urdaibai (Biscay), and Perejaume´s art (1957–) (St. Pol de Mar, Catalonia). The *Biblioteca del Bosque* (Forest Library) by Miguel Angel Blanco (1958–) could be considered as an earlier example of ecological art. Blanco started to recollect specimens from Fuenfría (Guadarrama Mountains, Madrid). It now has more than one thousand box-books where natural materials, such as minerals or leaves, are kept like poetic examples of an ecosystem. This project is still in progress.
10. There are hardly any examples which were supported by local governments or town councils. In the 1990s one can find the first projects such as the one by Luis Ortega and Iraida Cano made for Valdelamusa, Huelva, an old loading bay for pyrites which now is placed in the middle of the urban area of the town, separating three different neighborhoods. The high level of contamination in the soil forced the artists to come up with a different solution than growing green plants. So, in the design of the park that they presented as a project, they made good use of the remains of the old Pyrite mine and reused them aesthetically to take advantage of the colorful oxide minerals remaining all over the place (Raquejo 90). The visual improvement joined the practical use of the area: as a public space for leisure and walking around from one neighborhood to the other. The project was designed at the request of the government of Andalucía in 1997 but unfortunately it has not been carried out yet.
11. The photographs of one of the exhibits by Carma Casulá can be found in the Autumn issue of *Ecozon@* 4.2 (2013): 206–18.
12. For three years, the young artists selected by the Campo Adentro Committee, have been spending time at different rural villages through a resident program where they have developed an artistic project regarding the social and cultural peculiarities and site-specificity of the place. To see the projects: Campo Adentro Web site.
13. For other projects in Spain where artistic actions has taken in order to defend irrigated land against real estate speculation see, for instance, *Estar en la punta*

an exhibition which shows years of activism (1994–2000) of La Unificadora, an association to preserve the vegetable garden in La Punta, Valencia, see Albelda, Collette, and Lorenzo.

14. Research group founded in 2012 and funded by the Ministry (see note 1).

Works Cited

Albelda, José, N. Collette, and P. Lorenzo, eds. *Estar en la Punta*. Valencia: Universidad Politécnica, 2000.

Albelda, José, and José Saborit, *La construcción de la naturaleza*. Valencia: Valencia Generalitat, 1997.

Araujo, Joaquín. *La Cultura Ecológica*, Lanzarote, Teguise: Fundacion Cesar Manrique, 1995.

Araujo, Joaquín. *Bosque de Bosques*. Dir. Oscar Miranda. Vistapalabra. 2010. Web. 8 Aug. 2013.

Ardillo, José. *El salario del gigante*. Logroño: Pepitas de calabaza, 2011.

Arribas, Diego. *Minas de Ojos negros. Un filón por explotar*. Calamocha: Centro de Estudios Jiloca, 1999.

Arribas, Fernando. "El arte de la siega." *Naturalmente artificial. El arte español y la naturaleza 1968–2006*. Ed. J.M. Parreño. Segovia: Museo de Arte Contemporáneo Esteban Vicente Segovia. 2006. 93–98.

Arte y Ecología. *Arte y Ecología*. 2014. Web. 7 June 2014.

Barella, Julia. *Praderas de posidonia*. Madrid: Huerga y Fierro, 2013.

Barry, Peter. *Beginning Theory: An Introduction to Literary and Cultural Theory*. 3rd ed. Manchester: Manchester University Press, 2009.

Basurama. *Basurama*. 2014. Web. 7 June 2014.

______. *Basura y espacio público en Latinoamérica 2008–2010*. Salamanca: Editorial Delirio, 2011.

Binns, Niall. *¿Callejón sin salida? La crisis ecológica en la poesía hispanoamericana*. Zaragoza: Prensas Universitarias de Zaragoza, 2004.

Campo Adentro. *Campo Adentro*. 2014. Web. 12 May 2013.

Cancho, Protasia, and Agripino Terrón y Holga Méndez, *La misma vida como*. Exhibition catalogue. Jaén: Universidad de Jaén, Vicerrectorado de Extensión Universitaria, 2012.

Carretero González, Margarita. "El 'ecologismo reaccionario' de José María de Pereda en *Peñas arriba*". *Ixquic* 2 (2000): 96–112.

Casal, Paula. *Martina y el mar*. Madrid: Hiperión, 2007.

Casulá, Carma. "In the Mouth of Filomena." *Ecozon@: European Journal of Literature, Culture, and Environment*. 4.2 (2013): 206–18. Web. 7 June 2014.

Centro de Arte y Naturaleza. *Arte y Naturaleza*. 2014. Web. 7 June 2014.

Cesar Manrique Foundation. *Cesar Manrique*. 2014. Web. 24 Feb. 2013.

Chirbes, Rafael. *Crematorio*. Barcelona: Anagrama, 2007.

Cobos Wilkins, Juan. *El Corazón de la tierra*. Barcelona: Random House Mondadori, 2001.

Combalia, V. *La poética de lo neutro*. Barcelona: Seix Barral. 1975.

Cuadri, Antonio, dir. *El Corazón de la tierra*. 2007. Film.

Culturhaza. *Culturhaza*. 2014. Web. 7 June 2014.

Delibes, Miguel. "Discurso de entrada a la RAE." *Educación Ambiental*. 1975. Web. 20 May 2013.

"Ecofeminismo/s: mujeres y naturaleza" *Feminismo/s*22 (Dec. 2013).

Fergusson, Bruce W. "El Matam El Mish-Masry (The non-Egyptian restaurant)." Generación 2013. *Proyectos de arte Caja Madrid*. Madrid: Fundación Caja Madrid, 2013. 101–104.

Fernández, Joaquín. *El Ecologismo español*. Madrid: Alianza Editorial, 1999.

Fernández Polanco A., M. Mora, and C. Peñamarin, eds. *Construcción y destrucción del territorio. Memoria de lugares españoles*. 4 vols. Madrid: Universidad Complutense. 2009.

Ferrer, Virginia. *Recuerda Mundo. Novela ecológica en la tierra de Oliva Sabuco*. Barcelona: Sirpus, 2008.

Fluxá, Bárbara. *Bárbara Fluxá: Artista* Plástica. 2013. Web. 13 Aug. 2013.

Flys Junquera, Carmen. "Blanca Bosco's 'loving eye': An Ecocritical Approach to Juan Cobos Wilkins' *El Corazón de la tierra*." *Literature, Ecology, and Ethics: Recent Trends in Ecocriticism*. Eds. Timo Müller and Michael Sauter. Heidelberg: Universitatverlag, 2012. 171–87.

Flys Junquera, Carmen, Jose Manuel Marrero Henríquez, and Julia Barella, eds. *Ecocríticas: Literatura y medio ambiente*. Madrid: Iberoamericana Vervuert, 2010.

Flys, Carmen, Irene Sanz, Monserrat López, Esther Laso y León, eds. *Actas del III Congreso de EASLCE. "Paisajes culturales: herencia y conservación."* Alcalá de Henares: Servicio de Publicaciones de la Universidad de Alcalá, 2009.

GIECO. "GIECO: Grupo de investigación en ecocrítica." *Instituto Franklin de Investigación en Estudios Norteamericanos*. 2014. Web. 7 June 2014.

González Calvo, Olga. *The Land Before Them: Three Women's Perspectives of the Frontier Experience in North America*. Ph.D Diss. University of Alcalá, 2010.

Hernando Carrasco, J. "Visiones de la naturaleza: el arte y la sensibilidad ecológica." *Tendencias del arte, arte de tendencias a principios del siglo XXI*. Eds. J.A.Ramirez and Carrillo, J. Madrid: Cátedra, 2004. 53–84.

Huerta, Rocío. "Cuando la ciudad no es para el arte" *El País Cultural*, 9 Aug. 2013. Web. 7 June 2014.

Idensitat. *Idensitat*. 2013. Web. 7 June 2014.

López, Marta. "Julio Llamazares: dándole voz a la naturaleza en *La lluvia amarilla*." *Ixquic* 2 (2000): 113–23.

López Múgica, Montserrat. *Lectura ecocrítica de la obra de Charles-Ferdinand Ramuz.* Ph.D. Diss. UNED, 2006.

López Pacheco, Jesús. *Ecologas y urbanas.* Vitoria-Gasteiz: Bassarai, 1996.

Lootz, Eva, *Dis-cursos de Agua.* Exhibition catalogue. Burgos: Centro de Arte Caja Burgos, 2012.

Luzinterruptus. *Luzinterruptus.* 2013. Web. 6 May 2013.

Llamazares, Julio. *La lluvia amarilla.* Barcelona: Seix Barral, 1988.

Maderuelo, Javier, ed. *Actas. Arte y Naturaleza.* 5 vols. Huesca: Diputación Provincial de Huesca. 1995–1999.

_____. *Pensar el paisaje.* Huesca: 5 vols. Diputación Provincial de Huesca. 2006–2010.

Marchan Fiz, S. *Del arte objetual al arte de concepto.* Madrid: Alberto Corazón. 1972.

Marqués, Josep Vicent. *Ecología y lucha de clases.* Madrid: Ediciones Zero, 1978.

Marrero Henríquez, José Manuel, ed. *Pasajes y paisajes: Espacios de vida, espacios de cultura.* Las Palmas de Gran Canaria: Servicio de Publicaciones de la Universidad de Las Palmas de Gran Canaria-Gabinete Literario, 2006.

_____."Ecocrítica e hispanismo." *Ecocríticas. Literatura y medio ambient. Ecocríticas: Literatura y medio ambiente.* Eds. Carmen Flys Junquera, Jose Manuel Marrero Henríquez, and Julia Barella. Madrid: Iberoamericana Vervuert, 2010.: 193–217.

_____, ed. *Literatura y sostenibilidad en la era del antropoceno.* Las Palmas de Gran Canaria: Mapfre Guanarteme, 2011.

_____."De ecocrítica e hispanismo y de Francisco González Díaz, apóstol del arbolado modernista". *Tropelías* 12–14 (2001–2003): 293–312.

Matos Romero, G. *Intervenciones artísticas en espacios naturales. España (1970–2006).* Ph.D. Diss. Madrid: Universidad Complutense de Madrid, 2008.

_____. "Un posible recorrido por espacios naturales a través de intervenciones artísticas." *Naturalmente artificial. El arte español y la naturaleza 1968–2006.* Ed. J.M. Parreño. Segovia: Museo de Arte Contemporáneo Esteban Vicente. 2006. 15–73.

Miro, Neus. "Naturalezas perversas." *Montserrat Soto.* 2013. Web. 4 July 2013.

Montero, Rosa. *Lágrimas en la lluvia.* Barcelona: Seix Barral, 2011.

Nerter 15–16. "Ecocrítica." Verano-Otoño 2010.

Novo, María. "Ecoarte" *Ecoarte.* 2013. Web. 13 Jul. 2013.

Oliva, Juan Ignacio, ed. *Realidad y simbología de la montaña.* Alcalá de Henares: Universidad de Alcalá, 2012.

"Observatorio." *Observatori del Paisatge.* 2013. Web. 12 July 2013.

Palacios, Manuela. "Landscape in Irish and Iberian Galician Poetry by Women Authors." *Comparative Literature and Culture* 13.5 (2011): Art. 7.

_____. "The Course of Nature. An Ecofeminist Reading of Irish and Galician Women Poets." *Writing Bonds: Irish and Galician Contemporary Women Poets.* Ed. M. Palacios and L. Lojo. Oxford: Peter Lang, 2009. 77–96.

Paredes, Jorge, and Benjamin McLean. "Hacia una tipología de la literatura ecologista en el mundo hispano." *Ixquic* 2 (2000): 1–37.

Parreño, J.M., *Naturalmente artificial. El arte español y la naturaleza 1968-2006*. Segovia: Museo de Arte Contemporáneo Esteban Vicente, 2006.

_____. "Silencio Blanco/Canción del agua." *Dis-cursos de agua*. Exhibition catalogue. Burgos: Centro de Arte Caja Burgos, 2012. 69–77.

Pedrosa, Adriano. "Asunción Molinos Gordo, Townhall." *Art Forum* (May 2011): 306.

Pérez Abad, Miguel. "Miguel Delibes: ¿el primer verde? Una lectura ecocrítica de su obra." *Ixquic* 2 (2000): 124–46.

Prádanos, Luis. "Decrecimiento o barbarie: ecocrítica y capitalismo global en la novela futurista española reciente." *Ecozon@: European Journal of Literature, Culture, and Environment* 3.2 (2012): 74–92. Web. 7 June 2014.

_____. "Toward a Euro-Mediterranean Socioenvironmental Perspective: The Case for a Spanish Ecocriticism." *Ecozon@: European Journal of Literature, Culture, and Environment*. 4.2 (2013): 30–48. Web. 7 June 2014.

Pratt, Dale, and Barbara Gordon. "The Environment and Nineteenth- and Twentieth-Century Spanish Literature." *Literature of Nature: An International Sourcebook*. Ed. Patrick D. Murphy. Chicago: Fitzroy Dearborn, 1998. 248–56.

Raquejo, Tonia. *Land Art*. Madrid: Nerea, 1998.

Rhei, Sofia. *Bestiario microscópico*. Sportula, 2012.

Riechmann, Jorge. *Con los ojos abiertos. Ecopoemas 1985–2006*. Tenerife: Ediciones Baile del Sol, 2007.

_____. *Un mundo vulnerable: ensayos sobre ecología, ética y tecnociencia*. 2nd ed. Madrid: Cataráta, 2005.

Rueckert, William. "Literature and Environment. An Experiment in Ecocriticism." *The Ecocriticism Reader. Landmarks in Literary Ecology*. Eds. Cheryll Glotfelty and Harold Fromm. Athens: University of Georgia Press, 1996. 105–23.

Sampedro, José Luis, and Olga Lucas. *Cuarteto para un solista*. Barcelona: Plaza Janés, 2011.

Sanz Alonso, Irene. *Redefining Humanity in Science Fiction: The Alien from an Ecofeminist Perspective*. Ph.D. Diss. University of Alcalá, 2014.

SDF. *Sindomiciliofijo*. 2014. Web. 7 June 2014.

Tafalla, Marta. *La Biblioteca de Noé*. Barcelona: Herder, 2006.

Tusset, Pablo. *Oxford 7*. Barcelona: Destino, 2011.

Vázquez-Figueroa, Alberto. *El mar en llamas*. Madrid: Martínez Roca, 2011.

◆ 2

Nunca Máis: Ecological Collectivism and the *Prestige* Disaster

John H. Trevathan

> Las casas antiguas tienen un pozo negro, hemos aprendido que las democracias tienen alcantarillas y los Estados en general tienen una fosa en algún lugar apartado para los residuos tóxicos. Galicia es la fosa donde están los residuos tóxicos resultantes de la extraña química que precipitó la democracia española.
>
> —"Esperpentos en la niebla," Suso de Toro

> (Old houses have septic tanks; we have learned that democracies have sewers and States in general have pits in secluded places for toxic waste. Galicia is one pit where there is waste resulting from the strange chemistry that created Spanish democracy.)

The renewed interest in contemporary Galician cultural studies has paid close attention to the diverse meanings of the term *galeguidade* (Galician-ness) within the politically and culturally pluralist climate of post-Franco Spain (Hooper). While many have contextualized Galician identity globally through the region's waves of emigration, literary innovations and popular culture, little work has been done to situate Galicia through the region's recent ecological crises caused, in large part, by its close geographic contact with the Atlantic Ocean and international shipping routes. Isabel Castro Vázquez has begun to consider the important role of ecology in *Reexistencias* (2007), a dissertation on Galician author Manuel Rivas's ecological approach to language, bodies, and environments in his work.[1] María do Cebreiro Rábade Villar has also observed that several protest and cultural movements in Galicia have transformed nature into a productive political sphere. She considers how the natural sphere as a space of resistance begins to construct what she describes as "critical ecology." This resistance develops a new alliance between human life and the surrounding world.

She observes: "Its fortune resides in the fact that it is a faithful image of the terrain without being completely unfaithful to the rootlessness which involves the discovery of a—sometimes irretrievable—distance between subject and world" (91). Such approaches offer a new understanding of how local identity in the twenty-first century relates to the global networks of environmentalism and capitalism.

The 2002 *Prestige* oil spill is the largest environmental disaster in Spanish history and also the largest maritime oil spill in Western European history. As 125 tons of heavy petroleum washed ashore every day after the *Prestige* tanker sank 275 kilometers off the coast of Galicia, multiple levels of the Spanish government refused to acknowledge the extent of the unfolding disaster, as we shall soon examine in greater detail. This event is not an isolated incident. In fact, there have been eight major ecological accidents in the region since the 1970s, including the mass dumping of nuclear waste and several oil spills.[2] It is conceivable to connect these major environmental events to the (mis)function of global capital and its inherent risks. It has become clear that few will be held responsible for the environmental and economic devastation and that little has been done to prevent another disaster. In November 2013, the Audiencia Provincial de A Coruña handed down a judgment of impunity for the mere three individuals taken to trial over the accident. Only the hard-of-hearing, elderly Greek captain was sentenced to jail time (which he will not serve) while all prospects of reparations and prosecutions were dismissed due to "lack of evidence" of culpability. After eleven years, what became clear in the court's ruling is that little has been done to prevent another disaster of equal or greater magnitude and that, as Galician author Manuel Rivas quipped shortly after the ruling, "El *Prestige* se hundió porque quiso" (Rivas) (The *Prestige* sank because it wanted to). Environmental justice, on the view of this court, is a virtually nonexistent prospect. Impunity, as the Spanish environmentalist Juan López Uralde pointed out, begets future crime ("Otro *Prestige*").

In light of this ruling, one wonders what more needs to be said about the *Prestige* case, other than marking it as a major failure of the environmental justice movement in Spain. Reflecting on the term *Unzeitgemäss* (untimeliness), Nietzsche describes a present-day project only to be properly understood in the future. He writes: "for as long as that for which it is always time, and which the present time has more need of than ever, continues to count as untimely—I mean: telling the truth" (55). Considering the long-awaited outcome of this case, one might consider the importance of environmental justice in Spain as a critical set of questions we inevitably postpone: the timeliness of untimeliness. The observer must ask, then, what questions does the *Prestige* raise for environmentalism in Spanish, European, and global contexts? What truths, if any, does it reveal?

The wake of the *Prestige* event created two opposing narratives about the spill. An official response of denial and censorship and a second known as *Nunca Máis*, a democratic protest movement that gathered popular support to defend the ecological fragility of the region and, in so doing, demonstrated an alternative version of Galician identity in light of being entangled in this ecological disaster. Surveying the journalistic writing of Manuel Rivas, I contend that studying these two narratives in juxtaposition is an exposure of the fault-lines at work in the Spanish political system, which themselves are indicative of problematic relationships to global capital.[3] We shall see, then, two glimpses into the mechanisms that undergird the global movements of capital and how they affect the vulnerable maritime ecosystem of Galicia's *Costa da Morte.*

The official response, alluded to above, carried out by the right-wing Aznar and Fraga governments, chose to avoid and cover up the profundity of the crisis on regional and national nevels. The government never used the words *marea negra* (oil spill). In fact, Spain's current prime minister, Mariano Rajoy (at the time deputy prime minister) officially denied that an oil spill had occurred and has never retracted the statement.[4] Instead of referring to spilled oil as spilled oil, Rajoy metaphorically described it as "hilillos de plastilina" (little trails of clay); in effect, this verbally transforms a disaster with weighty ramifications into a minor incident (Moreno). Other public figures, such as the former Minister of Public Works and Transport Francisco Álvarez-Cascos, were simply absent. (Álvarez-Cascos had gone hunting because he deemed his participation as "frivolous.") Furthermore, many larger media corporations and interest groups, most overtly TVE, aided this denial by not airing footage of the coastlines as oil washed up in the *Costa da Morte*; the shadowy right-wing group *Manos Limpias* accused the protest movement of misusing public funds and embezzlement.[5] The subtext of this official narrative maintains a sense of normalcy; that is to say, it denies that Galicia's ecosystems and economy have been damaged by a human-induced disaster. While this denial is an obvious blunder against democratic transparency, the government's position also fails to highlight the slow violence that will continue to unfold for decades after the spill.

As one might glean from Rajoy's metaphor, this narrative has a decisively literary component. In one essay dedicated to the *Prestige* crisis, Rivas refers to North American journalist Greg Palast's story from the 1989 *Exxon Valdez* spill about the "miracle barrel" (*A cuerpo* 268). Palast discovered that a "barrel" of clean water had been substituted for the contaminated water sampled from the spill site, rendering the test results clean when they should have shown oil contaminants. The purified water-text took the place of the contaminated one. "The facts" of the spill actively obstructed any revelation about the ecological state of marine life and local communities. Insofar as politicians did not feel obliged to resort to proof and instead simply denied the incident, there was not an exact parallel event to Palast's miracle barrel in Galicia. Yet Rivas contends that Galicians were *inside* a miracle barrel. Public officials and media outlets discursively replaced the real contaminated beaches with imaginary clean ones. The obvious effect of this narrative is the shrouding or silencing of facts; that is to say, the iconic pastoral image of a rural Galicia obscures the truth that its ecosystems and communities were in a vulnerable position vis-à-vis international shipping routes of the petrol industry. Furthermore, the persistence of the miracle barrel narrative did not dissipate even as petroleum washed up on the beaches of *Costa da Morte*.

The ineptitude of the government's response also carried over into contemporaneous politics around cultural policies created by the former Galician President Manuel Fraga and his ideology, what came to be known as *fraguismo*. From 1989 until 2005, the government of the Franco-era minister Fraga dominated the region's politics.[6] One goal of *fraguismo* was to support certain kinds of ideologically "neutral" cultural production, specifically avoiding any references to Galician nationalism. As soon as Fraga was elected as President of the Xunta of Galicia, he found himself challenged by, among others, the editors at the cultural magazine *Tempos novos*, who accused Fraga of neutralizing Galician culture. In part, *fraguismo* perpetuated what the magazine called *autocompracencia esencialista* (essentialist self-satisfaction), which represents rural lifestyle akin to what is described above through some forms of *galeguidade*. María Reimóndez contends that Fraga's "*galeguidade*" did not advance an international consciousness that Galicia was (and is) an autonomous community within Spain (Hooper 196). On the contrary, its presentation of Galicia was a perpetuation of Spain's rural backwoods, open for tourism but devoid of its own cultural space.

One observes a similar strategy in the government's response to the oil spill, which leaves these sinister aspects of the region's unique relation to the dark side of international capital buried underneath the Galician verdant pastures and iconic seascapes. This ideological technique can be better comprehended through what poet Joyelle McSweeney has called the necropastoral: a literary genre that exposes the peaceful as the sinister or the verdant as the polluted. In a blog post entitled "Necropastoral, or, Normal Love," McSweeney writes: "The Pastoral, like the occult, has always been a fraud, a counterfeit, an invention, an anachronism. However, as with the occult, and as with Art itself, the fraudulence of the pastoral is in direct proportion to its uncanny powers." Instead of glorifying the problematic mechanisms of the global economy, the necropastoral genre seeks to obscure them by, for instance, refusing to address the presence of petroleum on Galician beaches. What is the "power" or effect of this obfuscation? On the one hand, the pastoral has the capacity to bury and obscure the physical and psychological wounds of ecological trauma. Nature becomes ideologically mobilized as a weapon of disguise and deceit, created in order to avoid or divert blame for the fallout. The necropastoral, then, minimizes the perception that *we* manufactured the risks that created the disaster. This utilization of the pastoral deemphasizes the idyllic image's relationship to actual landscapes and communities. In the *Prestige* case, the government's denial of the oil spill, never using the words *marea negra*, is an attempt to bury the necropolis within the pastoral, the great womb of Mother Nature. The pastoral here indicates a desire to reassert a pristine, unblemished, and unchanging nature *even in the midst of modernity's collapse*. The point is, ultimately, that the pastoral cannot conceal modernity's toxicity. It inevitably seeps out.

To understand the "uncanny power" of the necropastoral, it is useful to consider the risks that such an ideological strategy seeks to obscure. Environmental disasters require an important distinction between what Anthony Giddens has described as "external" and "manufactured" risks. The external risk of disaster is "experienced as coming from the outside" of a given community (26). A hurricane, for instance, is perceived as "natural" or "external" because it is a possible outcome of living near a coastline. A plague is likewise viewed as a consequence of coming into contact with contaminated animals or goods (or, for others, caused by divine intervention). No matter the exact source, external risk dictates that we cite the blame as stemming from outside of human communities. Conversely, manufactured risk originates from human activity. Giddens explains that recently, "we started worrying less about what nature can do to us, and more about what we have done to nature" (27). For example, on March 11, 2011, the Tohoku earthquake and tsunami triggered major equipment failures at the Fukushima Daiichi nuclear plants. As if the "external" disaster were not enough, the installation of nuclear facilities in such a location created "manufactured" risk far beyond what the force of a massive wave could do on its own. Indeed, the incurred damage to Fukushima Daiichi *entangled* different communities across the world as radioactive contaminants spread.[7] This sort of entanglement is increasingly global in scale. Manufactured risks are not merely dangerous for one particular location, but rather constitute dangers for entire ecological networks. As Giddens himself points out, events such as hurricanes, plagues, and tsunamis underscore the fact that these risk outcomes do not strictly affect nature or human communities, but rather the shared space of humans and nonhumans.

The *Prestige* incident qualifies as a discernible manufactured-risk situation. According the U.S.-based Ship Structure Committee, the *Prestige* was a single hulled oil tanker damaged in rough seas ("Ship Structure"). After controversially being refused safe harbor in Spain, France, and Portugal, the ship was towed toward open sea on November 13, 2002, in the hopes of avoiding a spill near the Spanish coast (García-Olivares 537). On November 19, the craft split in two and sank about 275 kilometers west of Vigo in international waters. Subsequently, an estimated 61,600 tons of oil spilled into the sea (eighty percent of the ship's 77,000 tons). The cause of the initial damage is not known, nor is it likely to be known because the *Prestige* is submerged in two miles of water (García-Olivares 537). Yet the U.S.-based Ship Structure Committee report suggests that the initial flooding may have been due to "fatigue of welded plates" or "sustained side shell damage." The ship, however, was perfectly within industry norms and should not be conceived as a rogue object or a limit case. Instead, the *Prestige* should be considered risky insofar as it is a typical, everyday object. While such objects typically remain invisible and silent, occasionally they speak through their unstable aspects. What this spill reveals is the dark side of these everyday risk objects and our intimacy with them. Bruno Latour argues that these *are* risk objects precisely because they are deeply entangled in our economies and communities (22–23).

In the end, over a hundred miles of Spanish, French, and Portuguese coastlines were contaminated with highly toxic oil and an estimated cleanup cost of €2.5 billion (the financial equivalent of the 1989 *Exxon Valdez* spill in Prince William Sound, Alaska) (Garcia 19). Moreover, analysts have suggested that the spill cost the costal economy about €71 million (García-Olivares 533).

The *Prestige* is also emblematic of the common practice known as "flags of convenience." The 26-year-old ship was built by Hitachi Shipbuilding and Engineering in Japan and sailed under a Bahamas flag. It was registered in the port of Nassau and owned by a Liberian corperation (Mare Shipping), managed by the Greek company Universe Maritime and insured by the London Steamship Owners Mutual Insurance. This litany of facts expresses the almost absurd sense in which the *Prestige* is an object *inscribed* with global capital. While its port of call is multifarious, the ship's true home seems to be nowhere, circulating along the network of global oil trade. The *Prestige*, then, becomes paradigmatic of systemic risk within late capital: its danger is from nowhere potentially affecting anywhere.

Challenging the official story, *Nunca Máis* sought to draw out Galicia's entanglement with risk objects like the *Prestige* and, in so doing, garnered support across Iberia, Europe, and the world. Two days after the *Prestige* sank, the political coalition BNG (*Bloque Nacionalista Galego*) and many intellectuals set up what would be come the first meetings of *Nunca Máis*.[8] Rivas himself sums up the spirit of the movement: "La sociedad civil reclama su derecho a ser ciudadanos, no meros votantes. El problema es que el PP teme a los ciudadanos. *Nunca Máis* es un rechazo a esa concepción" (Lobo) (Civil society reclaims its right to be citizens, not merely voters. The problem is that the *PP* fears citizens. *Nunca Máis* is a rejection of this concept). Beginning as a grassroots reaction to the oil spill and the awkwardly slow response by the PP government and international agencies, *Nunca Máis* united people towards a greater awareness of the fragility of environments and surrounding communities. This globalized protest of the official silence sought to decry, denounce and cleanup large amounts of *chapapote* washing onto Galician shorelines. The term *chapapote* took on special meaning in 2002. The word itself is a corruption of the Náhuatl word *tzaucpopochtli*, meaning "god of the roads and pathways." While the word generally refers to any kind of asphalt or oil in a Mexican context, the term gained much currency in the Iberian context during the Galician oil spill specifically meaning oil spill due to negligence or criminal behavior. As one clean-up volunteer noted, a Google search today of "*chapapote*" still yields a series of results about the *Prestige* spill. The appropriation of this term runs parallel to the *Prestige* as another globalized object, this time linguistically absorbed from one context to reflect the concerns of another place. Indeed, *chapapote* is indicative of a larger critical response that became a way to recompose Galician identity within a fragile nexus of the twenty-first-century global economy.

This risk-turned-disaster situation altered the perception of Galician landscapes in a way similar to Rábade Villar's cultural identity project in her article "Spectres of the Nation: Forms of Resistance to Literary Nationalism." She begins her essay citing Samuel Beckett's phrase "unspeakable homes," to argue that Galicia, as a peripheral cultural identity, should be a project of rendering what is unspeakable into the speakable because it should ally itself with those "who have received the greatest injustice" (Rábade Villar).[9] Cultural and linguistic identity, then, precludes any unified and stable project, based on a codified linguistic foundation. She notes that instead of understanding Galicia as a codified Heideggerian house of language, as did older forms of *galeguidade*, it is more productive to analyze the cracks or fissures in the cultural project because such analytical attention allows for a greater focus on the heterogeneity and openness of Galician identity. In other words, *galeguidade* shoud concern itself with those that have no voice within the nepotism of political regimes like *fraguismo*. Indeed, considering the unspeakable home, or better, the *oikos*, of Galicia in a moment of environmental disaster draws our attention to these fissures and our exposure to them. Specifically, the political problems plaguing Fraga's Galicia came to light during the *Prestige* crisis. Nationalism, then, becomes less of a singular unity and more about "distance, disaffection, and exile" (245). These terms, according to Rábade Villar, ally cultural identity with "residual" forms of subjectivity, a category that I will now extend to nonhuman and nonlife forms of being: the creatures and objects that compose Galicia's maritime culture and ecosystems.[10]

Rivas allows us to consider cultural strategies and styles as they attempt to grasp the scope and impact of the *Prestige* crisis. The author has published many essays on the subject of *Nunca Máis*, arguing that the democratic movement not only united a diverse group of interests and ideologies, but also responds to nonhuman cries for help. Such ecological concern, I find, also carries over into his fiction and poetry.[11] Rivas attributes the first cry of *Nunca Máis* to the sea itself (254). His claim suggests that ecological mobilization certainly involves nonhuman alliances just as much as it involves human collaboration. *Nunca Máis*, then, galvanizes Galicia's human and nonhuman inhabitants in a bid to reevaluate the region's local identity in the midst of a crisis truly brought on by the global forces surrounding the *Prestige*. I will examine how this insight further augments the necropastoral through an exploration of petro-melancholia, which not only underscores the fragility of ecological systems but also the root causes in modernity that cuase disaters such as the *Prestige*.

Nunca Máis became the space for communicating the screaming voices of the human and nonhuman collective. In order to evaluate this movement, it is important to first examine the flux of speech and silence at work in Rivas's "viaje de periodismo *indie*," in *A cuerpo abierto* (2008), which develops a "des-lectura" and allows for a critique of these facts and fictions. Rivas's "unreading" (*des-lectura*) of the facts becomes a means to understand how *Nunca Máis* reformulated Galician culture as a response to ecological crisis. *A cuerpo abierto* is a compiled text based on different newspaper columns written in the late 1990s and early 2000s. There are at least two ways of reading the work. On the one hand, this compilation might be dealt with simply as a compendium of nonfiction opinion pieces. Reading these texts in one place allows us to follow a trajectory of his thought. Yet the text is not arranged chronologically; if we view it thematically, the content is almost sporadic, jumping back and forth among a series of interconnected themes. It rapidly shifts through a set of postures on contemporary Spanish culture and politics from the debate about historical memory to Spain's involvement in the American-led Iraq War and sundry environmental issues. The text is divided into the following sections: "La mirada indie," "La amnesia retrógrada," "La revolución del mar," and "Re-existencias," each entirely composed of these short fragments, ranging from half a page to seven pages. What ties these sporadic themes together is the attitude of each section. The first section, "La mirada indie," encourages a "des-lectura" (an unreading) of a hypothetical apocalypse ("el Estado de Apocalipsis Permanente") after a fictitious presidential win of the conservative politician Mariano Rajoy in 2004. This ironic posture de-naturalizes the genre of journalism as a collection of facts and encourages the reader to slow down and question the reading of these "facts." Speech slows down to allow the reader to reflect on what truth it *does not say.* The second set of texts deals with a "retrograde amnesia," which wants to denounce a collective forgetting of the horrific events of the twentieth century, in Spain's particular case: the atrocities committed under Francoism. Part III, which concerns us here, is entitled "La revolución del mar" (The revolution of the sea). This short set of essays is an interconnected reflection piece on increased environmental awareness in Spain before and after the 2002 *Prestige* oil spill. Part IV is a short set of essays called "Re-existencias" that deals with the future of politics and culture in Spain. Rivas's injection of fiction into these journalistic texts urges ecological reconsideration after this crisis, suggesting that *Nunca Máis* is not simply an inflammatory reaction, but rather a different type of politics sensitive to the fragility of human and nonhuman systems. Though the third section predominantly deals with *Nunca Máis*, Rivas also works to connect this event to other environmental disasters in the region, such as the dumping of nuclear waste in the Atlantic Ocean and the Urquiola oil spill of 1976.

If Rivas's irony denaturalizes the "journalistic integrity" of these pieces through fictionalizing historic events, how are we to read them? To understand the significance of this turn towards fiction, it is important to underscore the use of irony in these texts. The meta-manifesto "Estamos hartos," that appears in Part I, "La mirada *indie*," helps explain the sort of irony that Rivas has in mind:

> Ya era hora de que estallara una nueva vanguardia. [. . .] "¡Estoy harto de Picasso!", proclama Antonio López. Y al día siguiente, Francisco Ayala va todavía más allá: "¡Estoy harto de Francisco Ayala!" Así se empiezan los auténticos manifiestos. . . . (37)

> (It's past time that a new avant-garde began. "I am tired of Picasso," exclaims Antonio López. And the next day, Francisco Ayala goes further still: "I am tired of Francisco Ayala!" This is how authentic manifestos begin . . .)

This "*nueva vanguardia*" is *hartismo*: *un hartazgo artístico* (an artistic satiety). Manifestos often begin with negations, setting apart new work from the old, the energetic from the tired and spent. Rivas claims that (h)*artismo* is not simply a new artistic form but also a collective discontent about "truth" in contemporary political discourse. Why? There is an assumption here that contemporary discourse suffers from excessive hyperactivity. Due to the velocity, volume and complexity of information in a global society, it has become more difficult to synthesize raw data and construct narratives. Truth, however, if it is to be a part of these discourses, must be built slowly and, paradoxically, contain fictions that point to the real uses of denial and mendacity at work in political systems (such as the Spanish government's denial of the spill). *Hartismo,* then, pinpoints the crux of what Rivas considers to be avant-garde protest pieces, pointing to truths latent or absent in political discourse. Rivas writes that "El *hartismo* es la verdad abriéndose paso laboriosamente en una realidad empachada" (37) (*Hartismo* is truth making way laboriously in an impeding reality). The adverb "laboriosamente" is a key insight into the irony of *hartismo.* It is a proclamation about the imperative to slow down our observations and to gaze into the features, structures, and disasters of what Rob Nixon has called slow violence: invisible and silent destruction of landscapes often occurring on the peripheries of nation-states or other global actors.[12] It is less about the supposedly cathartic or graphic kinds of violence often represented in the media and more about ecological absorption and decay, inside, outside, and around human communities. *Hartista* irony, then, slows down narratives to contest what we consider to be the factual status quo.

Rivas also underscores a kind of slow and silent decimation. His writing suggests that our politics of truth is saturated with lies and noise; our landscapes are decaying from ecological corruption; the planetary atmosphere thickens with carbon dioxide. These ambient metaphors are not easily detectible in daily speech because they are background noise to our everyday activity based on assumed facts. *Hartismo* moves against this velocity, impatience, and inflation of "facts" and presses us to observe how this slow, silent violence affects ecosystems and communities. It encourages us to look into the spatial backdrops of our real and imagined communities and begins to construct new narratives based on what has been buried in material and discursive sludge. In this way, *hartismo* is an ironic manifesto asking us to slow down our reading in a globalized world because hyperconsciouness composes a reality that is *empachada*—it is all stopped up and impeded.

For Rivas, the experimental irony of *hartismo* was at work in the way that the *Nunca Máis* protests transformed everyday iconic objects connected to Galician identity. Indeed, *hartismo* gives us two ways to read the narrative of *Nunca Máis*. First, *hartismo* is a literary diagnosis of the popular sentiment at the time of the spill, which was not only fed up by the disaster itself but also by an overtone of political malaise with Fraga. Second, *hartismo* presses us to consider the "facts" and "fictions" spewing out of the mouth, that is, the hull of the *Prestige*. In other words, Rivas presses us not only to look at the hope and disillusionment of the volunteers cleaning up the spilled oil, but also to the global mechanisms that helped produce the spill and the political mechanisms that tried to cover up the ecological fallout.

In this way, *hartismo* operates in an opposing manner to the journalistic process of catchy headlines and short and often misleading stories. And there is much need for pause. One aspect of ecological crises such as the *Prestige* spill or the 2010 BP Macondo blowout in the Gulf of Mexico is the fast-paced narrative surrounding the event of the crisis. For instance, throughout the BP disaster, there was an inundation of remedies and solutions in the media. Words like "kill switch" or "kill mud" became main characters in a silent drama unfolding on a live stream broadcast from more than 35,000 feet underwater. Yet below this drama is a myriad of losses with deaths of birds and fish and sea turtles in addition to its impacts on human economies.

Ecological slow violence produces what Stephanie LeMenager has called petro-melancholia: a critical attitude toward modernity after the realization that it is failing. *Hartismo* wants to slow down the narrative of the *Prestige* oil spill and help us reflect on the ecological toll of the crisis, what its causes are, and how we might avoid such crises in the future. The prescience of petro-melancholia requires an examination of the invisible and often silent structures and mechanisms at work in today's global economy. Several parallels between the *Prestige* case and the BP oil spill in 2010 become instructive on this point. Much like Galicia's case, the U.S. Gulf Coast and more generally the southern United States has remained underdeveloped when compared to its northern counterpart. Modernity, for a state like Louisiana, has largely consisted of a patronage system with big oil companies and an acceptance of the health risks associated with large-scale oil refining operations.[13] Writing on Hurricane Katrina and the BP blowout of 2010 in the Gulf of Mexico, LeManager argues that environmental consciousness and activism display a melancholia concerning the "industrial-era infrastructures" that are ultimately destructive to human and nonhuman ecosystems. She writes:

> We learn from these two events on the Gulf Coast of the southern United States not only that modernity and ecology are entangled objects, to paraphrase Bruno Latour, but also that the melancholia for a given Nature which has supposedly characterized modern environmentalism might be eclipsed, in the twenty-first century by an unresolvable grieving of modernity itself, as it begins to fail. (27)

LeMenager contends that new forms of environmentalism also mourn the failures of modernity in relationship to the machinery of a global economy. In this sense, petro-melancholia is not only the mourning of environmental violence but also a grieving due to the impacts of modernity's inherently manufactured risks. In Galicia's case, petro-melancholia embodies what Rábade Villar described above as disaffection, which begins the examination and protest of the causes of the crisis. Rivas captures this perfectly with the phrase "*corazón chapapoteado*" (we might loosely translate this powerful phrase as "petro-melancholic heart"), which conveys the relationality between the internal and external wounds caused by the disaster (260). Moreover, according to our above analysis of the *Prestige* and the word *chapapote*, these wounds come from nowhere but affect everywhere.

What is most striking about the Galician case is where this melancholia and dissent led. In sum, *Nunca Máis* was composed of over 300 labor, cultural, and civil organizations and, at certain peak points, over 250,000 protesters ("A cuerpo abierto" 255). Yet the movement also has its nonhuman actants. As Rivas points out in *A cuerpo abierto*, it is true that this movement was born out of a union of people, but poetically, it was born from a call of the sea itself. His claim suggests that human mobilization certainly involves nonhuman alliances, many of which were "typical" symbols of Galician identity, such as umbrellas, bagpipes, conch shells, and suitcases. As we will later discuss, these Galician cultural objects become symbols of defiant protest in the *Nunca Máis* movement. On the other hand, the immediate and systemic causes of the spill also become visible through an interrogation with more threatening objects involved, such as the ship itself.

This sinister aspect of the *Prestige* appears in the section entitled "Mayday, Mayday, Mayday," from the environmental section "La revolución del mar." In the fragment "La boca del *Prestige*," Rivas reflects on Xurxo Lobato's famous picture of the sinking ship, attributing "a dramatic expression" to its final grimace (257). The *Prestige*, he claims, speaks in this photograph. It "vomits" or spews the anthropomorphic gesture of desperation. It is also as if the ship were looking at us, making visible what is invisible: its embodiment of a manufactured risk of global economic movement. What do these human-like qualities do to the image of the boat? I submit that our detection of anthropomorphism is a sign that something sinister and "alive" is at work in the *Prestige* itself. In contrast to an anthropocentric perspective, we are forced to reconcile that a nonliving, nonhuman object is *speaking* to us, that it is speaking to us about its silent yet volatile capacity. Rivas writes:

> [La foto] [l]lama a la humanidad, como si el mar, con el entendimiento de ese arquero zen que es el fotógrafo, lo mantuviese a flote el tiempo necesario para que el nervio óptico transmita un dardo estremecedor a la glándula de la vergüenza de los dioses impacientes, los de la Era del Petróleo. (257–58)

> (The photo calls to humanity, as if the sea, along with the zen archer understanding of the photographer, kept the boat afloat the necessary time so that the opitical nerve could transmit a shocking dart to the shame gland of the impatient gods, those of the Era of Petroleum.)

La boca del Prestige marks the vessel as an actant because it appears to communicate with us. The *Prestige*, as an instantiation of our maritime technology, speaks to the underside of its invention: the accident.

Understanding these entangled conditions requires risks to be connected with the "accident." In *The Original Accident* (2005), Paul Virilio argues that any technological invention is also the invention of its eventual failure, malfunction or disaster. The invention of the airplane, for instance, inversely carries with it the phenomenon of the plane crash. The *Titanic* summoned icebergs as villainous points of collision. The nuclear power plant already contains the seeds for a meltdown. The accident (*accidens*) provides insight into "what is beneath" (*substare*) any of these tools: their inevitable failure (10). In the twentieth century, the amount of accidents began to pile up at an alarming rate. Political ecology, as Virilio calls it, must now transform politics to draw attention to these accidents.

On this score, *political ecology* cannot long go on sweeping under the carpet the *eschatological* dimension of the calamities caused by the positivist ideology of Progress (11). Virilio's accidents are another way of considering the kinds of manufactured risks inherent in human-designed technology. These broken projects, so often swept under the carpet (of the pastoral), are the scraps of both broken and working equipment that constitute the machinery of the petroleum industry. Political ecology, for Virilio, must take into account the impact of this invisibility, a task aided by Rivas's literary project of *hartismo*.

The sea is another powerful actant in Rivas's text. As mentioned above, Rivas poetically observes that the sea began the *Nunca Máis* movement. In the second fragment of "Mayday, Mayday, Mayday," the event is born out of silence: "Primero fue el silencio" (254). But this is not an example of mute silence but rather of a fecund sort: "Ese silencio inquietante que precede a las tormentas y a los seísmos. Luego, el estupor. El choque, el impacto, el dolor. Otra clase de silencio: trágico, hamletiano, de inseguridad, miedo y duda" (254) (That disquieting silence that happends before storms and earthquakes. Then, stupor. The shock, the impact, the pain. Another kind of silence: tragic, Hamletian, from insecurity, fear, and doubt). This is an uncanny silence that paradoxically works as a medium for the speech of the sea: "El hecho de que el mar hable: ¡Yo acuso! ¡Yo denuncio!" (255) (The fact that the sea speaks: I accuse! I denounce!).[14] The human silence gives way to the oxymoronic deafening (and damning) silence of the sea. We should read the sea's silence as a kind of uncanny speech that marks the beginning of a dynamic protest movement.

The sea becomes the catalyzing actant in this democratic movement precisely because it is so intertwined with Galician identity. Rivas presses us on this point, arguing that life on the Galician coast attunes one to the rhythms and tones of the sea. Just as a farmer is said to be in tune with the cultivated soil, so too is the seafarer to the sea. Indeed, he jokes that sailors have physically mutated because of their intimacy with the sea. They have "the largest ears of all" (254). This joke also points to the intimate connection between the seascape and the cultural identity in Galicia. Rivas begins to sketch a vernacular landscape counterpoised to the official silence of the Fraga government. The language of the vernacular landscape begins with the shockingly voiced silence of the sea itself. Speaking out is a form of channelling the language of Galicia's vernacular landscape and, furthermore, transforming how it appears in political discourse.

The sea is not the only local idiosyncrasy of Galician identity to show up in the movement's images. Rivas considers the recycling of iconic Galician symbols to be instances of avant-garde protest art. Due to Galicia's famously rainy weather, protesters showed up in massive numbers with umbrellas in tow, objects that Rivas describes as other appendages for Galicians (261). Rivas also references the off-kilter Galician joke that normally is meant to indicate the "servitude" of Galicians: "mean por nosotros y decimos que llueve" (261) (Piss on us and we will say it is raining). For Rivas, the mobilization of umbrellas as protest objects marks an ironic (*hartista*) transformation of the umbrella into an instrument of liberation: "Soltaron la pesadumbre de la espalda y la abrieron en forma de alas" (261) (They shrugged off their grief and opened the umbrellas as if they were wings). Another important example is *La Marcha de las Maletas*, in which more than a hundred thousand showed up in A Coruña on February 3, 2003, with suitcases in hand. The suitcase as a protest-object also counteracts the old adage that Galicians do not protest; they emigrate (262). In opposition to this stereotype, the *maleta* was meant to cast off and say goodbye to the destructive "fatalism" of the *Prestige* crisis. Similar instances occurred with bagpipes (*la "marea gaiteira"*), conch shells (*los gramófonos del mar*), musical concerts, and soccer matches. This ironic play unbinds these icons from their essentializing connotations about Galicia. That is to say, these objects of Galician identity became *hartista* protest pieces that fundamentally altered the public discourse of the *Prestige*.

Moreover, these protests were not simply local to Galicia. There were protests and shows of solidarity in Switzerland, Madrid and Brussels, to name a few. Such acts not only re-affirm local identity but also translate it into a different register, one sensitively grafted onto global concerns of ecological fragility. In *Españoles todos* (2004), a series of essays dedicated to Spanish social and political issues, Galician author Suso de Toro argues that *Nunca Máis* is a model for a democracy to come insofar as the platform re-composed traditional icons into symbols of environmental protest and constructed important alliances around important environmentalist issues. Whereas some forms of identity tend to insulate a world and gaze inward, *Nunca Máis* became what Toro describes as transnational and open through the common cause of environmental concern. As a platform for political action, *Nunca Máis* became inclusive, uniting old school nationalists with the left against ineptitude and silence. "Nunca Máis es un movimiento trasnacional, horizontal, abierto, adherible, de regeneración democrática, de exigencia de decencia a una administración inepta, mentirosa y culpable" (124–25) (*Nunca Máis* is a transnational movement, horizontal, open, adherable, built on democratic regeneration and the demand for decency from an inept, mendacious, and culpable administration). Looking into human responses through the posture of *el hartismo*, Virilio's political ecology needs literary and cultural explorations because it is not simply about being "fed up," but rather offers different ways of paying attention (to an ecological crisis) and to what we mean by "the facts."

Regional identity read through an environmental disaster becomes an example of an open and porous set of alliances between animals, rocks, humans, umbrellas, suitcases, land and water, all woven together in a heterogeneous cultural object. This would allow for a version of identity that at once moves away from clichés and creates awareness about Galicia as a cultural center in Spain. Moreover, this transformation, then, embodies what Rábade Villar described above as a "speakable" home. The *oikos* of Galicia, often remaining invisible and "unspeakable," entered the forefront of political discourse as a fragile composite of humans and nonhumans. This sense of interconnection creates a new politics because we begin to realize that we are terrifyingly dependent on fragile equilibriums on global and local levels.

As I noted in my introduction, a Spanish trial concerning culpability for the spill has recently concluded, virtually dismissing any responsibility or negligence on the part of the Spanish government or the ship's crew. Due to the complexity of international standards, the Spanish courts have only taken to trial four individuals. The trial has dragged on for a total of ten years: eight years of investigation and an additional two years mired in the legal complexities of international law. Moreover, the trial has become comically tragic, with, for example, media attention focused on the ship's ailing 77-year old Greek captain, Apóstolos Mangouras, who not only requires translation services, but also is hard of hearing (Ordaz). The trial has also brought to light many contradictions inherent in the case, with diverging accounts of how damaged the ship was before it entered rough seas and why the ship was refused safe harbor. Testimonies from the likes of the ex-Director of the Merchant Navy José Luis López-Sors, Mangouras and the ship's chief of machinery and Nikolaos Argyropoulos all seem to contradict one another and assign blame elsewhere (Obelleiro). Added to this, a fourth plaintiff in the trial, Filipino First Officer Ireneo Maloto, has not made an appearance in court and seems to have disappeared entirely. In terms of policy, it appears that the trial and the politics surrounding the case will only affect few, if any, changes to international standards for shipping petroleum.[15]

Reflecting on *Nunca Máis* more than ten years after the spill, I submit that the movement has much to tell us about future formations of identity as alliances between life and nonlife, between the human and nonhuman and between multiple communities. Txetxu Aguado has argued that the values of *Nunca Máis* should inform new models for thinking about national identity in the twenty-first century. Indeed, as I have outlined here, the movement was not only about environmental disaster but also about political disgust with the regional and national governments' continual unwillingness to act with transparency on behalf of its citizens. More broadly, the movement rejects Galician identity as stereotypically backward and instead becomes culturally transformative. Such a moment requires a certain amount of reflection that I will use to conclude. Aguado observes that "el primer gesto consitirá en el básico de volver a mirarnos y decidir qué se ha sido y qué se quiere ser" ("Suso de Toro") (The first gesture is one of returning to self-reflection and deciding what has been and what we want to be). As I have argued here, what is significant about this paradigm of identity is that it is not *merely* inwardly drawn toward an internal recognition of a single group, but rather as a community in relation to its surrounding ecosystems. Ecological fragility, then, comes to involve a pluralism of identities and habitats, sharing in collective strengths and vulnerabilities. It is this idea that is, in my view, so significant for rethinking the question of life in the Iberian Peninsula. While *Nunca Máis* certainly expresses particularities unique to Galicia, its appearance contains the seeds of how Iberian identity might be re-formulated in a time of increasing ecological awareness and globalized risks. After all, Galicia was not the only region affected by the spill; other contaminated coasts include those of Portugal, the Basque Country, and France. Aguado meditates on this kind of pluralism in two questions:

> ¿Es posible desde mi identidad llegar a conocer las otras locales? E incluso añadiría, ¿puede la propuesta local relacionarse con lo universal? En la primera pregunta, mi identidad local quiere ser consciente de lo que ocurre a su alrededor. No renuncia a participar en ese acervo común que compartimos como seres humanos. ("Suso de Toro")
>
> (Is it possible, from my identity, to come to know others? And I would also add, can a local proposal relate to the universal? In the first question, my local identity wants to be aware of its surroundings. It does not waiver in participating in this common good that we share as human beings.)

This *acervo común* points to a common environment that, when threatened and destroyed, as we saw in the government's necropastoral narrative and Rivas's petro-melancholia, propelled a democratic resurgence in response. The *Prestige* crisis functions as a catalyst to experience our connection to these ecosystems. Moreover, the above analysis has indicated that culture itself is not simply related to its surrounding ecosystems but dramatically dependent on their wellbeing. Various manufactured risks of a globalized economy expose the interconnectedness of local cultures via the fragility of ecosystems. Instead of denying these factors, local cultural identity becomes a way to speak with or for the *oikos*: to render the unspeakable speakable.

I indicated above that there were two competing narratives of what happened in the *Prestige* case. On the one hand, many forces quickly mobilized to cover up and obscure the impact of the spill, generating a narrative that Rivas describes as "the miracle barrel." Meanwhile, the democratic surge of *Nunca Máis* filled in this vacuum of public discourse and united in the absence of a governmental response. Juxtaposing the two narratives of the *Prestige* not only exposes the ecological fissures in the official narrative, but also offers a new articulation of Galician identity as a diverse ecological collective of human and nonhuman elements. The invisible wounds from the *Prestige* disaster come to light through a process of juxtaposition and extraction inherent in the necropastoral.

By introducing Manuel Rivas's discussion of *hartismo*, we observed a strategy of *des-lectura*, of slowing down and challenging many of the fictions that are often rendered truths in the hyperconsciousness of today's world. I pointed out that such reading is a key for understanding the impact of ecological violence often unfolding in the periphery or background of our world. This strategy urges us to pay closer attention to the behavior and "speech" of objects and environments, insofar as they will be the first to indicate a moment of crisis. By slowing down our reading of environments, petro-melacholia becomes a mode of disaffection during an ecological crisis, which is not merely the mourning of a lost nature but also the lamentation that modernity has failed us, insofar as its mechanisms and behavior are slowly decimating the biosphere and human communities. In turn, the production and study of Galician culture gains visibility and complexity through an invigorated ecological consciousness, creating a nexus of human and nonhuman voices to combat the threat of disappearance.

Notes

1. Castro Vasquéz observes that thinking ecologically about culture encourages plurality: "Esta ecología busca preservar la variedad y la riqueza de significados inherentes a todas las lenguas así como confrontar la contaminación con que amenaza el discurso del poder y las fosilizaciones de los significados que imponen categorías fijas sin dejar lugar a la pluralidad" (6) (This ecology seeks to preserve the variety and richness of the signified inherent in all languages as well as confronting the threatening contamination from the discourse of power and the fossilization of the signified that impose fixed categories without leaving room for plurality).
2. Eight of the most notable environmental disasters in the region include four other oil spills: the *Polycommander* (1970), *Urquiola* (1976), *Andros Patria* (1978), Good Lion (1983), Cason (1987), *Mar Egeo* (1992). As the independent report made by the Barcelona-based Institut de Ciències del Mar (ICM-CSIC) observes: "[E]very 7 years on average an oil spill accident occurs off the Galician coast, and we should expect

the next to occur in the next decade" (533). Other major problems include various European nations dumping nuclear waste into the Atlantic Ocean and several major forest fires.

3. Manuel Rivas (1957) is a preeminent Galician novelist, poet and journalist. Though he is a major representative of literature published in Galician, he also writes regularly for the Spanish-speaking newspaper *El País*. His work has received international recognition, with two film adaptations of the Spanish Civil War novel *El lapiz del carpintero* (1998) and the short story collection *¿Qué me quieres, amor?* (1995), which includes "La lengua de las mariposas," a story later turned into an internationally successful film. He was also a founding member of Greenpeace España and a central voice of the *Nunca Máis* movement.
4. On the occasion of the spill, Rajoy remarked: "Afecta a una parte importante de A Coruña, pero no es una marea negra" ("El juicio del *Prestige*").
5. *Manos Limpias* has gained notoriety over the years as a far-right interest group intent on impeding peripheral nationalist movements in Galicia, the Basque Country, and Catalonia. The group was also instrumental in bringing charges against Baltasar Garzón in his attempts to open up investigations surrounding the Law of Historical Memory. For more on their accustation against *Nunca Máis* see "Archivan la querella de la manos limpias contra nunca máis" ("Archivan la querella").
6. Fraga, a career politician, eventually fell out of favor with the PP and ended up in Galicia. As Suso de Toro has noted, he has infamously spoken out in favor of the Chilean dictator Augosto Pinochet and condemned the actions of Judge Baltasar Garzón ("*Españoles todos*" 55–57).
7. Likewise, debris from coastal Japanese towns continues to appear along the western coasts of the United States.
8. In fact, *Nunca Máis* had been used once before after the protests of the 1992 *Mar Egeo* oil spill.
9. Rábade Villar's reference is to W.G. Sebald. Literature can be "the memory of those to whom the greatest injustice was done."
10. There is an important parallel here between Jonathan Mayhew's classification of Riechmann's work as the poetry of difference and Rábade's emphasis on "residual subjectivity." Indeed, as many animal studies scholars point out, what category could be more marginalized than nonhumans? (Cf. Wolfe, "Human, All Too Human.")
11. While my discussion here is only limited to Rivas's essays and poetry, animals have a predominant role as main characters in several of his novels, including *En salvaje compañía* (1994) and *Los libros arden mal* (2007).
12. Rob Nixon describes these rates of disappeared forests and people as slow violence, which overwhelmingly involve peripheral, minority groups precisely because they also reside in the very *sites* of capitalist production, in this case, *A Costa da Morte*.
13. Robert Bullard notes that, in spite of the "constant" media coverage of the oil spill,

few noted how BP disposed of the waste from the clean-up efforts, most of which ended up near minority communities in southern Louisiana.

14. One is reminded here of Émile Zola's open letter "J'accuse" published in *L'Aurore* in response to the Dreyfus Affair.
15. For some time, there has been a Madrid-based proposal to change shipping routes to further avoid another *Prestige* case, but the initiative has been stalled for some time now (Moreno).

Works Cited

Aguado, Txetxu. "Suso de Toro y una identidad gallega: desde local a lo global." *Dissidences: Hispanic Journal of Theory and Criticism*. May 2010. Web. 5 Aug. 2013.

"Archivan la querella de 'Manos Limpias' contra 'Nunca Máis.'" *La Vanguardia*. 15 Apr. 2003. Web. 5 Aug. 2013.

Bullard, Robert. "Environmental justice communities bear brunt of BP's oil spill waste disposal." Apr. 2011. Web. 5 Aug. 2013. *Communication Power*. Oxford: Oxford University Press, 2011.

Castro Vázquez, Isabel. *Reexistencia/ Reexistence: A Obra De Manuel*. Vigo: Edicións Xerais de Galicia, 2013.

Garcia, Raúl. *The Prestige: one year on, a continuing disaster*. WWF-Spain. Nov. 2003. Web. 5 Aug. 2013.

Giddens, Anthony. *Runaway World: How Globalization Is Reshaping Our Lives*. New York: Routledge, 1999.

Hooper, Kirsty and Manuel Puga Moruxa. *Contemporary Galician Cultural Studies: Between the Local and the Global*. New York: MLA, 2011.

"El juicio del 'Prestige' comenzará diez años después del naufragio." *Público*. Web. 19 June 2012. Web. 5 Aug. 2013.

Latour, Bruno. *Politics of Nature: How to Bring the Sciences into Democracy*. Cambridge, MA: Harvard University Press, 2004.

LeMenager, Stephanie. "Petro-Melancholia: The BP Blowout and the Arts of Grief." *Qui Parle* (Spring/Summer 2011): 25–55.

Lobo, Ramón. "'Nunca Máis', una burla negra." *El País*. 4 Jan. 2003. Web. 5 Aug. 2013.

López Uralde, Juan. "Otro *Prestige* es posible." *El Diario*. 13 Nov. 2013.Web. 29 Apr. 2014.

McSweeney, Joyelle. "Necropastoral, or, 'Normal Love.'" 13 Jan. 2011. Web. 2 May 2013.

Moreno, Mariola. "A estas alturas nadie espera nada del juicio del "Prestige."" *Público*. 16 Oct. 2012. Web. 5 Aug. 2013.

Nixon, Rob. *Slow Violence and the Environmentalism of the Poor*. Cambridge, MA: Harvard University Press, 2011.

Obelleiro, Paula. "'Prestige,' enigmas sin resolver." *El País*. 22 Dec. 2012. Web. 29 Apr. 2014.

Ordaz, Pablo. "Estaba en el puente y oí una explosión." *El País*. 27 Nov. 2002. Web. 5 Aug. 2013.

Rábade Villar, María. "The Concept of Resistance in Contemporary Galician Culture: Towards a Poetic Ecology." *Cosmos and History: The Journal of Natural and Social Philosophy* 6.2 (2010): 82–92.

_____. "Spectres of the Nation: Forms of Resistance to Literary Nationalism." *The Bulletin of Hispanic Studies* 86.2 (2009): 231–47.

Rivas, Manuel. *A cuerpo abierto*. Madrid: Alfaguara, 2008.

_____. Entrevista digital. 5 Dec. 2012. Web. 2 May 2013.

_____. "Otro *Prestige* es posible: Reflexiones para evitar otra marea negra." Greenpeace España. Nov. 2012. Web. 5 Aug. 2013.

Ship Structure Committee. "Prestige." Ship Structure Committee. Web. 2 May 2013.

Toro, Suso de. *Españoles todos*. Barcelona: Península, 2004.

Virilio, Paul. *The Original Accident*. Cambridge: Polity, 2007.

PART II

Ecological Crisis and the Neoliberal Appropriation of Public Space

3

Tourism and "Quality of Life" at the End of Franco's Dictatorship

Eugenia Afinoguénova

Utilizable
será la cualidad absoluta de la vida
y las vidas, utilizables en un lugar
y el acto de utilizar en un lugar
será el tiempo. . . .

(Utility
shall be the fundamental quality of life
and lives, utility in a place
and the act of utilizing in a place
shall be called time. . . .)

—Manuel Vázquez Montalbán,
"Poema de Dardé" 1969/1986

In the Spanish Constitution, "quality of life" appears several times as a fundamental value related to the economic and cultural wellbeing of citizens, protection of the environment, and participative democracy (León Jiménez). Yet the term is, in fact, an international conceptual chameleon that is found in contexts ranging from philosophy texts to the summaries of social indicators. The Canadian philosopher Storrs McCall, who was among the ones who noticed it early on, lamented in 1980:

> Not only do we not know what quality of life is, we don't even know what category of thing it is. Is quality of life a state of mind or a state of society? Does its defini-

> tion vary from individual to individual, from culture to culture, from geographical area to geographical area, or is it the same for all people everywhere? Is quality of life measurable, and if so why do there continue to be profound differences of opinion over which social indicators are relevant to its determination? (qtd. in Draper and Thomson 649)

The notion of quality of life acquired international popularity in 1972–1975 against the backdrop of the Western economic crisis and soon moved on to become a staple of biopolitics. The delegates to the first international meetings defined this new standard in relation to growth and the responsible use of resources. Yet in Spain, both concepts were deeply compromised. Growth had been the cornerstone of the 1960s development strategy known as the *desarrollismo* that the Falangistas—the members of the Spanish fascist party supporting a "social state"—were now trying to combat. The resource management crisis (a product of the uncontrolled tourism boom of the 1960s) was the Falangistas' main argument, but many supporters and opponents of Franco's regime also backed this criticism. The essay analyzes how the "quality of life" debate unfolded during the final years of the regime and how that debate (and its resolutions and lack of resolutions) eventually made the category of "quality of life" central to Spain's transition to democracy after Franco's death in 1975.

Tourism and the Domestication of Global Capital

In Spain, discussions about "quality of life" were initially limited to the branch of industry that had produced the 1960s "*desarrollista*" boom: tourism. This gave the whole debate a paradoxical twist: in economically developed nations, the dream of a new life standard involved the very activity that Spaniards considered their major pollutant. In the end, tourism had to be rehabilitated because it was, after all, a biopolitical force necessary for the smooth operation of Spain's dictatorial regime in the new global economy. Since 1961,[1] the Spanish government had been making conscious efforts to "sell" tourism to the domestic population, mobilizing Spaniards as economically active yet politically idle subjects, as hosts and occasional guests (when there were not enough proverbial "Swedish girls [*suecas*]" to fill the bill). The hospitality business, erected into the nation's homemade "Marshall Plan," channeled the government's desire to find a neoliberal path to "defascistization" (Payne 363–98): an economic liberation without political or personal liberties (Crumbaugh 15–39). Hence, in practice, tourism served as a vehicle or a pretext for an array of neoliberal policies ranging from redistribution of land to promotion of unfair taxation schemes. While the rich were getting richer and the regional inequalities sharper (Martín Aceña and Martínez Ruiz), officials used Spaniards' growing purchasing power—called "nivel de vida" (level of life)—as a new way of legitimating the regime. So at the time my story begins, tourism was cherished as the cornerstone of a dictatorship whose claims to legitimacy were grounded in its services to Spaniards, "and not in its origins" (Sánchez Biosca and Rodríguez Tranche 422).

The "monoculture of sun" (as tourism was sometimes called), however, had powerful opponents in the government, and by 1969, Spain's "European vacation" was over (Pack). Yet, even then, on the surface, things continued the same: Spain was still moving on to become a world leader in cement production, newspapers were replete with photographs of construction sites, and new airports popped up faster than new highways. In Spanish politics, this was the period when two conservative cabinets, led by the so-called technocrats that had closed the *desarrollista* "aperture" of 1961–1969, began to draft their plans for Francoism after Franco. Denouncing the "ideological pluralism on which party democracy is based," Franco's final years featured, in Paul Preston's apt wording (48), "monochrome" governments: first grey (for the 1969–June 1973 cabinet) and later (June 1973–1974) blue.[2] Neither of the two, however, pretended to invoke the colors of the Mediterranean, but rather reflected the grey metaphorical "bunker" from which the ministers were preparing to defend the regime and the blue uniform of the Falange that was coming back to rule. In December 1973, ETA's assassination of Admiral Carrero Blanco, the president of both governments and the designer of this "posthumous" Francoism project (Preston 48), would add the color of blood to the mix and usher in Franco's last (1974–1975), still unreformed, cabinet (Suárez Fernández 423). In spring 1972, the police fired against the strikers at the shipyards in El Ferrol, Franco's birthplace, and when the government threatened to apply military law, workers all over Galicia, the Basque Country, and Catalonia went on solidarity strikes (Preston 44). Ecology seemed the least of the government-sanctioned concerns; and the ones that were not sanctioned by the government did not enjoy easy coverage from the publishers scared by copious fines from censors. And yet, all of a sudden, in 1972–1973, sustainability and the environment became the talk of the day. The timing seemed right: Europe was beginning its long engagement with the "quality of life." But here was a mystery: the government ministers were coming and going, Spain's intellectuals in opposition and in internal exile were learning how to work around censorship, the clandestine militancy was on the rise, and yet what everyone in Spain and in the ideological underground had to say about the environment and quality of life sounded the same. Or did it?

In what follows, I will argue that these seemingly uniform statements reflected a debate about the forms that neoliberalism was taking under a regime trying to strike a hard balance between a fascist-type social ("national-syndicalist") state and an open-market global economy. The meaning of "quality of life" becomes much clearer if one approaches early 1970s Spain as a territory where global capital had to enter into contact with a nationally-bound political power that, in its turn, was pressured by its own regions seeking national sovereignty, semi-official and clandestine groups in opposition to its existence, and, finally, interest groups fighting for political influence: the technocrats linked to Opus Dei (a conservative Catholic organization) and the Falange. Since the 1960s, international investors into Spain had been mastering the lingo of tourism, Spain's first global industry. By the 1970s, however, it became clear to wider sectors within Spanish society that the government had been acting as these investors' direct partner, seeking occasional support from local bureaucrats and excluding national oligarchs and entrepreneurs. A large part of the 1970s debates about "quality of life" was these groups' way of demanding greater participation in this system, not to be confused with political opposition and the defense of the natural environment.

Roberto Roccu's recent work on globalization in national contexts, grounded in David Harvey's interpretation of neoliberalism, is relevant for understanding what was really at stake in the "quality of life" controversy. According to Roccu, for global capital to become acceptable in the territories that are on the receiving end, international financial operations have to find support among national oligarchs and adjust to their ways of doing business. Roccu distinguishes two stages in this process—"hybridization" and "domestication" (431–32)—whereby global capital acquires the forms and shapes recognizable to locals. While this mutual adjustment is taking place, and also after it is achieved, national authorities remain complicit with the policies of "dispossession" by which global capital expands: privatization of public assets, manipulation of crises, and state redistribution (Harvey, *History* 159–65, qtd. in Roccu 427). On the surface, the anti-tourism thread in Spanish discussions about "quality of life" could be explained by the growing awareness of these inequalities among different social groups: what Harvey calls a "broad alliance of the discontented, the alienated, the deprived and the dispossessed" (Harvey, *Capital* 240, qtd. in Roccu 424). However, although Spain's debut on the global arena indeed produced an alliance broad enough to cut through the harshest ideological differences and the opposing party affiliations, in my story, the discontented misunderstood dispossession and, with counted exceptions, forgot alienation. Tourism, chastised as a pollutant, was soon back to being disguised as "leisure," one of the most seductive biopolitical illusions that alienated humanity has produced to console itself, to its governors' delight (Vilarós; Minca and Oakes 12–37). Spain is still under the spell of that leisure dream.

"Quality of Life": A Concept and a Tool of Policy-Making

At present, the undetermined meaning of the phrase has not prevented "quality of life" from becoming an economic and social indicator and a branch of research in disciplines as diverse as philosophy, public policy, health, and food sciences (Nussbaum and Sen; Phillips; Michalos; *Quality of Life*). Since 1991, the World Health Organization has been developing instruments to assess quality of life, which could allow researchers to compare the term's connotations across cultures (WHOQOL). The World Database of Happiness, directed by Ruut Veenhoven, of Happiness Economics Research Organization from Erasmus University in Rotterdam, names eleven academic journals dedicated to this and related concepts, and the number is growing (*Journals on Happiness and Wider Quality of Life*). Peter Draper and David Thomson (649) have counted that, in medical literature alone, the number of papers that use this term has doubled every five years since 1976; yet they found only two papers that used this phrase between 1965 and 1970. Researchers and politicians in Spain have remained conversant with "quality of life" research and have contributed their measurements to the European and global indexes (Moreno Jiménez and Ximénez Gómez; Casas; Fernández López et al.; Anderson). Since 2010, The Unión General de Trabajadores (UGT) publishes a multi-lingual compendium on "quality of life" in the workplace from Europe and America (*Anuario internacional*). At a time when economists predict that the generations to come will be poorer and less educated than their parents, this seems to be the era of "quality of life." How did we get here?

Politically, the term emerged at the end of the 1960s in the context of the arms race. In 1969, while a draft of the disarmament treaty was debated in Helsinki, the head of the British delegation, Lord Chalfont, announced to the U.N. General Assembly that what was really at stake were the "quality of life in the future" and the question of whether humanity could prevent a nuclear disaster ("Segunda sesión de trabajo"). In March 1972, the unveiling of the results of MIT's *Limits to Growth* project (Meadows) commissioned by the Club of Rome unleashed a public debate about the overuse of resources and the impact of the environment on "quality of life." The issue quickly moved into international policy-making (Chomsky 118–25). In June of the same year, the Declaration of the United Nations Conference on the Human Environment in Stockholm identified the "wise" use of "man's capacity to transform his surroundings" as a path toward improving what was now officially called "quality of life" and traced the first blueprint for creating "good life" compatible with "the enhancement of environmental quality" (United Nations Environmental Program Preamble 3, 6). The term "sustainable development" had not yet emerged (Giddens 61), and the dilemma formulated in the Club of Rome report sounded like a choice between growth, on one hand, and "quality of life," on the other. For developing economies, that simply meant that the richer nations were attempting to infringe on their rights to develop. The Stockholm declaration reflected this unease, as well as the difficult compromise that was achieved: "Economic and social development is essential for ensuring a favorable living and working environment for man and for creating conditions on earth that are necessary for the improvement of the quality of life." Such an ambiguous balance between expansion and wellbeing offered no guidance for a country such as Spain, whose politicians wanted to fit with their counterparts from the more-developed nations without ceasing to promote extensive development. This was the beginning of Spain's long and complicated romance with "quality of life."

In the following months, the definitions given to "quality of life" in the international arena split. In the European political lexicon, it became synonymous with environmental protection and care for resources. As Spanish observers reported from the European summit held in October 1972, the idea continued the region's self-styling as, in the words of British Prime Minister Edward Heath, "una Europa para el pueblo, preocupándonos por la calidad de la vida y por el medio ambiente en que vivimos y trabajamos" (qtd. in "Hay, verdaderamente") (A Europe for its people that cares about quality of life and the environment in which we live and work). At that meeting, a decision was made to adopt a European Community policy for the protection of the environment, leading to the first Environment Action Plan in 1973 (Scheuer 18).

At the same very time, however, work was underway in the United States focused on broadening the definition of "quality of life," purging it from straightforward environmentalist connotations and transforming it into a set of measurable policy outcomes. In August 1972, a group of researchers and experts from the United States Environmental Protection Agency and the Booz-Allen Public Administration Services gathered in the Airlie House in Warrenton, Virginia, with a specific purpose: "disentangle the *Quality of Life* concept to the point where it has decision making relevance" (Environmental Protection Agency iii, italics in original). The symposium papers, published in 1973, contained a roadmap for transforming an individual's sense of happiness into a productive tool that policymakers could use. "Quality of life" was about to become a system of tasks and assessment instruments. The age of biopolitics had officially begun.

What meaning, then, did the Spanish regime support? Given the term's neoliberal origins and "undoubtedly 'bourgeois' flavor" (Harvey, "Marxism"), it is hardly surprising that Franco's government, still enjoying its honeymoon on the international stage, joined the "quality of life" choir. "Ni la expansión de la economía ni la lucha contra la contaminación son valores absolutos. No podemos olvidar que, por encima de cualquier otro, el objetivo general de la lucha contra la contaminación es salvaguardar la salud del hombre y mejorar su bienestar" (Neither economic expansion, nor the fight against contamination are absolute values. We cannot forget that, above any other target, the general aim of the fight against contamination is to protect human health and improve wellbeing), declared Laureliano López Rodo, a technocrat who had been in charge of Spain's Development Plans since the late 1950s ("Destacada Intervención"), while addressing the 1972 Stockholm Conference. The striking part of López Rodó's declarations was his attempt to represent development and environmental protection as relative values, subject to the needs of improving living conditions. This brought the Spanish government's position close to that of the developing nations that were defending their right to expand. What did, then, "quality of life" mean for Spain? In Stockholm, López Rodó did not even use the term, mentioning instead the concept of "wellbeing." Though appearing synonymous, "wellbeing" was a staple of the *desarrollista* Newspeak of the 1960s, of which López Rodó was the prime designer; it referred to Spaniards' growing purchasing power. Hence, for the domestic audience, López Rodó's position promised no change of course.

Several months later, however, priorities shifted. Prior to the Stockholm meeting, Spanish experts had drafted a report for the Common European Market on the protection of the resources of the Mediterranean, and in November 1972, a conference on "Tourism and Ecology in [the] Western Mediterranean" was held in Madrid, in which López Rodó made closing remarks. *Then* it became clear that "quality of life" would figure firmly in Spain's political future, with its meaning aptly adjusted to the government's new needs. This was a double-faced "quality." On the one hand, Spain was recognizing the European definition rooted in environmental responsibility. On the other hand, however, its leaders redefined "environment" to steer it away from controlled use of natural resources. López Rodó's speech at that meeting demonstrated how easy it was to pay lip service to Spain's neighbors' concern for nature, while veiling the technocrats' interest in continued expansion:

> El tema del turismo y la calidad de vida es de los que encierran mayor fueza ética.
>
> Al abordar este tema, la primera dificultad con que nos enconramos es la de clarificar este concepto difuso y a veces, confuso de medio ambiente, es decir, todo ese conjunto de factores físicos, químicos, bilógicos, estéticos y sociológicos que afectan el desenvolvimiento físico, mental y social del hombre.
>
> Entiendo que la política para la mejora del medio ambiente ha de tener un carácter profundamente humanista. (*Congreso de ecología y turismo del Mediterráneo* 38)
>
> (The issue of tourism and quality of life is one of those harboring the greatest ethical force. The first difficulty that one faces approaching this issue is the need to clarify the vague and sometimes confusing concept of environment, that is, the whole set of physical, chemical, biological, esthetic, and sociological factors affecting man's physical, mental, and social development. I understand that the politics aimed at improving the environment ought to be profoundly humanistic.)

The Minister's rhetoric, in reality, only meant to explain that contamination "no constituye, en la mayor parte de los países del mundo, un problema proritario" (was not a primary problem for most countries) while development was, and that any environmental policy had to include tasks unrelated to nature, such as "un conjunto de bienes y servicios que deben ponerse a disposición de la comunidad para que ésta alcance el indispensable nivel de vida" (38) (providing goods and services to allow the community to reach necessary life standards).

The idea that combining development with quality of life was a difficult, albeit mandatory, task became a formula used to insert the technocratic agenda into the post-Stockholm international climate without sacrificing growth. López Rodó summed it up as the need to "armonizar en una síntesis difícil, pero no imposible, el crecimiento económico con la calidad de la vida humana" (harmonize, in a difficult, yet not impossible, synthesis, economic growth and the quality of human life) at an Ibero-American summit in May 1973 ("El príncipe de España presidió"). Yet in order to signal a change from the previous philosophy focused solely on development and expansion, the technocrats had to explain how concern for the environment fit with the 1960s ideas of development and "wellbeing." While the differences of opinion about the best way to proceed were admittedly strong enough to spur changes in the Development Plan Committee, López Rodó, the Committee's leader, remained in place and found an ingenious formula to explain the innovation: "Queremos que prime sobre el puro crecimiento cuantitativo la mejora efectiva de la calidad de la vida" ("La política, de semana a semana") (We want to put an effective improvement of quality of life above mere quantitative growth). There was still no explanation of what "quality of life" meant.

Amidst these uncertainties, tourism, which during the technocrats' roaring 1960s had been conceptualized as a sure road to wellbeing, was called upon to show how development could be compatible with "quality of life." Without questioning the positive monetary impact of tourism, Alfredo Sánchez Bella, Tourism and Information minister (1969–1973), now recast this industry as a new ecology. Although Sánchez Bella replaced in 1969 the main author of the tourism boom, Manuel Fraga Iribarne, he was also closer to the Opus Dei technocrats than to the Falange. His reframing of the "tourism or ecology" dilemma as "tourism *and* ecology," where each part would support, and not contradict, the other, reflected an interpretation of "quality of life" that everyone could live with. Sánchez Bella's solution to the tourist/environmentalist conundrum was ingenious: the industry that constituted the greatest threat to the environment was recast as the one most interested in environment protection. Greeting the delegates to the same 1972 "Tourism and Ecology" meeting in Madrid where López Rodó spoke, Sánchez Bella came close to a formula that could tie development with environmentalism:

> De la misma forma que las aguas de la ribera mediterránea se van contaminando, el litoral se va cubriendo paulatinamente de un muro de cemento que amenaza con privarnos de la vista al mar. . . . Y un elemental estudio sociológico del fenómeno turístico nos descubrirá, como una de sus motivaciones inmediatas, el anhelo de evasión de la opresión ciudadana: del cemento, hierro y asfalto, y de la búsqueda de espacios abiertos, horizontes distantes, panoramas verdes y reposados. En suma, el afanoso deseo de un horizonte distinto a aquel en que se desarrolla la vida ciudadana. Y lo que estamos en peligro de descubrir es esta diferencia, ya que el desorden en la construcción de instalaciones para recibir a las masas que se desplazan constituye un factor desequilibrador del medio ambiente que tiene (sic) a igualar el paisaje de destino con el de origen . . . (*Congreso de ecología y turismo del Mediterráneo* 13)

> (Just as the waters of the Mediterranean shores are getting contaminated, the coast gradually becomes covered by a wall of cement threatening to deprive us of the view of the sea. . . . Even the most elementary sociological study of the tourism phenomenon would tell us that the strongest motivation for tourism is the desire

> for an escape from urban oppression—from cement, iron, and asphalt—and the quest for open spaces, distant horizons, green and quiet panoramas. In sum, an intense desire for a horizon different from one in which urban life develops. And we may learn about this [diminishing] difference the hard way, because the disorderly construction of facilities for receiving the moving masses constitutes an unsettling factor for the environment, making the landscape in the destination indistinguishable from one's point of departure.)

Yet the Minister's awareness of the human longing for "open spaces and distant horizons" did not translate into protectionism. Instead, his approach to nature sounded like advertising for a hair product: "Humanicemos el paisaje de nuevo con un adecuado uso del color, del volumen, del material" (Let us re-humanize the landscape with an adequate use of color, volume, and material), he exhorted his listeners (14). His words drew the future of environmental protection as a new branch of industry that would turn the cause of destruction into its remedy: "El progreso técnico, la máquina, son grandes responsables de la degradación. Utilicemos ambos para recuperar el control de la biosfera" (12) (Technological progress and the machine are the two forces responsible for degradation. Let us use both to recover the biosphere). With a postmodern *simulacrum ex machina* twist, the Spanish government's environmental concerns only promised a return to machine-made nature.

This is how the environmentalist slogans were co-opted and convoluted by the government, which began to claim that the remedy to overusing the seaside could come in the form of more bucolic abuses such as hunting, fishing, and camping. Thus paradoxically, having described the undesirable consequences of tourism for the environment at the National Tourism Awards meeting in early 1973, Sánchez Bella went on to exalt the multimillion-peseta potential of "el turismo 'au plen (sic) air', de la acampada, del 'camping' y del 'turismo de los lagos'" ('plein air' tourism, staying in tents, 'camping,' and 'lake tourism'). Having achieved the limit of coastal development, Spain was simply forced to unpack its wilderness to increase its tourism capacity. Tourism may well have been recast as an essential ingredient of "wellbeing" in peace with nature and without a view of the cement walls, yet its numbers-driven character was not about to change. In haste, Spain's advertising partners abroad commissioned articles on Spanish wild animals, fishing, and other activities beyond the sun and beach. In the next few months, the Dirección General de la Promoción del Turismo (Tourism Promotion Department) produced a series of posters under a common title "Environment Protection" showing the vistas of rivers and nondescript woods, sunsets, and autumn leaves. Although the series was obviously made for expanding the list of Spanish destinations, it was also summoned to propagate the new meaning of tourism among the domestic audiences.

Newspapers lauded the program's "preocupación trascendental y modernísima" (transcendental and utterly modern preoccupation) with exalting Spain's beauty "en armonía con el medio ambiente y el hábitat turístico" ("Nueva serie de carteles") (in harmony with the environment and tourism habitat).[3] Yet what did "tourism habitat" really mean? Initially, it sounded like an oxymoron pointing at the need of naturalizing or humanizing tourism. Indeed, since 1972 politicians began to insert the words "human" and "humanistic" randomly into texts about tourism. At the congress on the Mediterranean, Gregorio López Bravo, Spanish Minister of Foreign Affairs, called tourism a dangerous "human phenomenon" that can alter the "profoundly human system of life that we have created" (8) (un fenómeno humano . . . que debemos cuidar . . . para evitar que pueda constituir un elemento de alteración del sistema de vida profundamente humano que simpre hemos procurado). Yet new language was already being forged that would soon make the word "tourism" obsolete altogether. Its environmentally friendly substitute was to be called "leisure": "El turismo no es un muro de cemento, es la arquitectura del ocio" (tourism is not a cement wall, it is the architecture of leisure), Sánchez Bella stated at the November 1972 Mediterranean meeting. This playful formula anticipated the future evolution of Spanish tourism that rebranded itself as the "leisure industry" conceptually in 1973 and factually in the late 1980s (Granados Cabezas). Crucially, by positioning tourism as leisure, Sánchez Bella made it clear that tourism itself set the standards for "quality of life" and could therefore not be discouraged or restricted.

In the final balance, in the Spain of 1972–1973, the idea of "quality of life" had become shorthand for a new program of "humanizing" development by redirecting it to previously neglected markets. Clearly, such an official environmentalism figured prominently in the courting ceremonies between "prepostfrancoist" Spain (to use Eduardo Mendoza's coinage) and its global investors. The wedding, however, would have to wait and would not be yet officially announced. Tourism, the flagship of Spain's development, now became a testing ground for the new approach, charged with offering productive models for a "quality" development. Suddenly, newspapers were filled with classist remarks about the "scum of Europe" bringing Spanish beaches down to their low standards of hygiene. Already in January 1973, in preparation for the meeting of British Tour Agencies in Mallorca scheduled for November, officials decided to shed the tired sun and beach panorama. The correspondents of *La Vanguardia* didn't fail to notice the irony:

> nada de playas pobladas de bañistas y sombrillas, nada de hoteles mastodónticos, nada que pueda suponer aglomeraciones y tumulto. Hay que presentar la Mallorca bucólica, apacible, tranquila e ideal para un descanso en cualquier estación del año. Hay que presentar rebaños de ovejas pastando en el verde campo mallorquín, mostrar los olivos centenarios y las calas solitarias que todavía quedan. ("Los agentes de viaje británicos")
>
> (no beaches filled with bathers and umbrellas, no mastodontic hotels, nothing that could suggest agglomeration and disturbance. Mallorca should appear bucolic, peaceful, tranquil, and ideal for relaxation in any season. Flocks of sheep should ramble in the green Mallorcan valleys, centenary olive trees and the few remaining solitary coves should be in full view.)

Yet when the tour agents and British dignitaries eventually got to Mallorca, the leadership of Spain's Tourism Ministry had changed: in June 1973, the Tourism Minister Sánchez Bella who had envisioned the "leisure industry" model as a way of continuing expansion, was sacked. The turnover was peaceful, and the reasons for the firing were not announced. On the surface, the official language of tourism and its policies stayed the same as they had been under Sánchez Bella. However, extensive debates about "quality of life" were about to emerge, changing the course that Spanish environmentalism would take.

The Discontented and the Domesticated

Interestingly, even though there had always been opposition to tourism-driven development in Spain, nothing really changed until foreign investors began to suffer the consequences of the fuel crisis in 1972. Their dwindling support made the 1960s model of tourism, one that had allowed Spanish officials to work directly with foreign tour-operators that were funding (in amounts that were kept secret by the government) the construction of large hotels in exchange for future reservations, unviable. Stephen Danos of Centrehurst Limited (a British public relations agency), who was in charge of promoting Spain in the United Kingdom starting in 1963, remembered that in the 1960s "the most senior officials of the Ministry were available to the [tour] operators, either as a group, or individually. Various promises were made, which were hard to fulfill" (Danos, Hopkinson, et al.). In a proposal for a new public relations campaign drafted in October 1973, Danos shed light on the timing and the reasons for the changes that took place between the 1960s and 1973. In the summer of 1973, the economic indicators made it clear that tour operators were no longer reliable. Unavoidably, the model was no longer viable, and the international "quality of life" buzz—a product of the same economic crisis that was hurting Spanish tourism—facilitated the shift. With the official meaning of "quality" still fluctuating, it was a time when different concepts began to clash around, and be tested on, the tourism industry. In hindsight, it is clear that the debates served a twofold purpose: the economic need to "domesticate" global investment and the political need to promote the Falangista-inspired social agenda against the tradition of *desarrollismo*. And even though there were plenty of people in the opposition or in the underground parties who talked about "quality of life," they simply went for a neoliberal ride that the Spanish "discontented" were starting to enjoy, as we will see below.

Prior to 1973, Spanish tourism was organized in a way that worked well for large foreign investors, Spain's landed aristocracy, banks, and the officials who used the boom for personal enrichment. Not only those Spaniards who did not take vacations or who would have been denied admission to the resorts, but also large numbers of business owners, felt excluded and united their voices against the hyper-development of coastal areas. In an article eloquently titled, "Spain: The Nightmare of Tourism," the German magazine *Der Spiegel* explained that Spain's tax system was giving incentives to the owners of large estates to lease or sell to foreign companies for building resorts. In Cerrado de Carderón, in the Málaga province, a Belgian-Dutch trust was developing the lands of the Count de Larios; in the Canary Islands, the lands of the Counts de la Vega Grande were leased to a German company. Banks from Spain's industrial cities such as Banco de Bilbao or Banco Coca from Madrid also entered the development game. And there was definitely no shortage of trafficking of influence and corrupt officials. *Der Spiegel* cited an example of the Hotel "El Tope" in Puerto de la Cruz (Tenerife), built in a protected green area because the mayor, who also happened to be the owner of that hotel, reclassified the terrains. Basically, the boom entailed a de facto redistribution of land without a corresponding sharing of benefits. Unsurprisingly, then, the continuing over-development created growing discontent. Surprisingly, though, newspapers in Spain and abroad began to voice this discontent in unison precisely at the time when foreign tour operators failed.

To understand the ways in which the government used this opposition and the new meanings of environmentalism that emerged as the result, we must know more about Spain's new business partners and their plans. The proposals drafted by Stephen Danos, the British public relations expert mentioned earlier, were clear: "Very often large numbers are synonymous with low prices" (Danos,

Hopkinson, et al.). The imperative was to provide a product whose higher costs could be explained to and rationalized by the consumer. And this is why environment became a crucial factor. The campaign that Danos designed in 1973 revolved around the notion of "good product." In part, Spain's advantage hinged on the problems of Spain's neighbors, France (where Danos emphasized "sea pollution") and Italy (whose reputation was hindered by a recent "heavy cholera outbreak"). Among the factors negatively affecting the image of Spain in Britain, Danos mentioned Spain's similar issues such as "a) Medical. Cholera, typhoid and paratyphoid on several occasions; b) Organizational" (Danos, Hopkinson, et al.). Putting the plan into action, the following year Danos either commissioned articles or consulted their authors about Spain's lesser-known areas and activities unrelated to sun and beach. Attuned to the interests of a middle-class audience wishing to move beyond the beach, he crafted a hedonistic image of Spain that combined pleasant climate and old culture, beaches and woods, urban anonymity and the promise of rural seclusion. Still, when it came to Spaniards themselves, Danos was an obvious supporter of the 1960s boom and considered it beneficial to all Spaniards. He wrote in a letter to his counterpart in Spain's Tourism Ministry:

> From time to time Spanish tourist authorities are attacked in the newspapers for the amount and quality of development which has been allowed to go on along various Mediterranean Coastlines.
>
> It would be most helpful to me if I had available a counter argument to the critics who, if truth be known, think back nostalgically to the days when all the peasants were walking around barefoot, the people were poor and it was nice for the Englishmen who could buy a bottle of brandy for a few pennies. (Danos)

While he agreed with the technocrats and only opposed their attraction to the numbers of visitors instead of the quality of product, politically, Danos did not envision a system capable of facing this challenge as anything but a strong centralized state supporting niche development. As his proposal demonstrated, the guarantees of the "high quality" of the Spanish tourism product were "Control by the Spanish tourist authorities" and "Concentration of resorts making for easier control and transportation." In other words, no political change or even economic liberalization was necessary to ensure a quality tourism product. The Falange politicians who thought that it was their moment to implement, finally, a "social state" could go along with that. However, both the foreign agents like Danos and the Falangistas underestimated the movement of people conjured by the recent economic development who no longer wished to be mere consumers of government-made bounty and wanted to come closer to the source of it all.

The last two years of Franco's dictatorship were not a time of public agreement. Yet when it came to "quality of life," almost everyone seemed to agree. The reason was simple: this blanket notion allowed the Falangista revivalists in the government to steer the environment debate in the direction that disarmed all possible opposition. Spanish entrepreneurs, of course, wanted the liberalization of economic activity—the opposite of the strong fascist "social state" providing benefits to its citizens. Those in communist and other similar opposition groups had their own ideal of "social state"—but they did not want Falange in it. The entrepreneurs demanded access to the power over resources and therefore criticized the government's irresponsibility. The opposition criticized the government, citing the misuse of the environment as merely one of the regime's thousand crimes against the Spanish people. So, official and dissident press alike ran articles predicting a tourism-borne ecological catastrophe. The dissident journal *Cuadernos para el diálogo* dedicated a special supplement to the problems of tourism in the Balearic Islands with a telling title: *El turismo: ¿un falso boom?* (Tourism, a Fake Boom?), written by the Guillermo Díaz-Plaja, a distinguished intellectual who has just become the dean at Barcelona's Escuela Universitaria de Estudios Empresariales. The no-less-critical *Cambio 16* featured a series of articles by a communist, "Felipe" (member of FLP: Frente de Liberación Nacional (National Liberation Front)), Carlos de Zayas,[4] focusing on the disastrous consequences of tourism. One of them announced its contents: "¡Vaya pulpo (el turismo)!" (What a flop [Tourism]!)."

When foreign correspondents joined the choir denouncing the excesses of Spanish tourism policy, they found equally critical interlocutors among the mayors of coastal towns. José María Revente, the mayor of Benidorm, Emilio Puignau Nogier, the mayor of Cadaqués, and Cayetano Utrera Ravassa, the mayor of Málaga gave their testimonies to Henry Giniger for an article, "Spaniards question mass tourism," for the *Herald Tribune* (1973) expressing their concern for the environment or denouncing the disastrous consequences caused by tourism. As Cayetano Utrera Ravassa, the mayor of Malaga, said, "the blue Mediterranean is becoming increasingly less blue off the bathing beaches." Local business owners also joined in. Josep Ensesa Gubern, a Barcelona oligarch, called the Spain of excessive tourism development "the playground of Europe's rabble" (Giniger). *Der Spiegel* quoted the President of Málaga's Chamber of Commerce: "We cannot continue using tourism as a cover for the enormous boiling pot of our underdevelopment" ("Spanien: Alptraum Tourismus").

The list of critics would be incomplete without the economists, scholars, and social scientists who also revealed the glaring deficiencies of Spain's tourism planning and its ecological consequences. The article in *Der Spiegel* mentioned above cited a whole series of studies, ranging from the economics of tourism in Málaga by Juan Joné Solá to the work on the destruction of agriculture in the Canary Islands by Jeronimo Saavedra, professor of labor law. Adolfo Drigani, the Argentinian journalist, cited "expertos y observadores" (experts and observers) who were afraid that many people would abstain from coming to Spain "por falta o inexistencia de una infraestructura adecuada (agua, saneamientos, sanitarios, etc.) en algunos lugares; [. . .] o por el creciente deterioro del medio, agresión urbanística, expolio y suciedad de playas y paisajes" (Drigani 28) (because adequate infrastructure [water, sanitation facilities, toilets] were, in some places, scarce or nonexistent [. . .] or because of the deteriorating environment, aggressive urbanization, plunder, and dirty beaches and surroundings).

It would appear that Spain's "discontented" were joining the environmentalist movement. Yet the real interests of these critics were various, and not all of them were "green" or even in solidarity with the "dispossessed." While the intellectuals opposing the regime wanted to lay bare its complicity with foreign capital, business owners interested in bringing foreign capital home demanded that the government authorize gambling. Differences of opinion due to class and ideological clashes were also matched by the diversity of regional claims: while all local authorities demanded a compensation for the export of their resources, the sense of dispossession was higher in Catalonia, where the tensions caused by tourism became a part of national self-determination. Hence, though everyone concurred that irresponsible development was bringing Spain to a catastrophe, not many critics would agree about the remedies.

Amidst this surprising accord between the "discontented" and the few intellectuals and administrators speaking on behalf of the "dispossessed," almost nothing was being said about alienation as the root of what was happening in Spain. A rare shot of skepticism came from Manuel Vázquez Montalbán, the writer who dismissed the whole debate about both nature and tourism. For Vázquez Montalbán (47–49), this was a pseudo-choice offered up by financiers in order to distract humanity from the real problem that was none other than the capital furnished by the financiers itself, fueled by the "irracionalismo patológico que explica la mística y la urgencia del week-end, del 'retorno a la naturaleza', de las vacaciones" (pathological irrationalism that explains the mystique and the necessity of a weekend, of "return to nature," of vacations). From that standpoint, the concerns about nature conservancy were from the outset compromised by the capitalist origins of the concept of "nature" itself and were therefore to be rejected as a mere "ideología de la tregua que el hombre lobo concede a sus semejantes, igualmente hombre lobos" (47–49) (ideology of breather that a wolfman concedes to his fellows, who are wolfmen as well).

Voices of discord of a different kind came from the activists of animal protection—all of them women. As soon as the government officials began to mention nature in their speeches, Dolores Comas Marsans, then the vice president of the International Federation of Animal Protection, wrote a series of letters to the Tourism Propaganda officials, telling them about her work in Spain and abroad. Without asking for the officials' support, she tried to educate the government that had just discovered environmentalism, introducing different definitions of nature and the environment and engaging the authorities in a dialogue (Comas Marsans). Carmen Alonso, writer and activist from Oviedo, was far less diplomatic in her attitude toward the official environmentalism that ignored the needs of animal protection: "Hoy los Gobiernos de todas las naciones promueven y apoyan campañas en pro de la naturaleza; en estas campañas no se incluye al animal doméstico (como si este no fuese también naturaleza)" (Alonso) (today the governments of all nations promote and support campaigns in defense of nature; in these campaigns domestic animals are not included [as if they were not a part of nature]). Had they been heard, these women could tell their "discontented" male comrades a few things about the alienation of humanity from nature.

The Triumph of "Quality of Life"

Once the question of who controlled the use of resources was taken out of the equation, very few could dispute the need for a development consistent with an improving "quality of life." It was the Falange that dominated that debate. The Falangista intellectuals had kept a varying distance from things that Franco was doing in the 1960s. Some, like Dionisio Ridruejo, passed on into the opposition and found themselves in jail. Others stayed with the regime and became its commentators and occasional critics. Since the technocrats fell out of favor in 1969, the Falange's role in the government was on the rise, and so was the importance of their opinions. In the early 1970s, one of the Falangistas' most important causes became the defense of "quality of life" against the "quantity of possessions" ("Calidad de la vida"). Thanks to their efforts, the environment-based European definition of "quality of life" that initially triumphed in Spain was supplanted by a new, synthetic definition: an effect of a guided state policy, elaborated in the United States.

In the summer of 1972, while the delegates of the Stockholm Conference were debating the future of development, Franco's old-time literary adviser Manuel Aznar sensed that a paradigmatic shift was approaching: "se presiente que los pueblos rectores de la evolución del mundo quieren organizar y aplicar todas las fuerzas de la ciencia y de la técnica de manera que sirvan de acrecentamiento de la 'calidad de vivir'" (18) (one can feel that the nations who lead the world evolution want to organize and apply all forces of science and technology to raise the "quality of life"). Aznar wrote this in a preface to an ambitious publication: a coffee-table-sized album of bird's-eye views of Spain with texts collected from Spanish writers, some of whom had only recently been officially rehabilitated. Although published by the Dirección General de Promoción del Turismo, this was, however, a volume in Spanish for the domestic market; after publication, all newspapers had to run a press release introducing the book and glorifying its relevance. Its main objective was to explain the 1970s version of tourism to the new generation of Spaniards, and Aznar was charged with finding, in the incoherent array of posttechnocratic policies, a unifying direction—or inventing it. He did so using the rancid colonial metaphor of a feminized land awaiting a conqueror: "La 'calidad de vida' ha de ser obra vuestra, españoles jóvenes. Ahí tenéis a España, aguardándoos. Miradla desde el tejado de las nubes, desde el alto cielo, y llegaréis mejor al entendimiento de vuestra empresa: más claramente alcanzaréis el horizonte de vuestra misión" (20) ("Quality of life" has to be yours, young Spaniards. Here is Spain, waiting for you. Look at her from the platform of clouds, from the high sky, to better grasp your task: to see clearly your mission's horizon).

If the technocrats were trying to work around the European environmentalism in their version of "quality of life," the Falangista thinkers such as Aznar relied on the visions of society coming from America. So Aznar's prospects for Spain were grounded in the work of the futurists Herman Kahn, Anthony J. Wiener (1967), and Alvin Toffler (1970), who predicted that a technological revolution would spur a radical rupture with the past (18). The conclusions that Aznar derived from this fact, however, were closer to the 1930s dusty eschatology than to science fiction. If change was coming, he claimed, then the dream of a "social" state could come true because the growing economic prosperity would make spiritual pursuits relevant again:

> El grito será, si no lo es ya, "conquistemos una vida de mayor plenitud". Lo cual no equivale a pedir "mayor comodidad", ni "más amplio confort", ni siquiera "una riqueza más abundante" para un bienestar material más completo, sino que se pretende añadir a la vida valores más altos. "La calidad de vida" implica el retorno

> a estimativas mucho más exigentes y a la reinvención de estímulos de cultura, de anhelos espirituales, de satisfacciones de los sentidos y de incitaciones morales que proyectándose sobre la personalidad del hombre lo exalten y enriquezcan. (18)
>
> (The war cry will be, if not yet, "let us conquer a life of greater plenitude." It is not the same as asking for "enhanced convenience," or "more comfort," or even for a "greater prosperity" to achieve a more complete material wellbeing. It is, rather, about adding higher values to life. "Quality of life" implies a return to applying much stricter estimates and to reinventing cultural stimuli, spiritual quests, and a satisfaction of moral senses and challenges that would glorify and enrich the human personality onto which they are projected.)

Hence while Aznar's version of "quality of life" polemically targeted the technocratic materialism, it was, nevertheless, no more nature-friendly than the machine simulated "biosphere" that tourism officials like Sanchez Bella were envisioning at that time. The only difference was that Aznar was essentially proposing a return to the Falangista 1930s ideal of a "new man."

The Falangistas did not reject economic growth any more than their technocratic opponents: their anti-prosperity language simply reflected a need to harvest political benefits from the uneven distribution of income during the *desarrollismo*. A spiritually-enhanced concept of "quality of life" was a tool for achieving this goal. The interesting part of this story, however, is that the "quality of life" that the Falangistas put on Spain's banners was consistent with the "eight social areas" of the policy elaborated by U.S. experts at the end of 1972: "health; education; employment; income; housing and physical environment; leisure and recreation; public safety and legal justice; and population" (*Quality of Life* 20). The Falangista debates took shape in the press at the same time when these tenets were being elaborated—the end of 1972—making it difficult to trace a direct connection. Yet the similarities between the "quality of life" as a bundle of quantifiable variables of relevance for policymakers and the Falangista idea of "quality" as a set of social programs are striking. So, too, are the differences between this line of thought and the European definition of "quality of life" that the Spanish technocrats were using. While European politicians defined "quality of life" as synonymous with a well-preserved environment, the new multi-factor definition reversed this relationship, making the human understanding of the environment inseparable from how "quality of life" was defined. The reports of the 1972 meeting in Virginia directly stated that "a wide range of policy issues, both domestic and international, that relate to the environment may well be determined by considerations based on what man considers an adequate level of *quality of life*" (*Quality of Life* II–10, italics in original).

For the Falangista thinkers, this also became the foundational concept on which to ground their project of a "social state." Jesús Vasallo, a writer who at the time directed the Prensa del Movimiento, aptly summed up this position when he stated that neither "per capita rent," nor "el aumento de las unidades de consumo nos dirán nada en el orden del espíritu, si no mejoramos también la calidad de la vida del hombre. La educación, la sanidad, la cultura, el ocio, la formación profesional y un ambiente sano han de estar al alcance de todos. De lo contrario, podríamos dejar de llamarnos ya sociales" ("El reto del desarrollo regional") (the rising units of consumption would [not] tell us anything about the spiritual realm if we do not also improve the quality of human life, education, healthcare, culture, leisure, professional formation, and healthy environment have to be at everyone's reach. Otherwise, we can stop calling ourselves "social [thinkers and politicians]"). With the rise of the new language, the technocrat tourism ministers found themselves alone in their attempts to reconcile development with the responsible use of resources because, outside of the tourism context, the environment did not seem to be an issue. "Quality of life," the triumphant concept, dominated Spain's 1972 Economic Development Plan, gave sense to regional planning, and even merited its own convention: a National Biennial Quality Congress.

The same synthetic formula of "una justa distribución de la riqueza y del bienestar, de la cultura y del ocio, de la salud y de la calidad de la vida (a fair distribution of prosperity and wellbeing, of culture and leisure, of health and quality of life)," as José Utrera Molina, the Falangista President del Movimiento in Franco's last (1974) government, had it ("Vivimos en una época de subversion total"), became the slogan of the dictatorship's final years. In December 1974, looking at Spain's Development Plan for 1975, Ramiro Campos Nordmann, a professor of economics who used to support technocratic expansion, concluded that "quality of life" defeated the *desarrollismo* when he stated that Spanish planners acquired a vision of "mayor bienestar con independencia del desarrollo por el desarrollo. . . . Hoy ya afortunadamente en nuestro país preocupa el medio ambiente sereno y la calidad de la vida, y en algunos casos, priman sobre otras consideraciones" (5) (a greater well-being independent of development for development's sake. . . . Thankfully, today in our country tranquil environment and quality of life are what we care about—in some cases, above and beyond all other considerations). Campos Nordmann's optimistic conclusion referred directly to the social policy of well-being coming from the United States: Paul Samuelson's theory of "net economic product," which Campos Nordmann characteristically translated as the "Bienestar Económico Neto" (net economic well-being).

In Franco's Spain, the winning concept of "quality of life" was, therefore, a flexible notion that combined the Falangista ideal of social justice with the American approach to happiness as a social policy. It no longer had as its cornerstone the European standard of responsible use of environmental resources. The same strikingly ubiquitous development, however, can be traced in the rest of Europe and on all sides of the ideological spectrum. In 1974, French Socialist President Valéry Giscard d'Estaing created a post for a Minister of Quality of Life, a fact widely commented on in the Spanish press. More surprisingly, considering Spain's specific historical conditions, the old-school antibourgeois thrust of the Falangista thinking stroke a note similar to the spirit of the communist opposition. In the Spanish underground, the Partido Socialista Unificado de Cataluña, the Catalan Communist Party that had not rejected Marxism-Leninism or followed Santiago Carrillo on the path to Eurocommunism, used very similar language to justify the need for a social revolution. The only difference was that Catalan communists had their own alienation to fight—they claimed national sovereignty for Catalonia:

> La supresión de los vestigios feudales, la consecución de las libertades nacionales de Cataluña, la consolidación de la democracia, una profunda reforma de la enseñanza, de la sanidad, la mejora de la calidad de la vida, son objetivos urgentes que sólo podrán ser alcanzados mediante una revolución social que ponga las bases del socialismo, régimen basado en la socialización de los principales medios de producción y de cambio que permite abolir la explotación del hombre por el hombre. ("La situación política y las tareas del P.S.U.C.")
>
> (The suppression of the remains of feudalism, the achievement of national liberties for Cataluña, the consolidation of democracy, a deep reform of education and health care, the improvement of quality of life are aims that can be achieved only through a social revolution establishing the foundations of socialism, a regime grounded in the socialization of the main means of production and exchange, that would end the exploitation of man by man.)

Indeed, not even the most radical ideological opposition could offer an alternative to the "quality of life" politics.

How did these changes affect tourism, the sector where the environmentalist definition of "quality of life" initially took off? There, too, the term was quickly recast as a social issue only marginally related to the environment. The ability to offer pristine landscapes was, of course, important because it resulted in a more expensive tourism product. Yet the definition of "tourism" was departing from the meaning it had in the 1960s (the hospitality business) to become something that Spaniards could also consume: leisure. Fernando Liñán, who succeeded Sánchez Bella as Tourism Minister in 1973, defined new tourism as "la posibilidad de escoger y disfrutar efectivamente del tiempo libre" (the possibility to choose and effectively enjoy free time) and a principal component of "quality of life" ("El Ministro de Información y turismo inauguró"). As leisure, tourism became a matter of social policy. In October 1973, the Dirección General de la Promoción del Turismo was replaced by a new structure, the Dirección General de la Ordenación del Turismo, a name that heralded that planning ideal. On the surface, its program was not too far removed from the tasks of the previous Ministry: raising the quality of tourism product and expanding the activity away from the coasts. On a deeper level, however, it signaled a new approach: the major factor affecting the nation's wellbeing was now the tourism activity of the domestic population, rather than income from foreign visitors. "El desarrollo de un pueblo es paralelo al de su turismo interior (a nation's development is parallel to that of her interior tourism)," stated the new General Director, Francisco Javier Carvajal, invoking the idea of the "social state" and proclaiming the need for "planteamineto diferenciado del turismo interior por edades, profesiones o incluso 'deseabilidades'" (a differentiated approach to interior tourism by age, profession, and even "preference") that would ensure that "cada cual encuentre el turismo justo que necesite" ("Nueva Dirección General de Ordenación del Turismo") (everyone finds the type of tourism that they need) . Regional planning, niche tourism, and leisure would survive beyond that short-lived government and Francoism itself and would become the three foundational blocks of Spain's Transition.

Conclusions

The debates analyzed here had nothing to do with the "ecology of the poor" (coined by José Martínez Alier) or any other form of contestation normally associated with anti-capitalist activism. Rather, this was an official environmentalism that different pressure groups exploited for their own needs in order to secure their political or economic power while also prolonging Franco's dictatorship beyond the life of Franco himself. This story of Spain's environmentalism gradually subsumed under "quality of life" should therefore read as a cautionary tale about capitalism producing a puppet version of its enemy and a script according to which the threats to humanity can be cured with more capitalism, provided it includes the interests of local businesses. It was this context of a late dictatorship that entrusted its economic stability to tourism that makes it so difficult for a present-day reader to tell a dissident from a loyalist. Some of the opinions reflected the positions of the supporters of the 1961–1969 *desarrollismo*, losing power. Other views were backed by the social-corporate opposition to that experiment coming from the "camisas viejas." Only the third, at the time the least visible, group inspired by the Communist and Euro-Communist programs uncovered the complicity between the regime, capital, and Spain's territorial structure; but this group also used the policy outcomes coined in the United States.

Before Michel Foucault and his followers developed the notion of "biopolitics," it was easier to understand the alienating effects of the official environmentalism than to find fault with "quality of life." In Spain, it did not take long for the intellectuals to uncover environmentalism's neo-liberal or "neo-capitalist," as it was then called, thrust. One of the most radical denouncements was made in 1979. It belonged to Alfonso Pérez Agote Poveda (at present, professor of sociology at the Universidad Complutense de Madrid), who stated: "el peligro de una catástrofe producida por la ruptura del sistema ecológico desdibuja las contradicciones del sistema social. El enemigo—y el amigo a encontrar—es la Naturaleza. . . . No es la catástrofe futura, sino el peligro actual de la catástrofe futura lo que funciona como integrador del sistema social" (11) (The danger of a catastrophic rupture of the ecological system overshadows the contradictions of the social system. The enemy—and the future friend—is nature. . . . It is not a future catastrophe, but rather the present danger of a future catastrophe that works as the social system's integrating force).

As to "quality of life," after Franco's death it only acquired greater relevance. An early statement of the transitional government promised that:

> el Gobierno dirigirá su política hacia los grandes objetivos sociales y económicos de conseguir para todos los españoles, incluso los hoy emigrados de la Patria, un puesto de trabajo asegurado; el reparto de las cargas y sacrificios de forma equitativa, pero con especial atención a protección de los más débiles; y la consolidación y avance de los niveles de bienestar y calidad de vida. . . . ("Declaración del gobierno" 2)
>
> (the government will aim its policies at the grand social and economic objectives of achieving, for all Spaniards, even those who are today in immigration, a secure job; a distribution of obligations and sacrifices, paying special attention to

> protecting the weakest; and consolidation and improvement of the levels of well-being and quality of life. . . .)

The term's scope of usage also expanded. The Central Committee of Spain's newly-legalized Communist Party, the Partido Comunista Español (PCE), trumpeted the falling quality of life as a sign of an imminent social overturn; still, it used the language of the American policy indicators: "la enseñanza, la sanidad, la vivienda, el urbanismo, y, en general, la calidad de la vida" ("El PC rechaza el proyecto de pseudemocracia otorgada") (education, health care, housing, urban development, and quality of life in general). In the heat of the electoral campaign, PCE consistently cited "las mil deficiencias de la calidad de la vida bajo el capitalism" (the thousand and one deficiencies in quality of life under capitalism) to bring out the voters (Martín 2). But it was hard to win a battle whose slogans appeared on everyone's banners. The sub-secretary for the interior market aspired to "una sociedad nueva y distinta que desea para todos iguales posibilidades y, por otro lado, una calidad de los productos creciente que permita satisfacer las aspiraciones permanentes y crecientes de mejora en la calidad de la vida" ("Los consumidores") (a new and different society that seeks for everyone equal opportunities and, on the other hand, an improving quality of products that could satisfy permanent and growing aspirations for improving quality of life) . King Juan Carlos argued: "El verdadero progreso es irrealizable en la escala social, si no se perfecciona a calidad de la vida humana" ("Jornada de los reyes en Stuttgart") (the real progress is not realizable on a social scale if the quality of human life does not improve). Meanwhile, private companies and banks reformulated their statutes to add "quality of life" to their mission ("Hidroeléctica Española"). During the debates about the draft of the 1978 Constitution, the article on the environment, centered on the notion of quality of life, suffered many amendments ("Numerosas enmiendas"). It is only logical, then, that the concept appears three times in the Constitution's final version and still has many supporters and believers today.

The analysis here of the "quality of life" ideologies helps to elucidate the complexities of the dictatorship's endgame. Yet crucially, the dialectic of Marxism, neoliberalism, and postfascism underpinning these debates did not evaporate after Francoism. Revealing the early steps of neoliberalism in Spain, these debates are still affecting the present-day understanding of life and the environment. The 1972–1975 version of environmental protection as tourism quickly became obsolete. Still, rather than lose relevance, tourism recast as "quality of life" became a lasting part of the Francoist heritage—or, shall we say, baggage—responsible for many values that were eventually ratified during the

Transition and Democracy and heralded, especially, by the Partido Socialista Obrero Español (PSOE) trend of leftism. Among the values underpinned by such a pseudo-environmentalism we find, not only sustainability, but also and more ominously, beauty and leisure, which remained in the foundation of post-Franco politics of all political colors. We are still unwittingly suffering the consequences of that version of environmentalism.

Notes

The research for this article was funded through Marquette University Regular Research Grant (December 2012-January 2013). The author wishes to thank Sally Anna Boyle and the editors of this volume for their comments and thorough corrections.

1. The year when Manuel Fraga Iribarne was appointed the Minister of Information and Tourism.
2. The Words about pluralism belong to the Torcuato Fernández Miranda, vice president of the 1973–1974 cabinet.
3. Conversely, the editor in Tenerife did not even understand the concept of the "defensa del entorno" and the advertising appeared under the military-sounding title "Turismo y la defensa del contorno" ("Más de siete millones de folletos"). Tenerife suffered from unsustainable tourism development, and the pun could have been intended.
4. Historians of the 1960s and early 1970s list Carlos de Zayas, the future PSOE Deputy for the Balearic Islands, as a "young socialist" and a PCE-infiltrated member of FLP (McDermott, 966, and Chust Calero and Broseta, 146).

Works Cited

Alonso, Carmen. "El maravilloso mundo de los animales. El absurdo complejo de inferioridad." *La Nueva España* (Oviedo), 21 January 1973.

Anderson, Robert, Hans Dubois, Tadas Leoncikas, et al. *Third European Quality of Life Survey: Quality of Life in Europe; Impacts of the Crisis.* Luxembourg: Publications Office of the European Union, 2012. Eurofound. Web. 14 Aug. 2013.

Anuario internacional sobre prevención de riesgos psicosociales y calidad de vida en el trabajo: regulación de los riesgos psicosociales en la sociedad global, una perspectiva entre Europa y América. Madrid: Secretaría de Salud Laboral UGT-CEC D. L, 2010. UGT. Web. 14 Mar. 2014.

Aznar, Manuel. "Prólogo." *Desde el techo de España.* Madrid: Ministerio de Información y Turismo; Dirección General de Promoción del Turismo, 1972. 9–20.

"Calidad de la vida." *Jaén*, 23 Aug. 23 1972. Reprinted in *Hoja del lunes* (Madrid), 28 Aug. 1972: 2. Web. 15 Aug. 2013.

Casas, Ferrán. "Calidad de vida y calidad humana." *Papeles del Psicólogo* 74 (1999). Web. 14 Mar. 2014.

Campos Nordmann, Ramiro. "El bienestar económico nacional." *El Mediterráneo* (Castellón de la Plana). 7 Nov. 1974: 5. Web. 15 Aug. 2013.

Chomsky, Noam. *Language and Politics*. Oakland, CA: AK Press, 2004.

Chust Calero, Manuel, and Salvador Broseta. *La pluma y el yunque: El socialismo en la historia valenciana*. Valencia: Universitat de Valencia, 2003.

Congreso de ecología y turismo del mediterráneo occidental. Textos y documentos. Congrés d'ecologie et tourisme de la méditerranée occidentale. Textes et documents. Madrid, 30 Octubre-3 Noviembre 1972. Madrid: Ministerio de Información y Turismo, 1972.

Crumbaugh, Justin. *Destination Dictatorship: The Spectacle of Spain's Tourist Boom and the Reinvention of Difference*. Albany, NY: SUNY Press, 2009.

Danos, Stephen. Letter to Francisco Perez Naranjo, Jefe Negociado Publicidad y Relaciones Públicas, Ministerio de Información y Turismo. 1 February 1973. Manuscript. Archivo General de Administración, Alcalá de Henares.

Danos, Stephen, C.D. Hopkinson, J. Schiefield, S. Forde, W.L. Lewis. *Informe y propuesta de relaciones públicas. Redactado para el Director de Oficina Nacional Española de Turismo en Londres. London, October 10, 1973*. Manuscript. Archivo General de Administración, Alcalá de Henares.

"Declaración del gobierno." *El Mediterráneo* (Castellón de la Plana). 16 Dec. 1975: 1–2. Web. 15 Aug. 2013.

"Declaration of the United Nations Conference on the Human Environment." Web. 14 Mar. 2014.

"Destacada intervención de Don Laureano López Rodo: 'Por encima de cualquier otro, el objetivo general de la lucha contra la contaminación es salvaguardar la salud del hombre', dijo el ministro español." *La Vanguardia* (Barcelona), 8 Jun. 1972: 19. Web. 14 Mar. 2014.

Díaz-Plaja, Guillermo Luis. *El turismo, ¿un falso «boom»?* Suplemento monográfico N° 29 de *Cuadernos para el Diálogo*. Madrid: Destino, 1972.

Draper, Peter, and David R. Thomson, "The Quality of Life: A Concept for Research and Practice." *Journal of Research in Nursing* 6: 3 (May 2001): 648–57. Web. 14 Mar. 2014.

Drigani, Adolfo. "El momento de escoger." *Visión* (Madrid), 22 Sep. 1973: 28–32.

"El Ministro de Información y Turismo inauguró anoche Hogarhotel 1973. En su discurso estimó compatibles el turismo de cantidad y de calidad." *El Mediterráneo* (Castellón de la Plana), 18 Nov. 1973: 21. Web. 15 Aug. 2013.

"El PC rechaza el proyecto de pseudemocracia otorgada." *Mundo Obrero Semanal* (Madrid), 24 Nov. 1976: 3–4. Web. 14 Mar. 2014.

"El príncipe de España presidio . . ." *El Mediterráneo* (Castellón de la Plana), 22 May 1973: 18. Web. 15 Aug. 2013.

Fernández-López, Juan Antonio, María Fernández-Fidalgo, Alarcos Cieza. "Los conceptos de calidad de vida, salud y bienestar analizados desde la perspectiva de la clasificación internacional del funcionamiento (CIF)." *Revista Española de Salud Pública* 84. 2 (2010): 169–84. Web. 14 Mar. 2014.

Giddens, Anthony. *The Politics of Climate Change*. New York: John Wiley & Sons, 2013.

Giniger, Henry. "Spaniards question mass tourism." *International Herald Tribune* (Paris), 27 Aug. 1973: 6. Web. 15 Aug. 2013.

Granados Cabezas, Vicente. "Nuevas formas de turismo: el turismo rural en Andalucía." *Boletín económico de Andalucía* 10 (1988): 35–50. Web. 15 Aug. 2013.

Harvey, David. "Marxism, Metaphors, and Ecological Politics." *Monthly Review* 49.11 (April 1998). Web. 14 Mar. 2014.

_____. *A Brief History of Neoliberalism*. Oxford: Oxford University Press, 2005.

_____. *The Enigma of Capital, and The Crises of Capitalism*. London: Profile, 2010.

"Hay, verdaderamente, una voluntad de hacer progresar las comunidades europeas."
El Mediterráneo (Castellón de la Plana), 22 Oct. 1972: 1. Web. 15 Aug. 2013.

"Hidroeléctica española; Junta General de Accionistas." *Hoja del lunes* (Madrid), 26 Jan. 1976: 26. Web. 15 Aug. 2013.

"Jornada de los reyes en Stuttgart." *El Mediterráneo* (Castellón de la Plana), 23 Apr. 1977: 1. Web. 15 Aug. 2013.

Journals on Happiness and Wider Quality of Life. <http://www1.eur.nl/fsw/happiness/journals.htm>. World Database of Happiness. Web. 14 Mar. 2014.

Kahn, Herman, and Anthony J. Wiener. *The Year 2000: A Framework for Speculation on the Next Thirty-Three Years*. New York: Macmillan, 1967.

"La política, de semana a semana." *El Mediterráneo* (Castellón de la Plana), 1 Oct. 1972: 8. Web. 15 Aug. 2013.

"La situación política y las tareas del P.S.U.C. Informe en el III Congreso del P.S.U.C., febrero 1973." *Treball* (Barcelona) 36: February 1 1973. s.p. Web. 14 Mar. 2014.

León Jiménez, Fernando. "Dimensiones del concepto constitucional de 'calidad de vida': especial referencia a la ambiental." Web. 14 Mar. 2014.

"Los agentes de viaje británicos desean cambiar la imagen turística de Mallorca. En el congreso, que celebrarán el próximo noviembre en Palma, estudiarán, con los elementos locales, la presentación de temas bucólicos y sedantes." *La Vanguardia* (Barcelona), 20 Jan. 1973: 4. Web. 14 Mar. 2014.

"Los consumidores podrán dialogar y trabajar con los responsables de legislar y ordenar el comercio." *ABC* (Madrid), 7 May 1977: 172. Web. 14 Mar. 2014.

Martín, Aurora. "¿Qué piensa el 'ama de casa'?" *Mundo Obrero* (Madrid), 11 Apr. 1977: 2–3. Web. 14 Mar. 2014.

Martín Aceña, Pablo, and Elena Martínez Ruiz. "The Golden Age of Spanish Capitalism: Economic Growth Without Political Freedom." *Spain Transformed: The Late Franco Dictatorship, 1959–1975*. Ed. Nigel Townson. New York: Palgrave Macmillan, 2007. 30–46.

"Más de siete millones de folletos y dos millones de carteles, para la promoción del turismo español y defensa del contorno." *El Día* (Tenerife), 14 April 1973.

McCall, Storrs. "What is quality of life?" *Philosophica* 25:.1 (1980): 5–14. Web. 14 Mar. 2014.

McDermott, Patrcia. "Gorkin y Cia: Una interrogación sobre la 'Conspiración liberal' a través de la srevista de exilio exterior e interior durate la guerra fría cultural." *Escritores, Editoriales y Revistas Del Exilio Republicano de 1939*. Ed. Manuel Aznar Soler. Valencia: Renacimiento, 959–69.

Meadows, Donella H, et al. *The Limits to Growth: A Report for the Club of Rome's Project on the Predicament of Mankind*. New York: Universe Books, 1972. Google Books. Web. 14 Mar. 2014.

Michalos, Alex C., ed. *Citation Classics from Social Indicators Research: The Most Cited Articles*. New York: Springer, 2006.

Minca, Claudio, and Tim Oakes, eds. *Real Tourism: Practice, Care, and Politics in Contemporary Travel Culture*. London: Routledge, 2012.

Moreno Jiménez, Bernardo, and Carmen Ximénez Gómez. "Evaluacion de la calidad de vida." *Manual de evaluación en Psicología Clínica y de la Salud*. Ed. Gualberto Buela Casal et al. Madrid: Siglo XXI, 1996, 1045–70. Google Books. Web. 14 Mar. 2014.

"Nueva Dirección General de Ordenación del Turismo." *SPIC* (Madrid), Oct. 1973: 29.

"Nueva serie de carteles." *El Alcázar* (Madrid), 16 Apr. 1973.

"Numerosas enmiendas." *El Mediterráneo* (Castellón de la Plana), 31 Aug. 31 1978: 9. Web. 15 Aug. 2013.

Nussbaum, Martha, and Amartya Sen, eds. *The Quality of Life*. Oxford: Clarendon Press, 1993.

Pack, Sasha D. *Tourism and Dictatorship: Europe's Peaceful Invasion of Franco's Spain*. New York: Palgrave Macmillan, 2006.

Payne, Stanley. *Fascism in Spain: 1923–1977*. Madison: University of Wisconsin Press, 1999.

Perez Agote, Alfonso. *Medio Ambiente e ideología en el capitalismo avanzado*. Madrid: Encuentro, 1979.

Phillips, David. *Quality of Life: Concept, Policy, and Practice*. 2nd ed. London: Routledge, 2012.

Preston, Paul. *The Triumph of Democracy in Spain*. London: Routledge, 1986.

Quality of Life Concept. A Potential New Tool for Decision-Makers. Washington, DC: Environmental Protection Agency. Office of Research and Monitoring. Environmental Studies Division, 1973. Google Books. Web. 14 Mar. 2014.

Quality of Life: A multidisciplinary journal of food science, environmental science and public health. Banja Luka: Pan-European University, Apeiron, 2010. Web. 14 Mar. 2014.

Roccu, Roberto. "David Harvey in Tahrir Square: The Dispossessed, the Discontented, and the Egyptian Revolution." *Third World Quarterly* 34.3 (2013): 423–40.

Sánchez Biosca, Vicente, and Rafael Rodríguez Tranche. No-Do: *El tiempo y la memoria*. Madrid: Cátedra, 2000.

"Segunda sesión de trabajo en la conferencia de Helsinki." *El Mediterráneo* (Castellón de la Plana), 21 Nov. 1969: 11. Web. 15 Aug. 2013.

Scheuer, Stefan, ed. *EU Environmental Policy Handbook. A Critical Analysis of EU Environmental Legislation.* Brussels: European Environment Bureau. Web. 26 Jul. 2013.

"Spanien: Alptraum Tourismus." *Der Spiegel*, 27 Aug. 1973 27: 72–87. Web. 15 Aug. 2013.

Toffler, Alvin. *Future Shock*. New York: Bantam Books, 1970.

Vasallo, Jesús. "El reto del desarrollo regional." *El Mediterráneo* (Castellón dela Plana), 4 Mar. 1972: 9. Web. 15 Aug. 2013.

Vázquez Montalbán, Manuel. "Los ritos de la fiesta o los estuches transparentes." *CAU*, Barcelona 4 (Oct./Nov. 1970): 47–49.

Vilarós, Teresa M. "Banalidad y biopolítica," *Desacuerdos* 2 (2005): 29–56.

"Vivimos en una época de subversión total y frente a la subversión no bastan ni represión ni reacción". *El Mediterráneo* (Castellón de la Plana), 22 Oct. 1974: 1, 13–14. Web. 15 Aug. 2013.

WHOQOL. Web. 15 Aug.2013.

Zayas, Carlos de. "¡Vaya pulpo! (turismo)." *Cambio 16*, 93, 27 Aug. 1973. 10–18.

_____. "*Turismo 1973* (I y II)." *Cambio 16*, 93, 27 Aug. 1973, 21–24; 94, 3 Sep. 1973. 18–22.

4

Die and Laugh in the Anthropocene: Disquieting Realism and Dark Humor in *Biutiful* and *Nocilla experience*

Katarzyna Olga Beilin

Scholars in the environmental humanities frequently debate which genres and moods are most appropriate for talking about the ecological crisis in order to awake the public and promote a change of attitudes. Fréderic Neyrat believes that "[w]e need to create a fiction of a catastrophe and act as if it was certain. [Because] we can, perhaps, avoid the worst if we believe that it is certain." Timothy Morton's concept of "dark ecology" follows a similar logic in describing a world in which the catastrophe has already begun to happen, having taken its course like a mortal sickness from which there is no redemption possible. In both there is an emphasis on incorporating the future into the present, detecting the roots of what will eventually happen in our everyday experience. This activity reminds me somewhat of looking for ghosts of the future, believing them to be as real as those of the past. Thinking of future as present requires an ontological turn. Morton's "dark ecology" is characterized by processes that occur silently and invisibly on a planetary scale; he calls these processes "hyperobjects." Hyperobjects are not only imperceptible but also hard to grasp with common sense; however, Morton insists that they are more real than what our common sense believes is "real" because they have enormous impact on our environment and on us, transforming generations of life to come. Global warming, nuclear waste contamination, and a sixth mass extinction are among the processes that humans have provoked and which will continue to transform Earth for thousands of years after our deaths. Morton argues that the uncanny feeling that has so often been rendered by contemporary art is caused by the subliminal awareness of the destructive character of the mutations that our bodies and our environments are and will be undergoing. The feeling of disquiet is due to the sensation that something that cannot be seen is happening around us and inside of us. Disquieting realism, a mood that appeared in Iberian literature during the 1980s in the context of *desencanto*, a feeling of dissatisfaction with the experience of newly achieved political and economic freedom as well as consumerism that turned to addiction, now, perhaps not disconnectedly, expresses concerns with environmental, social, and economic crises. The future from which we were borrowing, knowingly and unknowingly, has caught up with us, signaling that there is no more credit to be given and the time to pay has come. Instead of the excitement driven by disregard of the future, we now need to learn how to get excited by the idea of growing the future like a garden, taking care of it as if it were our child. But how to transplant these parenting and garden-growing emotions into the economy of motivation in the realms of production and pleasure?

Various environmental thinkers notice that the discourse of crisis and mourning, while deeply moving, may not be productive of the right motivation.[1] Ursula Heise expresses the need for finding alternative discourses and moods to substitute for the tone of mourning ubiquitous in environmental art. She quotes Joseph Meeker's *The Comedy of Survival* (1974), in which the author argues that comedy may be a more appropriate way to talk about environmental issues. While the discourse of doom and loss characteristic of tragedy is anthropocentric in its nature, comedy brings in interrelatedness and translates as change what a tragic author views as a loss. As a result, Meeker and Heise suggest that comedy may constitute a better tool for reflection and may be more likely to lead to a change in the public's attitude than doom. Humor, and particularly black humor, may indeed have an important role in subverting structures of thought which are responsible for environmental destruction because it ridicules habits responsible for thoughtless and egocentric behavior.

This essay focuses comparatively on two works employing different strategies of concern: doom and humor, both containing a mood of disquiet, that stems from the perception that, in the words of Gibbson, "[t]he future is already here—it's just not evenly distributed yet" (Kennedy, qtd. in Heise). In Alejandro González Iñárritu's *Biutiful* (2010), concerns with the invisible processes of mutation surrounding us are expressed through darkly disquieting sensations and nearly apocalyptic intuitions of the demise of human civilization due to self-destruction. Disquieting realism is also present, however, in a lighter mood, filled with irony, sarcasm, and various darker shades of humor in Agustín Fernández Mallo's *Nocilla* trilogy (2006–2009). The discourses that both these works adopt especially challenge those conceptual structures regulating our understanding of reality that are responsible for setting up a lowered value of life in the present economical and political system. Each possible scenario of demise and survival in these works brings its own ethical dilemmas and varying degrees of hope; and as a result each motivates different existential attitudes. The conclusion considers the sources of hope in the doom and reflects on the most important ethical challenges that condition it.

Biutiful's Clouds and Money

Biutiful (2010) is a Mexican-Spanish coproduction, whose action is set in Barcelona's darker side. In his critique of the film, Benjamin Fraser seeks to understand *Biutiful* within the frame of what he calls "urban cultural studies" and in dialogue with the urban theory of Henri Lefebvre and Manuel Delgado's studies of Barcelona (20). While "Barcelona is undeniably recognizable as the film's co-protagonist" (21), a similar story could be told about various contemporary cities, or even of less-urbanized areas which attract human migration from poorer lands and whose diverse forms of contamination and corruption, driven by need and greed, provoke human mental and bodily illness. From the perspective of environmental studies, Fraser's idea that the film represents the "human costs of modernity" (21) can be extended to the nonhuman framework of Morton's "dark ecology." All life (human and nonhuman) that appears in the film is negatively affected by human economy. This portrayal of life evokes the notion of the "Anthropocene," a crisis of human civilization due to the destruction of its own environment.

The Anthropocene, a concept coined by ecologist Eugene Sotermer and popularized by the environmental chemist and Nobel Prize winner Paul Crutzen, is that stage in Earth's evolution when humans become a quasi-geological force, their actions having a decisive impact on the transformation of planetary ecosystems. The beginnings of the Anthropocene can be traced to the eighteenth-century Industrial Revolution, but it is noteworthy that the term appears when the consciousness of transformation of the life on Earth becomes worrisome and when our civilization is perceived as destructive for human and nonhuman bodies and communities. The Anthropocene reveals an understanding of the history of modernity from the perspective of today's perils, but it also involves an understanding of the present from the perspective of the feared future. In this sense, Neyrat refers to climate change as "a moment of danger," that kind of moment of danger that for Walter Benjamin brings memory of the past in a flash, asking to be controlled for the purpose of the future. The Anthropocene contains the fear of the unknown consequences of the changes, which may be imperceptible until it is too late and can mount in human-built environments, including the interiors of our bodies, as a result of the general corruption imposed by the route our civilization has taken. The Anthropocene is, then, a term closely related to "dark ecology" and to the mood of disquieting realism, in which the familiarity of human reality is subverted by intuitions and nonsensory perceptions of sinister mutations occurring under the surface of the visible.

The buzzing sounds, the anxious movements of the flocks of dark birds on the sky, the ghostly shades, and the momentary presences of real ghosts with whom Uxbal is able to communicate, give *Biutiful* a feeling of disquiet. There is a mysterious, eerie correspondence between the progression of the fungus on the ceiling and the growth of abnormal cells in Uxbal's body, as if the human flesh and the matter of its immediate environment had communicated. These two processes of deterioration are signs of a deep cancer of the environment, caused by the poisonous effects of neoliberal politics, portrayed by the film as an omnipresence of rule by money. It manifests as moral and organic corruption, as physical and mental sickness, and as deprivation, poverty, and suffering. The morbid side of things seems to be creeping out into the surface of the city, like the worm moving through the alabaster surface of a tomb in one of the film's brief scenes. Although the most disquieting images, such as this one or those of the ghosts of the dead floating under the ceiling, last only a second, and their brevity is such that after a single viewing of the film the audience might not remember them, they strongly contribute to the general feeling of disquiet. The borderline visibility of these scenes directly affects one's mood, but not necessarily one's knowledge. It is as if the director were making a statement that there is quite a bit here that the human eye may not register, not only in the film but also in our everyday reality.

One of the most striking features of *Biutiful*'s montage is its persistent alternation between short, medium shots of bodies and interiors and long, panoramic shots of Barcelona, of its streets, sea, and sky, which point to their role in the drama represented in the film. The reflections and shadows of people and things, which also abound in the movie, oftentimes floating in the air as if it was water, point to how humans and objects "enter" the environment in which they move. The sequences of inside and outside shots often have a condensed metaphorical value, as in the case of the fish mosaic on the wall whose scales are made out of money or when the embrace of Uxbal and Marambra is compared to two chimneys of the Badalona factory, sadly throwing dark, poisonous smoke into the air. The CAT scan image of Uxbal's lower belly cannot but remind us of the growing fungus stain on the ceiling of his room. There are three shots of that decaying ceiling spot in the film, each larger and more ominous-looking, which mark three stages of Uxbal's sickness and the three parts of the film: recognition, endurance, and surrender. This putrefying ceiling affects not only Uxbal, whose worried gaze appears every time in the reverse camera angle, but also it seems to be driving the development of events in the story. It is as if the growth of the fungus on the ceiling, intimately connected to Uxbal's sickness, progresses in parallel to the decay of the interiors of the corrupted city. From the formal point of view there is not much new about it. Nineteenth-century realist novels often begin with long descriptions of interiors that metaphorically anticipate events to come. In *Biutiful*, however, the connections between the protagonists and the environment are organic. There is no clear separation between human and nonhuman matter and all matter is somehow animated, as in the newly popular vitalist theories, of which Jane Bennett's is possibly the most discussed. Bennett's *Vibrant Matter* (2007) argues that matter can be attributed agency and that we have an exaggerated notion of control that we exercise over the objects that surround us. The change of perspective that she proposes leads to a blurring of the binaries between life and matter, human and nonhuman, organic and inorganic, that points to the need to change the political and economical relationship with things. Benett coins the concept of "thing-power," which displays "the strange ability . . . to exceed their status as object and to manifest traces of independence or aliveness" (xvi). This secret activity of things not accounted for by humans is another factor of disquiet in *Biutiful*. The growing stain on the ceiling, the spoiled fish, the urine, and the smoke function as elements of an organic chain of decomposition, progressively darkening, where meaning is coined through sickness that transcends the body and the mind of the individual life, permeating the cultural-natural environment where the life decays. The viewers of *Biutiful* will remember Barcelona not as the artful, harmonious, and clean city from tourist postcards, but rather full of decaying matter, smelling of bloody urine, vomit, and of a fish that was frozen and defrosted four times as Marambra could not pay her bills and the electricity in her apartment was cut. In this famously sunny Mediterranean city, the sun is now scarce and it seems to be fighting to get through the dense clouds in order to nourish the remaining life. A caterpillar moving on a shaded branch seems to never get enough of it. But the "biutiful" of the title is not only a reference to the city but also about the life on Earth in general. "Biutiful's" mutation in spelling reflects the mutations carried by all the literal and metaphoric processes portrayed in this film. The cancer that corrodes Uxbal's body, the disequilibrium which tortures Marambra, and the corruption of democracy are in turn connected to the economic and environmental exploitation of the non-Western world, provoking "slow violence" (Robert Nixon), which condemns peoples to massive and often catastrophic migrations, followed by sicknesses, losses, and deaths. The film's mood is that of mourning for life given up in suffering.

The obsessive positioning of human characters in their environmental context in the film's montage is meant to signal that they are organically connected and that the story of human sickness is also the sickness of the world. In the Anthropocene it is more than ever "a human world," an intertwined and fused culture and nature, or, to use Donna Haraway's well-known concept, "natureculture" (16) which is all embracing. In the "natureculture" environment that *Biutiful* represents, the decisive element defining relationships between people and things, life and death by fixing values, is money.[2]

Consider the following sequence. After the detention of Ekwame, the illegal Senegalese street vendor, Uxbal, who had attempted to protect Ekwame and his companions by bribing a police officer, goes to Ekwame's wife's apartment and offers her money, which she initially refuses to take. She reproaches Uxbal angrily that he did not keep his word and failed to protect her husband. In the interior patio, lightened only by a few sun rays, with rough cellar-like walls and a strikingly blue bowl in the background, Ige washes clothes by hand, looking at both Uxbal and his money with distrust and bitterness. As he leaves some cash on a pile of folded clothes, the dirty, crumpled paper contrasts with nicely washed textile. Ige takes the money slowly with her hands, very clean and still covered with foam, as Uxbal walks out to the narrow street filled with African immigrants' faces and scaffolding. The next shot is a panoramic view of Barcelona with the Sagrada Familia, also surrounded with scaffolding and cranes against a cloudy sky, and as the camera moves inward through the window we see a medium shot of Uxbal's reflection in the tiles that cover the interior walls and then Uxbal himself wearing a hospital gown color of the tiles (the pale medical green of fading life), connected to an IV, a sort of crane itself, and looking out with his habitual gloomy expression. The suggested comparison between the scaffolding, crane, and IV puts on stage the complex relationship between technoscience embedded in our current economic system and the sacred vulnerability of life. Sagrada Familia cathedral, in words of Gaudí, was supposed to be "a mirror of its people" and to bolster the life of the family. Built since its nineteenth-century beginnings from people's donations, the church has never been finished even as its oldest parts have already required restoration. According to Gaudí, "in the Sagrada Família, everything is providential," and if his vision reflects correctly God's designs, donations will follow and the project will be accomplished (Moore). The precarious character of the economy of gift contrasts, however, with the aggressiveness of that of gain as, one hundred fifty years after its conception, the Cathedral is still a work in progress, while supermarkets take over cemeteries. As a result of this dynamic Uxbal's family life has decomposed.

In the next scene, back in Uxbal's shabby apartment, the camera zooms in on the money (again, piles of crumpled, dirty paper) that he takes out of a sock and counts for several seconds. As he does so we see his sickly face reflected again, fragmented and decomposed in a small mirror kept on the top of the dresser. Instantly we find ourselves in a street with a mural of a huge fish, likely a shark, with scales made of money, as Uxbal walks to buy heaters for Chinese immigrants. Below the mosaic, there is graffiti ironically calling for people to eat fruit and representing a homeless man pushing a shopping cart filled with garbage, which in a close-up turns out to be ruins of a city, right into the fish's mouth. Money devours everything and transforms it into garbage, which is fed to the fish. The camera travels slowly to a shop window filled with several TV screens showing dolphins pushed by waves to the beach and dying on the sand. This last image announces the future scene in which the Mediterranean Sea throws to the beach bodies of the Chinese immigrants killed by the accidental gas spill from the heaters that Uxbal is just about to buy. The camera zooms in on the prices of different heaters. Uxbal points towards the cheapest one and the shop assistant begins to load them to a car. As he watches him do so, Uxbal is taken by a sudden fit of vomiting, as if he were the fish being fed garbage. He bends down right by the wall with the mosaic and the graffiti.

In this relatively short sequence, lasting only four minutes, money is constantly present, both materially as crumpled paper bills, and symbolically as the mural of a city wall and as the price of heaters. Articles that are to provide for a better life end up killing due to the conditions of injustice in which they are produced. The homeless man with a cart filled with garbage which he seems to be confrontationally pushing to the open mouth of the money-fish on the mosaic manifests the dark side of the money economy, its side effects, or in environmental language its "externalities," contaminating waste, but also inequality, poverty, and injustice. The fish dying on the screens of the television sets, sold by the electronics store where Uxbal buys the heaters, announce the future as death and Uxbal's vomiting signals the poisoning already being undergone by his body suffering side effects of chemotherapy. In the subsequent sequence, Bea tells him that chemotherapy is not a cure but a poison and that there is no solution possible to cancer but to order one's things and prepare to die. This statement contextualizes the sequence where Barcelona, with its Sagrada Familia "in treatment," is compared by the camera to Uxbal undergoing chemo to signal the ultimate failure of medical science and of capitalism. Science and technology cannot bring the diseased of this world back to healthy lives because they were created in service of an economy that profits from destruction and death.

Money is a sort of a cultural DNA that carries information into organic matter, mutating it and transforming as in the street mural where a fish's scales are already made of bills. Life is redefined, restructured, and organically transformed in money's terms as the interactions of various forms of flesh are regulated by the relations of their monetary equivalences and as live bodies are poisoned and dead bodies are displaced by commercial pursuits. When Uxbal negotiates construction jobs for the Chinese immigrants with the manager Mendoza, the latter complains that the immigrants' poor state of health will slow them down and will not really allow him to lower his costs. In order to lower these costs the workers have to sleep in a cold basement, where their health worsens even more. Uxbal's attempt to provide them some warmth is also conditioned by the need to keep costs low, leading him to choose the cheapest heaters, which turn out to be defective and kill all the workers with a gas leak. The defects of various products featured in the film, like the purses sewn by Chinese immigrants which do not even open, are caused by the condition of the sick and overtired humans who produce them, as well as by a system in which the inferior quality and low price of merchandise is designed to force the customer to shop again and again. Thus profit as the supreme objective contributes to the overuse of resources, exploitation of the poor, and also to the destruction of life by lethal products. The economy of cheap excess is revealed as death-dealing.

The uncontaminated life that "used to be called nature" (Heise) appears in *Biutiful* only as a fantasy of a future and of the past.[3] It is an unfulfilled dream of a happy family excursion to the Pyrenees to see the snow. It is also a premonition of the other world into which Uxbal passes to meet his long-deceased father. Their meeting takes place in a silent forest slowly covered by snow. As the camera moves through the trees, it stops for a moment on the body of a dead owl that symbolically represents the dead Uxbal. While the snow falls silently, the dead father soothes his dead son by imitating the sounds of the primordial Ocean, from which, as we are reminded, everything emerged and to which, we can suppose, everything is slowly returning. The watery motives in various scenes of the film, such as the growing wet stain on the ceiling, the painting of waves in Marambra's apartment and a similar one in Uxbal's room, the fish in the aquarium and on the shower curtain, Uxbal's t-shirt which reads "Ocean city," ghosts floating in blue as if in the sea, are signs of the presence of the nearby future; this world is submerging back into the matrix of life.[4] This connection between the past and the future, so obvious for the dead father, brings a smile back to Uxbal's face. This is the quietest scene of the film, in which its sense of time is altered. "Sabes que hace poco aquí no había nada, sólo agua salada?" (Do you know that not long ago there was nothing here, only salted water?). This "not long ago" is a reminder of the relatively short time that human civilization has lasted, compared to the age of our planet. The fact that the film begins with this scene and also ends with it (although some angles are slightly different, the dialogue and the meaning are the same), places human life in the context of nonhuman life and individual life in the context of death. This vision of the Anthropocene as a crisis surrounded by a primeval ocean constitutes a consolation for an individual death, which then loses its dramatic significance. The sea, which appears in the film as a leitmotif, is a reminder of the relative brevity of human dominance and of our coming out from and, perhaps, going back to the salted water. The scene in which the Chinese workers' corpses are washed out to the beach can be interpreted as a dark allegory of this vision of the ending of humanity, this time rendered with dread and no palliation. In *Biutiful,* the balance of the matter, mutated by money, is precarious as Iñárritu multiplies the signs that signal the catastrophe.

Money can be seen as yet one more "hyperobject," in Morton's sense of an overwhelming presence and process which transcends and conditions life with an all-embracing dynamic that every other object depends upon. Life assimilates money, runs on money, as if it were an addictive stimulant, killing through overexertion. Money permeates love and, as in the famous song by Liza Minnelli, moves lives around the globe. Ultimately money leads to exhaustion, not only of individual organisms, but of the whole human civilization. When, after a fight, Marambra and Uxbal embrace each other in a moment of tenderness, she immediately asks him if he could give her some money, and then if he could give her more. Like Marambra's character, the system engineered by money and diverse human technologies is bipolar. The Disco's euphoria, fed by cocaine and blinding lights at moments in hellish deadly green, is a direct counterpart of the dark dungeons inhabited by the poor immigrants. These dark spaces are the side effects of the economy of excess, which produces all sorts of cancer. When Uxbal asks Bea why *this* is happening to him, why he got sick, if this is some sort of punishment, she does not answer. Yet a few minutes later, she reminds him that the gift of speaking with the dead, which they both possess, was given to them for free and that she had told him in the past that they should give it back in the same way. She implies that Uxbal's sickness may somehow be connected with his selling out his capacity to communicate with ghosts.

An illusion that they will be able to make money brings immigrants from faraway lands, and its scarcity condemns them to suffering from overcrowded apartments, cold basements, and illnesses; and this hunger for money which brought them overseas also produces their death. Money also shapes future and past time. When Uxbal is urged by his friend Bea to order his things before dying, he thinks of the money that he can leave for his children to ensure their survival. While Bea convinces him that the universe will take care of them, he skeptically answers that the universe will not pay the rent for their apartment. Preparing for death is counting money. Similarly, when the cemetery is liquidated to make space for a supermarket it is as if money has destroyed the memory of previous generations. An economy of life wherein money becomes the single significant value transforms the perception of time. It gets rid of history, which is not rentable, and transforms it to the eternal present of production and consumption. The experience of time in this present is asphyxiating. We see Uxbal walk quickly up and down the streets of Barcelona struggling to pick up children on time in the evening. The time of rest before going to sleep is marked by tiredness and tension. After all his efforts, exhausted and sick, Uxbal manages to save for a year's rent, which to him seems like a lot. The only quiet, slow time is after death, away from the civilization of money, perhaps a premonition of the future after the primeval sea covers the dirty cities. In the Anthropocene the ocean has already begun to enter the city through the clouds. There is no horizon on various panoramic shots of Barcelona with the sea, but rather the ocean and the sky blur.

Nocilla's Dark Humor

Nocilla experience (2008), the second novel in the *Nocilla* trilogy by Agustín Fernández Mallo, can be interpreted similarly to *Biutiful,* as virtual travel through an Anthropocene in crisis, whose mood is one of disquiet. Here also the mutation of life is a source of uncanny sensation, and all the novel's transformations seem to be marked by a tension between human greed, which leads to wars, madness and death. The novel conveys a darkly humorous conviction that the environment is silently waiting for human self-destruction to rebuild itself. As in the song "Los pueblos del mundo" (The Peoples of the World) by the Galician band Siniestro Total (Total Disaster), with which Fernández Mallo openly dialogues, the novel indulges in the descriptions of uncanny changes, anticipating a possibility of extinction of the human race. Yet, for Fernández Mallo's narrator and his protagonists, this is not a reason to stop enjoying life and art. They are the only resistance to the system that we can oppose. It is indeed with a dark humor that the novel suggests that, as in the song, we shall "extinguish ourselves with a smile."

As in *Biutiful*, the catastrophe has already begun to happen in *Nocilla's* world. Mark and Josecho, two of the novel's protagonists, withdraw from their reality, driven by a premonition of a future dystopia. Both live in shacks constructed from cardboard and packing materials on the roof terraces of the city skyscrapers and observe from above the constant mutation of life, as if they had moved already to a posthuman future. Another couple of connected characters, Antón and Antxo, compare Boris Sagan's *The Omega Man* (1971) with the catastrophe of the oil spill from the tanker *Prestige*, which sank close to the coast of Galicia, contaminating hundreds of kilometers of beaches and killing all coastal life. Jota, an artist who paints the chewing gum glued to London's pavements, is obsessed with Roberto Rossellini's *Journey to Italy* (1954), where Ingrid Bergman cries while looking at the bodies of lovers who had died in an embrace covered by the lava from Mount Vesuvius. The haunting presence of the bodies of Pompeii, preserved for thousands of years in their death expressions, like the bones of dinosaurs, remind the reader of the past catastrophes connecting to the fear of the future ones. A chain of catastrophic premonitions of the future is formed by the apocalyptic images of the *The Omega Man*, where the last man on Earth explores the ruins of the past civilization, and by the *anagnorisis* scene from Franklin Schaffner's *Planet of the Apes* (1968), when the tip of the Statue of Liberty sticking out of the sand appears. These past and future imagined catastrophes are contemplated as disquieting spectacles that can, however, be enjoyable if approached with the distance of humor.

This pleasure originated by the imagination of future catastrophes is, perhaps, the most questionable aspect of the discourse of *Nocilla experience*, although the surreal, dark humor often mixes aesthetic pleasure with revelations of the dark sides of our civilization. The most memorable allegory of the capitalistic thanatopolitics is a self-sustaining slaughterhouse, which is also a shelter for refugees displaced by the wars in the Middle East. It has been constructed by a young Armenian who, after having earned some money traveling, returns to his native country to build an eight-story building where 900 pigs are bred and slaughtered with the assistance of the political refugees from the regional wars. Twenty families of the exiles live on the upper four stories of the building, with the owner occupying the penthouse. The fourth floor is designed for porcine birth and it is where the mother pigs live with their little ones. As they grow, the animals are passed down to lower stories until they get to the bottom, where the slaughter takes place. The warmth produced by the animals is used to heat the building and their gases serve to heat the water and to generate electricity. This self-sustaining construction is a model of the hierarchy of exploitation. As in our modern societies, life is skillfully protected and nourished in order to be exploited, killed, and consumed by those administering it from their penthouses. The vertical ordering of the slaughterhouse, in which life's energy is literally sucked out of it by the system, and where organisms descend to the bottom of the hierarchy to die for the benefit of those that are still higher but will come down in their turn, brings to mind the perfect interchangeability of human and nonhuman bodies in this context. In fact, the four-story building, where animals occupy the first four stories and humans the higher four, but where all life (except the owner's, for the time being) moves downward, suggest that after all the pigs are killed the political refugees may need to substitute for them in order for the house to stay heated. The slaughterhouse parable does not generate pleasure but rather a feeling of disquiet and abomination.

The disquiet in the novel is also deeply connected to a scientific perspective on life, which leads to the blurring of a number of old dichotomies, such as nature and culture, living and dead matter, and even synthetic and organic, since on the molecular level all is made from the same set of particles. As Miguel Espigado notes, Fernández Mallo opens literature to the techno-scientific developments that the visual arts had reacted to earlier, challenging conceptual categories that had not been updated in the literary world. Espigado notices that the focus on the material reality of things imposes a preference for recycling. Literary recycling is performed in *Nocilla* through *bricolages* that open up a vision of a world that is a rhizomic mesh of connections.

As in *Biutiful,* in *Nocilla experience* nature and culture intertwine into a "natureculture," where human influences are omnipresent. As an example, "seguidores culturales" (cultural followers, 27) are animals, birds, insects, and microorganisms whose evolution has been determined by the relation that they have maintained with human societies. These forms of life developed their behavior in order to adapt to humans, placing themselves in a blurred space in between *oikos* and *techne*. "Transpoetic narrative" (57), which *Nocilla experience* exemplifies, is defined precisely by its focus on these hybrid artifacts: mixtures of human and nonhuman, living and dead, present and past, real and imagined, which all merge together as if morphed. The diverse forms of art that appear featured in the novel are also environmental. For example, Harold, who runs around the American continent without stopping because of the depression he suffers after his divorce, is featured as an example of land art. Jota, who paints all the chewing gum spit out on the pavements of London into different colors, is an ecologist who cures the proliferating cancer cells of the city, sublimating them into artistic expression. Thinking of the unity of reality on a molecular level provides a way out from the killing divisiveness of a system driven by wars and commercial brands.

The disquieting mood emerges in connection with all that science still cannot explain. The "horizon of events" (52), an imagined circumference that surrounds the Earth somewhere in the cosmos, separates us from what we can never know. Yet even more mysterious is the fact that we often know certain things although, rationally we have no possibility of knowing them. One of the brief stories in *Nocilla experience* is about how scientific tests failed to show what was immediately recognized by an expert: that an examined statue, supposedly Greek, was not authentic (156). Both the undocumented knowledge or, in other words, unknown known and the known unknowns are manifested by a random quality of life, also known as luck. For example, randomness in the evolutionary changes, which have occurred over millions of years, has obsessed biologists for many years.[5] While random genetic changes are a condition sine qua non of evolution, randomness as a characteristic of the environment has been unsettling for scientists who believe that "science is an enemy of chance" (Appleman 559) and that it needs to be freed from the element of chance to achieve the certainty to which it aspires. *Nocilla experience* moves towards a vision of complexity, suggesting that what is thought to be random may in fact be a very complex combination of various contingencies. In other words, "random" is in fact an unknown where a meaning can reside.

Literature and life stories can provide insights into scientific unknowns, for example, when a strange path leads a person looking for a loved one directly to him or her, as in the story of Ernesto's encounter with a lost friend from childhood, and in Chicho's story of a search for his daughter who had disappeared. One of the leitmotifs in the novel is an analysis of the topology of balls, inspired by a mathematical theory. It is an attempt to calculate the possible position of each ball in space—which is random—with the help of an abstract theory and without taking into consideration the contingencies of the real world. In the epilogue, the narrator explains the utopian character of this undertaking: "No es que la teoría y la vida estén mal, es que no tienen nada que ver, como tampoco tienen nada que ver los pensamientos del pez que baja aleteando con los del carbón que tira el anzuelo y arriba espera" (187) (I do not mean that theory or life is wrong. The thing is that they have nothing in common just as the thought of a fish which swims moving its fins have has nothing to do with the thought of the jerk who throws the bait and awaits above). The randomness of events that make it impossible for the science to interpret the world completely is in the novel a source of a humorous enjoyment of life and possibly a source of hope. The appreciation of random chance creates an illusion of a freedom from doom. This is why the game of Parcheesi, based on luck, obsesses various protagonists of the novel.

Both the unknown and the resulting mood of disquiet in the novel can be also seen as consequences of partial human blindness. A group of scientists at the University of Southern California implanted video cameras into the eyes of blind people. Initially they thought that they would need a resolution of one thousand pixels to make the blind see, but in the process of experimentation sixteen pixels turned out to be sufficient. *Nocilla's* narrator explains that the scientists did not take into consideration the blind spot through which we do not see, which our mind fills with imagined elements. In this way our previously acquired knowledge and imagination provide for a large part of what we see, often substituting our expectations formed by concepts we have previously learned for a hidden reality. The narrator comments that in every life there are dark spaces that were never seen, but which were later reconstructed with the memory that he calls an "artifact." A disquieting supposition follows: "Puede que en realidad estén ahí ocultas las otras dimensiones, fantasmas y espectros que no percibimos y que vagan por el planeta Tierra a la espera de emerger como consecuencia de que alguien edifique una metáfora en ese punto ciego" (64–65) (Perhaps, in fact there are other hidden dimensions, specters that we do not perceive and that wander through Earth waiting to emerge as a result of a metaphor that someone will build in the blind spot of his eye). If it is a lack of concepts that prevent us from seeing parts of the reality which, however, exist hidden from our half-blind eyes, supernatural experiences may be insights. The stories of the unexplained paranormal phenomenon of the faces of Bélmez and of the experience of the Jesuit exorcist who managed to free a child from a malignant possession connect the scientific awareness of the unknown to stories of ghosts, possessions, and other paranormal phenomena. The connection is conceptual; we are only finding what we are looking for, that is, entities for which we have reality-connoting-concepts, whose necessary existence we first theorize. For example, the Higgs boson was believed to be necessary; it took many years of searching, but a few years after the publication of *Nocilla experience*, which relates to the process of this search, it has finally been discovered. The story of the Higgs boson makes us realize that there may be many objects or organisms which have never been conceptualized and therefore have never been discovered. These entities may, however, be real and thus somehow affecting us, producing weird coincidences and paranormal phenomena.

One of the protagonists of the novel, Antón, who collects barnacles that live attached to the rocks in the water, throws computer hard drives into the ocean, hoping that the information contained in them would turn to "informatina," the substance carrying information pertaining to DNA, which will in turn transform the taste of barnacles. This act of surreal humor can be viewed as a hybrid between an artistic performance and a scientific experiment, or even as a metaphor for genetic engineering. It is also a fantasy of blurring the borders between information and things, abstract and material. If human language and its metaphors are "edificadas" (65) (built into) the human brain and thus substitute for the unseen parts of reality, and then they condition the transformation of the world by humans, it can be said that language transcends and becomes a part of the reality, shaping it. If this is so, then maybe the information encoded by humans in a computer's hard drive (computer language) could also possibly infiltrate organic matter, leading to unexpected changes in life. The fantasy of the "informatina" seeping from the computer hard drive into the DNA of barnacles can be also connected to those mysterious processes, which result in paranormal phenomena, such as the appearances of the faces in the floor of a house in Bélmez. If the information coded in a computer can turn into the DNA's information carrying substance (informatina), than perhaps informatina, which had belonged to the bodies, can form pictures on the walls of a house.

Of all the environmental contexts where the fusion of human and nonhuman, artificial and natural takes place, the human body seems privileged by the novel, as it is in *Biutiful*. One of the recurring images of *Nocilla experience*, that provides the opening frame for other stories, is that of a dead body floating in a lake, which, having absorbed water with all its organisms, has increased its volume and doubled up like a tea bag that swells up when dipped into a cup of water. As the narrator reminds us while reflecting on this image, the process of absorbing of the environment by the body takes place not only after death, but also during our lives. The corresponding scene in *Biutiful* shows Barcelona entering the hospital room and Uxbal's face transcending into the green tiles of his surroundings. Our bodies are in fact containers and carriers of different elements of the environment, and this bodily capacity is used by the international contraband of radioactive isotopes that are smuggled in the stomachs of children walking inside oil ducts across the borders of post-Soviet countries. On the other hand, the body extends itself beyond the body's limits. We leave marks and pieces of ourselves in our direct environment; the shape of the body remains on the bed sheets when we wake up. The skin is the border between the body and the environment, to which too much meaning may be given. An obsession with skin's illusory and mystifying character appears in various chapters of the novel as the narrator reflects on the fear of the students of medicine to cut through skin, and on the hybrid art of Bodyworks displaying human bodies without skin, as well as various forms of pseudo-skin like the lava covering the bodies in Pompeii or the cloth circling in the washing machine.

The most important source of the disquiet in *Nocilla experience* is, as in *Biutiful*, the perception of a human world progressing to its demise due to the self-destructiveness of human civilization. The mesh of connections between present, past and future woven all around the globe follows the path of catastrophes. Similarly as in Iñárritu's film, in the novel, this is a world in decay. In both cases, the framing scene and the most significant referent for all that happens is death. While in *Biutiful* reality is marked by sickness and madness, in *Nocilla experience* similarly the most frequently returning motif is that of the growing madness of a soldier in the Vietnam War, from Francis Ford Coppola's *Apocalypse Now* (1979), followed by the body as a tea beg. In both works a prominent place is occupied by cancer, which in *Nocilla* consumes one of the main protagonists of the novel and proliferates in the city spaces obsessively cured by Jota. The difference lies in the vision of a possibility of transcendence. In *Biutiful*, beyond human life there is the sublime re-encounter in a silent forest, slowly covered by snow, while in *Nocilla experience* there is only material consolation of all sorts, including art. Notably, while the apocalyptic mood of the film leads to a hope of a sort-of afterlife, the dark humor and sarcasm of the novel rob the reader of any possibility of otherworldly experiences, except from returning like a face from Bélmez desperately adhered to a cement wall.

Conclusion: Coexistence and Hope

In *Biutiful* prevails the doubt, or even disbelief that, given a political and economic system based on self-interested gain and measured with money that is now determining human behavior and regulating politics, there is any chance of preventing the final catastrophe. In this state of things, an empathetic coexistence, alleviating each other's suffering, is the main ethical imperative. In this sense Uxbal's figure, as an intermediary between the living and the dead, and helping both, is a priest-like or even Christ-like figure. All his activities, although mostly illegal, are focused on helping marginalized people to find work and avoid deportation and harassment. He maintains himself from the money that he receives as an intermediary in bribing authorities or in exchanging favors between leaders of different immigrant mafias, so he can also be seen as a parasite. But a parasite is precisely a master of coexistence as it usually preserves the bodies that it feeds on. Uxbal, the parasite, allows others to live on him as well. Requested by Ige, he lets her move into his apartment, even though sharing the very small space with her crying baby is hard for someone as sick as he already is at that time. His young son Mateo does not understand why his father lets Ige move in with them and Uxbal says that he helps her simply because she needs help. These words can be viewed as an ethical motto of the film. Applied universally, this principle could alleviate some suffering. Shortly after Ige's moving in, Uxbal loses his remaining strength and then Ige takes care of him, and she stays to take care of his children. In *Biutiful's* vision only empathy and painkillers can help humans on their way back to the primordial ocean.

This focus on the planetary mesh of symbiotic connections, whose randomness appears amazing and overwhelming, also in *Nocilla experience* leads to a focus on coexistence in both space and time. For example, the story about a town in Kalmykia, Ulan Erge, where the winter is so severe that buildings were covered with sheets recycled from Siberian concentration camps, can be read as one of the metaphors connecting the present, past, and future generations in an environmental sense of a mesh where everything matters. Those sheets stained with body fluids and graffiti, marked with the suffering of prisoners, were turned into a sort-of-skin of human houses after the camps were closed, as if the people who are alive were covered with the skin of the dead. When the spring comes, one old man gets so attached to this connection that he does not want the sheets to be removed from his house even though the cold has ended. He thinks that "si la gente se diera cuenta de que una de las moléculas de la piedra que lleva en el riñón perteneció quizá al ojo de algún dinosaurio, o de que ese fragmento de cinc que ahora corre pro el torrente sanguínea de su vecino estuvo en la orina de Alejandro Magno, o que una particular del suero de la leche que fermenta ahora en sus estómagos la mamó antes un cordero en las montañas de Sahara, no le quitarían a él su galería de huellas humanas" (147) (if people realized that one of the molecules of the stone that he has in a kidney belonged perhaps in an eye of a dinosaur, or that the fragment of zinc that now runs in the blood of his neighbor was once in the urine of Alexander the Great, or that one particle of the milk that ferments now in their stomachs was sucked before by a lamb in the Sahara mountains, they would not remove now this gallery of human marks). *Nocilla experience* shares with *Biutiful* the belief in the need of ethical coexistence. "The little goodness"[6] can mitigate human doom on the planet ridden by suffering, just as the American Marine does not throw the grenade inside the Iraqi home, but rather, moved by the gaze of a woman that he sees, lowers his arm and goes away. He then pursues her and becomes her husband and a father of her child and they are happy together. "The little goodness," or "little love" in this story, however, can do little when confronted with the global conditioning of the system. When the child of the American Marine and of the Iraqi woman is seven years old another Marine throws a grenade to their house and kills the whole family. The random coincidences which can cause doom, can be also, however, a source of hope, for example, when a group of prisoners of Auschwitz randomly survive the final killings in the camp while hidden in a dark corner playing the game of luck, Parcheesi.

Yet another way out of the general doom in *Nocilla experience* is hybrid art that queers the concepts supporting the deadly system and the dark humor arising from the knowledge of contingencies that subverts this system from within. In this sense, *Nocilla experience's* attitude towards today's capitalism is more playful than *Biutiful's*, although it is also very critical of it. Chapter 100 ends with a conclusion that coincides with the main thesis of the famous article by Adorno and Horkheimer on "The Culture Industry: The Culture as Mass Deception," stating that "la vida es un anuncio de teletienda al que le han eliminado el producto anunciado" (151) (life is a publicity ad whose product has been eliminated). Josecho, one of the main protagonists of the novel, whose character may constitute Fernández Mallo's self-reflective mocking, is a writer who in order to promote his novel suggests to his editor that publicity may be financed by a clothing company (Chanel is chosen) in whose clothing the author would pose for all the posters and advertisements. This constitutes a meta-reflection on the *Nocilla experience*, inviting us to think that the publication of Fernández Mallo's novel could have been partially financed by the producer of the popular chocolate Nocilla. Indirectly, literature is also partaking in the system of production and consumption and as such it is co-responsible for its casualties. In this sense, comparing Nocilla to a drug, as German Labrador does, makes sense. These are both perfect products of the system that lives at the expense of the bodies addicted to junk which they work to get. In his analysis of the video "Nocilla la merendilla" by Siniestro Total, Labrador writes about "una generación, que chocopasta mediante, sobrevivió al franquismo para morir de sobredosis en su entrada en la democracia" (A generation that survived Franco times thanks to chocolate paste in order to die of overdose when democracy started). This logic continues and our generation perception of the injustice is also muted by all sorts of sweet pastes that are fed to us via mouth and eyes. There have been various Nocilla generations and the novel contains a muted complaint that there is no way out of Nocilla.

This interpretation needs to be nuanced however. Both the title of the novel that the reader has in her hands, *Nocilla experience,* as well as the one that Josecho, one of the most important characters, writes, *Ayudando a los enfermos,* evoke other song titles of Siniestro Total. The last song on the 1982 album, *Ayudando a los enfermos,* "Purduy," ridicules the brainwashing manipulation of the Western cultural production, mocking the British TV series *The New Avengers,* where Purduy, named after the arm factory, was a female spy working for the British intelligence. The golden rules of Samurai life extracted by Fernández Mallo from Jim Jarmusch's *Ghost Dog* (1999) are presented in *Nocilla experience* in the same mocking tone of Siniestro Total's "Purduy" song. The most intense intertextual dialogue, however, is established with Siniestro Total's song from a different album, entitled "Pueblos del mundo extinguíos" (Peoples of the World Extinguish Yourselves) and whose words express the same kind of poetics of the Anthropocene that the novel expresses: "Pueblos del mundo, ¡extinguíos! / Dejad que continúe la evolución; / esterilizad a vuestros hijos, / juntos de la mano hacia la extinción. (Peoples of this world, extinguish yourselves let Evolution continue / sterilize your children / hand in hand towards the extinction let's go). Both the song and the novel stress the significance of companionship in the face of death, but also openly refuse tragic mourning. It is not an accident that one of the fans of Siniestro Total who made a video to accompany the song beginning with words "Pueblos del mundo extinguidos . . ." and placed it on YouTube used images which are central in *Nocilla experience*: a T-Rex skeleton in a museum, the bones of two lovers unearthed in an embrace, and the head and the hand of the Statue of Liberty sticking out of the sand from the last scene of the *Planet of the Apes.* The pleasure principle so important in *Nocilla experience* should be interpreted in the line carrying the central message of the song: "Sonríe cuando te vayas a fosilizar, / que no piensen luego que lo has pasado mal" (Smile when you fossilize / so that they won't think that you didn't have a good time). Having a good time does not imply a passive acceptance of the system, but rather it is an expression of a resistance towards the system that wants to make us suffer while it also provides energy to deconstruct it. *Nocilla experience* develops its own readings of various products in resistance to the meanings inscribed in them by the hegemonic discourses of the market. It is in fact about how to deconstruct Nocilla. For example the shacks built atop the skyscrapers where Marc and Josecho live are made out of containers full of commercial trademarks. (In one of the last passages Marc opens a hole in one of them as if reaching for an illusion of a space of freedom). In the post-modern cuisine in Brooklyn, plastic supermen and Polaroid photographs are diluted and served customers as a part of their dishes. These are ways in which hybrid art contributes to a queering of the concepts responsible for destruction. *Nocilla experience* can be viewed as an act of what Simon During calls "'a transgressive' undermining or 'festive' overturning of routines and hierarchies" (10). This transgressive undermining is accomplished through "ironical mimicry, symbolic inversion, orgiastic letting-go, even daydreaming . . ." (10). The significance of the construction of an alternative pleasure economy that is accomplished in this way cannot be underestimated, because pleasure is the main motivation for action. The argument that, by becoming a sort of best-seller, *Nocilla's* transgression becomes itself institutionalized, can be rebuked in Espigado's words: "a pesar de que la obra periférica triunfa en su revindicación estética y se desplaza hacia el centro, el apriopiacionismo no pierde su dimesión reivindicativa, pues en su naturaleza está la contradicción de criticar el sistema desde dentro del sistema" (In spite of the fact the a peripheral work triumphs in its esthetic vindication and thus moves towards the center, its strategy of appropriation does not lose its revolutionary dimension, because it is in its nature to criticize the system from inside).[7]

While queering of the concepts that structure action and political systems and the pleasure derived from it, is necessary to transform human civilization, the pleasure built around the apocalyptic scenarios in the novel may be counterproductive from the activism point of view. The chapter about "el tercer paisaje," (third nature) announces nature's capacity to revive after human demise. Third nature emerges in the spaces from which civilization has withdrawn. These are roadsides, riverbanks or sites abandoned by agriculture or industry where biodiversity spontaneously returns transformed. If we think of the third nature's parallelism with the "third state," which, as Fernández Mallo states, was a social class that had no right to vote in the ancient regime, we understand that third nature's lack of function in human economy pushes it out of the political map of life. Paradoxically, however, precisely in this way, belonging to the unmapped territory, freed from the cartographic control, a variety of forms of life escape biopolitics and can strengthen, flourish and revive. If we consider that Karl Marx called human civilization "second nature" (146), then "third nature" acquires an ominous meaning of the state of affairs after the end of the Anthropocene, as the life in the primeval ocean in *Biutiful.* This posthuman stage would emerge, however necessarily mediated by the destructiveness of human activities. Reading *Nocilla experience* with Richard Dawkins's suggestion in mind that our body is merely a vehicle for "selfish genes" to make more copies of themselves, we can think of human techno-science as a mere trick of nature to move itself onto a new level. It may be that it is not us who control nature but rather nature that will eventually take us in the direction in which it wants to go. The pleasure of consuming Nocilla, that stands for general overconsumption of junk, may be just viewed as a trick programmed wisely by nature so that, as Siniestro Total advises, peoples of the world extinguish themselves with a Nocilla smile and let other forms of life develop anew. Fernández Mallo's novel with its dark humor and sarcasm is thus not more hopeful for human future than *Biutiful*, but it opens readers' eyes to the spaces of randomness which provide for doom, but also for freedom and doubt, and argues for the need to take advantage of these spaces for artful resistance. If we take advantage of a possible good luck, we may randomly survive a little longer.

Notes

1. The concept of crisis can work both ways; it may be useful for mobilizing the public, but it may also be taken advantage of by economic institutions to justify intensification of the very measures responsible for the problems. One of the defining characteristics of "crisis" is that it implies the need for a change; however, it is debatable what kind of change needs to be implemented since the causes of the crises are only partially clear and there is limited consensus about how to address them. Crisis implies various kinds of threats and losses, but also possible trade-offs. Insufficient resources may lead to more or less justice as the scarce goods may be spread more equally or less equally as a response. Crisis may also seem too deep to resolve and so it may cause the feeling of doom and no attempts to change may be undertaken. Thus crisis may generate passivity, sadness, alienation and depression rather than being a challenge for transformation and adjustment.
2. Iñárritu's obsession with the dark side of money is as long-lasting as his film director career. In fact, his first half-length film, aired on the national TV, was entitled *Detrás del dinero* (Behind the Money, 1995).
3. In her review of Bill McKibben, *Rambuctious Garden,* entitled "Encounters with the things formerly known as nature," Heise writes: "used to call it nature: forests, lakes, foxes, butterflies, mosquitoes, dandelions. Soils and oceans. Seasonal cycles. Also floods and heat waves and the occasional hurricane. But no more."
4. I owe these details as well as the images to Jenny Jeong.
5. See, for example, John Tyler Bonner *Randomness in Evolution* (2013).

6. It is the term used by Emmanuel Levinas in his reflection on the book by Grossman depicting the Second World War's atrocities. During those terrible times only "a little goodness" of one person helping another could alleviate the general suffering inflicted by the terror.
7. See Vicent Moreno's article "Breaking the Code" for further discussion on relations between "Generation Nocilla" literature the market and the literary field.

Works Cited

Adorno, Theodor, and Marx Horkheimer. "The Culture Industry: Enlightenment as Mass Deception." *Dialectic of Enlightment*. New York: Continuum, 1972: 120–67.

Appleman Phillip, ed. *Darwin: A Norton Critical Edition*. New York: Norton, 1970.

Bennett, Jane. *Vibrant Materialism: A Political Ontology of Things*. Durham: Duke University Press, 2010.

Biutiful. Dir. Alejandro Gónzález Iñárritu. Menageatroz, 2010. Film.

Dawkins, Richard. *The Selfish Gene*. Oxford: Oxford University Press, 1989.

Detrás del dinero. Prod. Alejandro Gónzález Iñárritu. Televisa S.A. 1995. Film.

During, Simon. "Introduction." *The Cultural Studies Reader*. 2nd Ed. Ed. Simon During. London: Routledge, 1999.

Espigado, Miguel. "Agustín Fernández Mallo & *Nocilla experience* (1.0)." *Afterpost*. 9 Mar. 2008. Web. 15 May 2013.

Fernández Mallo, Agustín. *Nocilla experience*. Madrid: Santillana, 2010.

Fraser, Benjamin. "A *Biutiful* City: Alejandro González Iñárritu's Filmic Critique of the 'Barcelona Model.'" *Studies in Hispanic Cinemas* 9.1 (2012): 19–34.

Gónzález Iñárritu, Alejandro. Dir. *Biutiful*. Prod. Spain-Mexico: Menageatroz, 2010. Film.

_____. *Detrás del dinero*. Prod. Mexico: Televisa S.A. 1995. Film.

Haraway. Donna. *Simians, Cyborgs, and Women; The Reinvention of Nature*. New York: Routledge, 1991.

Heise, Ursula. "Lost Dogs, Last Birds, and Listed Species: Cultures of Extinction." *Configurations* 18 (2010). Web. 23 Nov. 2013.

_____. "Encounters with the things formerly known as nature." *Publicbooks.org*. Web. 13 Nov. 2013.

Labrador Méndez, German. "Nocilla tiene el nombre del hambre. El rock-memoria de Siniestro Total, memoria de cacao y sus fantasmas biopolíticos en la movida española." *Crítica latinoamericana*. 2012. Web. 20 May 2014.

Levinas, Emmanuel. *Totality and Infinity*. Pitsburgh: Duquesne University Press, 1969.

Marx, Karl. *Social and Political Thought*. Ed. Bob Jessop and Russel Wheatley. Bungey, Suffolk: Routledge, 1999.

Montero, Rosa. *Lágrimas en la lluvia*. Barcelona: Seix Barral, 2011.

Moore, Rowan. "Sagrada Familia is nearly done, but would he have liked it?" *The Observer*, 23 Apz. 2011. Web. 15 May 2014.

Moreno, Vicent. "Breaking the Code: *Generación Nocilla*, New Technologies, and the Marketing of Literature." *Hispanic Issues Online* 9 (2012). Web. 13 May 2013.

Morton, Timothy. *Ecology without Nature: Rethinking Environmental Aesthetics.* Cambridge, MA: Harvard University Press, 2007.

Neyrat. Fréderic. "*Oikos* and *Polis*. Ecopolitical Thought and the Avatars of Humanism," University of Wisconsin, Madison. 4 March 2011. Lecture.

Nixon, Robert. *Slow Violence and the Environmentalism of the Poor.* Boston: Harvard University Press, 2011.

Ortega Cerdá, Miguel. ed. *La deuda ecológica española; Impactos ecológicos y sociales de la economía española en el extranjero.* Sevilla: Universidad de Sevilla, 2012.

Tyler Bonner, John. *Randomness in Evolution.* Princeton: Princeton University Press, 2013.

◆ 5

Cultivating the Square: Trash, Recycling, and the Cultural Ecology of Post-Crisis Madrid

Matthew I. Feinberg and Susan Larson

On June 1, 2013, people walking through Madrid's historic center may have been surprised to find that the Puerta del Sol—ground zero for the 15-M uprising of the "Indignados" in May of 2011—seemed to have been renamed. The name the iconic central public square had held since it began to serve as one of the city's principal gates of passage in the fifteenth century was changed when the British mobile phone company Vodafone signed a contract with Madrid's regional government to change the landmark Sol metro station to "Vodafone Sol" for a period of three years in exchange for three million Euros (approximately 3.9 million U.S. dollars), a sum that would be used to maintain the economically-troubled metro system.[1] This incident is the latest in a series of reminders of how in Madrid, as in other large cities, the "right to the city" (to use a key phrase of Henri Lefebvre's) is falling more and more into the hands of the private sector. As David Harvey points out in *Rebel Cities*, "[t]he actually existing right to the city, as it is now constituted, is far too narrowly confined, in most cases, in the hands of a small political and economic elite who are in position to shape the city more and more after their own particular needs and hearts' desire" (24). In the case of the Puerta del Sol, who occupies public space and how are being contested as competing groups fight for control over the ability to maintain different histories and legacies of place. The name "Vodafone

Sol" now reminds Madrid's citizens every day of the commodification of public space in the age of economic austerity and how the Puerta del Sol's history has been marked by the imprint of a multinational corporation. This propensity for naming public urban spaces after corporations and notable proponents of privatization, union busting, and deregulation only promises to continue. On September 17, 2014, the ceremony to commemorate the naming of the Plaza Margaret Thatcher, located in the Salamanca neighborhood of Madrid, was a source of sharp criticism.

The culture of urban spaces is of great political importance. Words, sounds, and images exist in a reciprocal relationship to urban space, creating meanings that are constantly being renegotiated over time. Moreover, urban cultural forms are invariably tied up in the production of social space. As Álvaro Sevilla-Buitrago points out in "This Square Is Our Home!", the 15-M movement combined some of the already-existing political answers to Spain's problems from the left-wing parties making up Spain's parliamentary government with very different (and oftentimes conflicting) calls for self-government that came from more radical groups.[2] Even if only for a short period of time, on the municipal, national, and global scales the Puerta del Sol spatially embodied a set of radically new political possibilities. The renaming by the British telecommunications conglomerate Vodafone of the very place where the Spanish 15-M movement began in 2011 is yet another battle in the ongoing war over who controls the city.

Modern cities have always been sites of revolutionary politics and places where the struggles over who controls access to economic resources and who dictates the organization of daily life play themselves out. Cities are also where new cultural forms are constantly being created to reflect upon, respond to and shape urban politics. These new cultural forms don't merely exist in an abstract dialectic with urban politics, however. They respond directly to the urban space in which they "take place"—reimagining, rebuilding, and reshaping the built environment to correspond to new political, economic, and cultural circumstances. Two such examples of contestatory spatial responses are the focus of this essay, which considers how the 2008 economic crisis in Spain coupled with the aggressive privatization of public space have given rise to such highly visible reactions as the 15-M movement, but also resulted in a proliferation of philosophies of space as well as small-scale grassroots endeavors that work toward creating new ways of living, moving, and theorizing the city. Individually and collectively, these community-based projects emphasize what can be called a "new ecology of urban space."[3] The notion of ecology being used here should not be confused with what is typically considered urban ecology, a subdiscipline of ecology that focuses on the biological processes of the urban by studying the presence of nonhuman habitats within urban environments (i.e., peregrine falcons in New York City). Nor is this working concept of an "ecology of urban space" reducible to cultural ecology, a field dedicated to studying how human culture responds to and adapts to its physical geography and climate (i.e., habitat). Rather, in an attempt to rework our understandings of both ecology and the urban, this essay looks at some recent directions in urban studies that question urban / nature binaries and encourage us to think of the city as an *ecosystem* of cultural, political, and material relationships.[4]

There are a number of recent philosophical approaches to the intersection of the urban and nature in the fields of geography and urban studies. A 2012 book edited by Ingrid Leman Stefanovic and Bede Scharper entitled *The Natural City: Re-envisioning the Built Environment*, for example, relies on a phenomenological approach to the term "natural." For them, the "focus is on the diverse spaces that emerge between the 'natural' and the 'cultural'" (Leman Stefanovic 23). This approach sees the urban environment not merely as "habitat" (a fundamentally biological understanding), but more in terms of what Stefanovic—with a nod to Heidegger—has called "dwelling" in which "people and built spaces are more than mere objects" (Stefanovic 30). David Seamon develops this notion further in his contribution to the book by suggesting that striving for the "natural" city assumes that "a particular fit between people and world contributes to a larger environmental whole of interpersonal encounter, sociability and place attachment" (232). This means that a "natural" or an "ecological" concept of the city relies not only on biological relationships that produce less pollution and have more Nature (i.e., green space), but rather is also founded on a cultural experience of place built on more just and humane connections between people. The political vision behind these theories clearly prioritizes a cultural model based on collaboration in the hope that people who are more connected to place and to each other will create more livable and sustainable cities and, ultimately, a more livable and sustainable planet.

This philosophical approach to theorizing the "natural" city contrasts with the more materialist political ecology found in the work of geographers like Erik Swyngedouw, Nik Heynen, and Maria Kaika (2006). They, too, emphasize the importance of human culture in understanding the ecology of urban space, but they frame this intersection in terms of an urban political ecology that sees the urban as a "process of socio-ecological change" (Heynen et al. 2). Although this second group also moves beyond the nature-culture binary to see cities as intrinsically part of a global ecosystem (i.e., the environment), they understand the urban as "the pivotal embodiment of capitalist or 'modern' social relations, and, by implication, of the wider (and often global) socio-ecological relations through which modern life is produced, materially and culturally" (Heynen et al. 25). In short, they see the overlapping and intersecting process taking place between the biological and climatological as a key component of material conditions of humans and nonhumans alike.

Swyngedouw's more recent work goes even further beyond an understanding of the urban as a socio-ecological process by not only rejecting the idea that the material and biological are separate but by suggesting that Nature as a concept has become a rhetorical and ideological tool to promote a neoliberal spatial reorganization of the globe. He writes that when any connections to systems of political economy are obscured, Nature is constructed as a fetishized Other that is a fixed, external and ahistoric "thing."[5] Within this frame, the potential catastrophe of a warming planet is posed as a universal and totalizing threat to humanity and as such it is (somewhat paradoxically) depoliticized and turned over to nonelected technocrats to manage. This depoliticized discourse surrounding Nature "[p]ermite y sostiene un planeamiento post-político recorrido por el miedo y guiado por una preocupación por gestionar las cosas de modo que podamos preservar lo que poseemos en la actualidad" (Swyngedouw 51) (allows for and sustains a postpolitical planning bound by fear and guided by the preoccupation of managing things in order to be able to preserve what we currently have). For Swyngedouw, as long as the official political discourse of the State remains tied to an overly fetishized notion of Nature there can be no pursuit of solutions outside of a capitalist framework to resolve our growing ecological crisis. The only recourse, he contends, is to be found in "[r]eclamar una democracia plena, espacios públicos democráticos, espacios para la enunciación de la disputa agonista como cimiento para y condición de posibilidad de ordenes socio-ecológicos más igualitarios" (62) (reclaiming full democracy, democratic public spaces, spaces for the enunciation of spirited debate as a way to cement and create the conditions for possible socio-ecological orders that are more egalitarian [and] the formulation of socio-ecological futures). If one takes Swyngedouw's thesis to heart, then, a possible solution cannot be ecological in the sense of simply protecting Nature, but rather one that foregrounds the politics of space, by cultivating democracy through new understandings of both habitat and dwelling. Inspired by Swyngedouw's call for action through greater urban-ecological awareness and collaboration, this essay is about how a number of architectural manifestoes and philosophical essays foreground two grassroots movements in Madrid that cultivate urban spaces for human connection and debate—spaces where new socio-ecological orderings of space are sprouting out of the detritus of the nation's current economic crisis.

Spanish architects Iñaki Ábalos and Juan Herreros along with philosopher José Luis Pardo began to write about urban ecology before the crisis of 2008 and their work has only become more relevant in recent years. While the architects and the philosopher differ in disciplinary approach what they have in common is a concern with the spatial reordering of discarded, abandoned, or otherwise unused city spaces. All three argue that significant cultural renovation is occurring in places that financial capital seems to have "refused"—a term that refers to a "refusal" to spend capital, but also to the notion of city space itself as "refuse" or trash. Ábalos and Herreros's concept of recycling urban space through architecture and Pardo's theory of a culture of trash both propose new ways of thinking about urban responsibility and citizenship that rely on a fundamentally un-modern understanding of the relationship between nature and the built environment. They speak to some of the grassroots initiatives of urban collectives being organized throughout Madrid, and also engage in significant ways with some of the recent directions in urban theory mentioned earlier that question a number of basic assumptions about what a city is and whether or not we live in an "Urban Age" at all.[6]

Ábalos and Herreros talk specifically about Spain's capital as early as 2000 in their essay "Recycling Madrid," where they put the city in the position of "begging to be recycled with imagination and pragmatism, adapting to a culture whose sensibilities and ways of being have dramatically changed" (22). They use notably Deleuzian terms to describe Madrid as "a conglomeration of natural, material, and immaterial elements (fluxes), that are simultaneously porous and fibrous, with dense and stable areas, loaded with memories, and vast diluted expanses that are devoid of character and almost liquid, made up of antithetical elements that have broken with the precision of the traditional limits between the natural and the artificial" (22). Likewise, "A New Naturalism: 7 Micromanifestos" consists of a series of brief but conceptually dense philosophical approaches to the urban that appeared in a 2002 volume of the architectural magazine *2G* devoted to their early building projects.[7] As in "Recycling Madrid," they distance themselves from both the modern and the postmodern, proposing a new program consisting of what they call "[U]n atlas mental, una geografía imaginaria capaz de iluminar cada lugar desde la propia proyección (invirtiendo el contextualismo tradicional) y dando forma, paso a paso, a *un jardín latente*" (23) (a mental atlas, an imaginary geography capable of illuminating each location from the planning process itself [inverting traditional contextualism] and giving shape, step by step, to *a latent garden*).

This image of the "latent garden" as an organic space that is emergent, participatory, and open is full of social and aesthetic possibility. It is not an Edenic garden, nor a utopian space or perfect modern form that stems from the artistic genius of the architect. Rather, the architectonic vision of the "latent garden" hinges on what Ábalos and Herreros call "la fusión de naturaleza y artificio, la disolución de límites disciplinares entre arquitectura, arte, jardín y pensamiento, la organización de la experiencia como secuencias narrativas, la primacia de lo visual y el movimiento, también de lo immaterial e invisible; la construcción de un espacio público y una arquitectura reflejo de las nuevas posibilidades" (25) (the fusion of nature and artifice; the dissolution of disciplinary boundaries between architecture, art, garden and philosophy; the organization of experience into narrative sequences; the primacy of the visual and of movement, of the immaterial and the invisible too; the construction of a public space and an architecture reflecting new possibilities). The roles and responsibilities of the architect, posit Ábalos and Herreros, have changed because the world has changed, and they are trying to adapt to these new circumstances. For them, "el trabajo del arquitecto se confunde con el del jardinero: desbrozar, preparar el terreno, escoger las especies y sembrarlas de forma organizada, cuidando después de que el paso del tiempo haga bien su trabajo" (26) (the work of the architect is like that of the gardener: cutting back the undergrowth, preparing the ground, choosing the plant varieties and sowing them in an organized manner, taking care afterwards that the passage of time does its work well).

At the heart of Ábalos and Herreros's call for the architect to think like a gardener is the desire to rethink the natural/artificial opposition on every scale of the built environment. Modern urban planners oftentimes thought of the city in moral terms and sought neoliberal or reformist solutions to the city's problems. For Ábalos and Herreros, breaking with (post)modern tradition is only possible when we get beyond the opposition between the natural and the artificial. They put their hopes in what they call a sense of "ecomonumentalism," which is rooted in a sensitivity to both the short- and long-term impact of the construction of buildings and urban spaces. They find inspiration in the unexplored possibilities of the undetermined, empty or abandoned spaces of the city as well as the peripheral zones in flux on the outskirts of the historical city center, places with as yet undefined landscapes that aren't properly urban yet can't be called the countryside.

For Ábalos and Herreros, their "new naturalism" arises from the hybrid situation of nature as both the subject of knowledge and of aesthetic experience: a crossbred, humanized form fused with its former opposite, artifice. This takes place in the peripheral no man's land of the *descampados* or still undefined spaces of the city. This "hybrid" or "mestizo condition," they make clear in the last *micromanifiesto*, is the only one possible in a hyperconnected world in which "economía y ecología están asociados, no solo etimológicamente (*oikos*, la administración prudente de los recursos domésticos" sino también estéticamente" (28) (economy and ecology are linked, not only etymologically [*oikos*, the husbanding of domestic resources], but also aesthetically). Lastly, Ábalos and Herreros make a final point of distancing themselves from the discourse of globalization, something that they believe "solo parece darse en una dirección, la de George Bush: mera intensificación del dominio de los de siempre sobre otros" (28) (only seems to go in one direction, George Bush's, a mere intensification of the domination of the usual group of people over the other). They prefer to use the term "conectividad" (connectivity) of people and information on a global scale as something that flows both ways. For them the contemporary architect is someone who is required to transfer a universal beauty that crosses nation, race, and class identities. Ábalos and Herreros aim at nothing less than "esa arquitectura que traspasa el tiempo y el espacio, que nos sobrecoge por simple, universal, feliz, barata e intensa. Ese es el jardín que vemos desplegarse" (29) (an architecture which transcends time and space, which takes us by surprise by being simple, universal, felicitous, cheap, and intense. That's the garden we see spreading before us).

"Simple, universal, felicitous, cheap, and intense" are adjectives that can also be used to describe the vision of city life that one finds in the work of philosopher José Luis Pardo. Trash, recycling, crisis—some of the ideas of Ábalos and Herreros mentioned above seem tinged with an almost perverse pleasure—as if the authors were expressing some satisfaction from vicariously experiencing the world being destroyed by extremist politics, corruption, rapacious capitalism, runaway technology, and ecological disaster. They bring to mind the famous phrase of Walter Benjamin: "our self-alienation has reached such a degree that it can experience its own destruction as an aesthetic pleasure of the first order" (242). Pardo's reflections on the current situation in which urban citizens find themselves have a lot in common with but are ultimately more pragmatic than those of the architects, however. In his 2010 book, he proposes that, "'Nunca fue tan hermosa la basura' [. . .] y si lo que llamamos basura no lo fuera en realidad? Entonces no tendríamos que preocuparnos porque nos devorase, no nos sentiríamos asfixiados por los desperdicios si dejásemos de experimentarlos como desperdicios y los viviéramos como un nuevo *paisaje urbano*" (170; emphasis in original) ('Trash was never so beautiful.' [. . .] what if what we call trash really wasn't at all? Then we would not have to worry about it devouring us, we would not feel suffocated by waste if we stopped experiencing it as waste and made a *new urban landscape* out of it).[8] Pardo's basic premise is that for the first time in human history we are living in a period in which there is not enough space for the waste that the world's population produces. Historically, argues Pardo, the more refuse that modern societies created (in terms of both quality and quantity) the wealthier, bolder, and more energetic they were considered to be. Wealth meant squandering, wastefulness, and surplus. Modern societies need ever-growing spaces to deposit trash so they carry on with life and keep on wasting in the midst of their own waste. These nonplaces (to use Marc Augé's term) become necessary and by extension even give rise to social nonplaces, locations where the populations are transferred that the productive and consumptive systems cannot absorb (in the form of suburbs, shacks, *favelas*, ghettos, and camps, for example). "Trash" for Pardo is that which does not have a place, that which is misplaced and, therefore, that which has to be moved to another place hoping that it might disappear as trash there, that it might be reactivated, recycled, extinguished: trash is that which searches for another place.

Pardo's philosophy of trash and its implications for what he calls a "new urban landscape" invite us to focus not on the monumental public spaces of the historic city center (such as the Puerta del Sol) but on the nonplaces or junk-places that don't have a clear purpose in the modern, industrial or even postindustrial sense of the word. Like Augé and like Abalos and Herreros, Pardo is interested in the politics of "transit zones" (waiting rooms, airports, train and bus stations, transport vehicles, hotels) and stresses the cultural and social importance of urban spaces such as lots where buildings have recently been removed, abandoned districts, housing in decline, preindustrial suburbs, peripheral shacks and outlying areas in ruins. He is interested in the potentialities of spaces that are not designed to be permanently inhabited but only to be occupied provisionally, to be crossed or to facilitate passage from one place to another. If a modern utopia was one where all trash had its place, the contemporary world is one of permanent recycling. In a Nietzschian turn, Pardo sums up his thinking in the following way:

> No se puede decir de manera más clara: allí donde nada es basura, todo lo es. Y es el mismo Marc Augé quien se ha dado cuenta de que, de seguir así las cosas, nuestra civilización será la primera del mundo que no deje tras de sí esa clase especial de basura histórica que son las ruinas. La ciudad genérica (la *ciudad-basura*) no deja ruinas porque, cuando un edificio entra en estado de obsolescencia, se puede reconfigurar enteramente para un nuevo uso. (176)

(It cannot be stated more clearly: where nothing is trash, everything is. And it is Marc Augé himself who has realized that, if things continue like this, our civilization will be the first one in the world that does not leave behind any ruins, that special type of historical trash. The generic city [the *junk-city*] leaves no ruins because when a building is in a state of obsolescence, it can be entirely reconfigured in order to provide a new use.)

Unlike Ábalos and Herreros, Pardo is not celebrating the new propensity for recycling as much as he is describing our current era. As a philosopher he is primarily interested in articulating how this elastic, flexible and modular idea of subjectivity in the age of trash bumps up against individual human identity. The spatial and urban components of Pardo's theory of trash and recycling lie in the fact that, as he puts it, "es como si cada enclave edificado en las calles debiera ser, al mismo tiempo, una seña de identidad inconfundible y un espacio infinitamente remodelable, es decir, una zona cero" (it's as if each enclave built on the street should be an unmistakable identifying mark and, at the same time, an infinitely redesignable space, that is, a ground zero). The recycling paradigm that Pardo articulates goes hand in hand with the provisional, experimental, mobile, constantly changing quality of contemporary urban space that so interests Ábalos and Herreros.

Others engaged in the fields of urban studies and urban theory in Spain have also been looking for ways to rethink some basic working concepts in an effort to better reflect on the current economic and political situation in which cities find themselves. Fernando Roch Peña, José Miguel Fernández Güell and Álvaro Sevilla-Buitrago, for example, summarize the efforts of a dynamic group of academics devoted to urban issues in their 2011 editorial for the journal *Urban. Nueva Serie*. For these urban theorists, it is now a question of "conceiving a *new urban society* for the purpose of developing a growing independence" from the economic system within which social justice and true long-term progress are inconceivable (5).[9] On the theoretical level, the discourses of the modern and the postmodern don't seem to mesh with the contemporary city's problems and potentialities.[10] Radical changes in demographic trends, the failure of the neoliberal state to deal with the economic crisis, changes in the global division of labor, the cracks in globalization's myth of the mobility of people and goods, the increasing division of classes by gentrification, and last but certainly not least, the impending ecological threats to our current understanding of the city: these are the issues that have urban theorists in Spain and elsewhere working towards new solutions to the city's current problems and redefining what the city is and can be.

Part of our inquiry into the evolving concept of the urban to reflect new economic and ecological realities in Madrid forces us to pay attention not only to the places and spaces that have been produced by the consumer city, but also to investigate the ways in which urban space is discarded, left derelict, and then recycled in *lugares basuras* (places of trash) or *no lugares* (nonplaces), as Pardo would call them. This approach is somewhat paradoxical since it seems to simultaneously treat urban space as object and product to be discarded, while also accepting the theoretical position espoused by urban theorists like Henri Lefebvre, David Harvey, and Neil Brenner that the urban is not a product, but a process. To understand the urban landscape as a socio-spatial process whose derelict spaces (which are nonproductive in the capitalist sense) form an integral part of the overall whole of the urban environment is to treat urban space like an ecologist might treat a forest that has been clear cut or devastated by fire. The forest does not stop being forest because there are stumps or charred logs, rather it enters into a new stage of forest succession. Different plants begin to grow, downed logs become a micro-habitat for grubs, small plants and mammals, and slowly the process of being a forest continues.

The second half of this essay looks at two spaces in contemporary Madrid that have radically changed their social purpose amid shifting patterns of socio-spatial organization in the wake of the economic crisis. These formerly vacant cityspaces exemplify Ábalos and Herrero's "latent gardens" that contribute to what they call "public space and an architecture reflecting new possibilities" (25). These examples of participatory urbanism began like small pioneer plants that took hold in a devastated landscape, spreading their roots and reconstituting the cultural ecosystem of the city by envisioning and then using urban space in new ways. These modes can be called "ecological" not because they protect some fetishized concept of Nature, but rather because they cultivate connections and relationships and offer a deeper, richer, more heterogeneous definition of what it means to produce urban space. They are also ecological in a more conventional sense because their efforts rely on actual urban gardens, the repurposing of recycled materials, as well as a culturally-aware approach to cityspace that, we contend, responds to Pardo's entreaty to stop experiencing "trash" as waste and "make a *new urban landscape* out of it" (170). Pardo encourages us to think of vacant lots, abandoned sites for commercial centers, or underused indoor marketplaces as potential sites for new urban landscapes. These places that the neoliberal city has not yet been fully able to weave into the network of global capital or that have been set loose from this network since the collapse of the Spanish economy in 2008 are precisely where particularly creative urban projects are taking place. In the examples that follow we look at

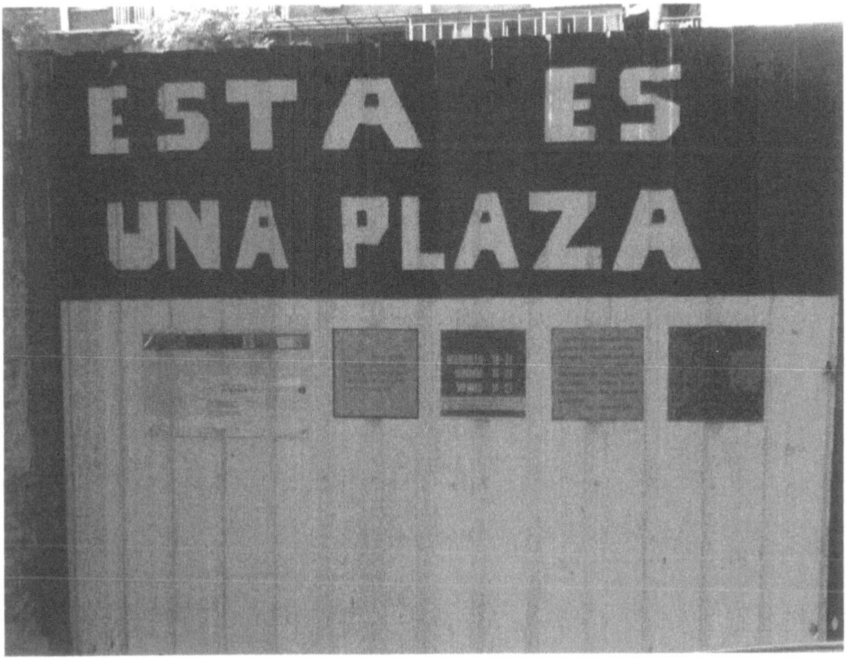

Fig. 1. Entrance to *Ésta es una plaza*. Photo by Matthew Feinberg.

how these "spaces of refuse"—in both senses of the word—have been reclaimed and renegotiated in ways that question some of our most basic assumptions about the (post)modern city and how it works.

¡Ésta es una plaza!

Halfway along the short stretch of Doctor Fourquet Street in Madrid's Lavapiés neighborhood one finds a half acre (1770.46 m^2) plot behind a long brick wall that contrasts sharply with its nearby neighbor, the iconic Reina Sofía Museum of Contemporary Art and its 30-million-dollar 2005 addition designed by French "star architect" Jean Nouvel. Behind this wall, this parcel remained cut off from the neighborhood's residents for approximately thirty years. Today the activity in this seemingly abandoned corner of the densely-populated neighborhood is announced by the brilliant colors and creative designs of graffiti art that covers the length of the wall and the large white painted letters painted on the black gate that read "ÉSTA ES UNA PLAZA" (THIS IS A CITY SQUARE).

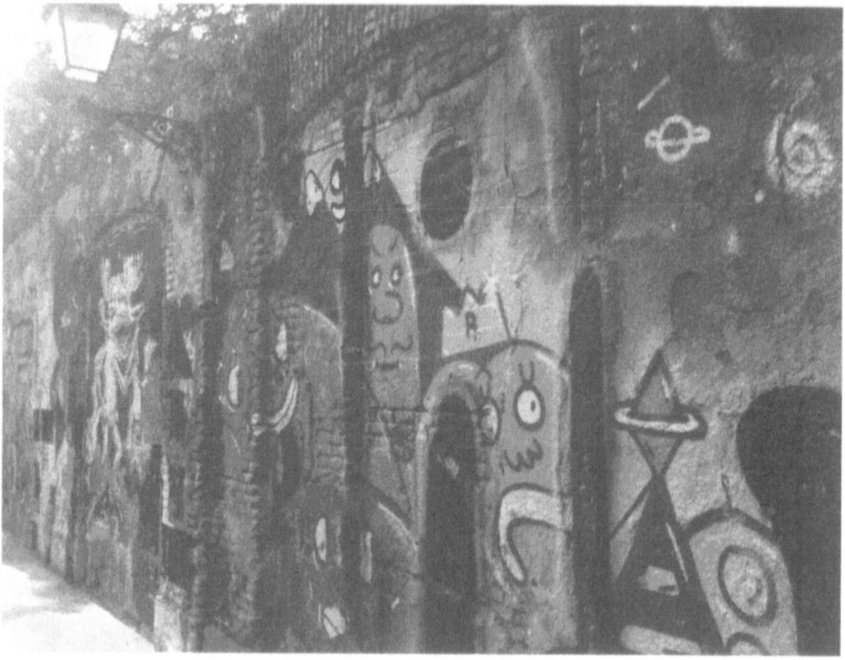

Fig. 2. Exterior wall of *Ésta es una plaza*. Photo by Matthew Feinberg.

Through its use of the demonstrative pronoun "this," the name is a proclamation that actively reconceptualizes what lies beyond the gate. Lacking an antecedent, the pronoun seeks out a referent and it is clear that it must be assigned to the parcel of land, which no longer linguistically an empty space, declares itself a plaza.

The term *plaza* derives from the vulgar Latin *plattea* meaning broad street or courtyard and carries with it a long cultural history of urban space as a point of encounter; a meaning revealed in the second definition offered by the Real Academia Española which defines a *plaza* as the place where "se tiene el trato común de los vecinos, y donde se celebran las ferias, los mercados y fiestas públicas" ("Plaza") (the common activities of neighbors are held as well as fairs, markets, and public festivals). These "common activities" are reflected in the smaller yellow sign on the gate that explains that it is a lot ceded by the municipal government "para convertirlo en una plaza pública" (for conversion into a public square). This focus on the public quality of the space is important because it reflects the overall objectives for the project that began in 2008. As part of a workshop on urban reclamation called the "Montaje de Acciones Urbanas"

(Montage of Urban Action) organized by La Casa Encendida, a social and cultural center located close to the Plaza, the Milanese firm Esterni received five thousand Euros to develop the parcel as an exhibition site of new models of urbanism.[11] The parcel was outfitted with spaces dedicated to theater, gardening, and general assemblies, among other uses. When the site was razed in May of 2009 by the municipal government, activists from the neighborhood and beyond established themselves as a nonprofit organization called the *Asociación Ésta es una Plaza* and petitioned the city's Department of Urban Planning and Housing for control of the space. Though part of the *Asociación's* interest in the lot was driven by a desire to preserve vegetated public space in a neighborhood where this was a rare commodity, their real objective was to "facilitar los procesos de ciudadanía activa, mediante la participación activa y consciente de los vecinos en la gestión de los espacios públicos del barrio" ("Proyecto Original") (facilitate the processes of active citizenship, by means of the active and conscious participation of the neighbors in the management of the neighborhood's public spaces). The "plaza" then is not merely a location, but rather a whole range of *spatial practices*—as Henri Lefebvre would call them (1991)—involved in the creation and management of the space.[12]

The most basic of these spatial practices, the occupation itself of the land, is important because of the ways that it problematizes notions of vacancy and emptiness by embracing an ecological understanding of the site. The problematic nature of the descriptor "vacant" can be seen quite clearly in images found on the website of *La Asociación Ésta es una Plaza* (www.estaesunaplaza.blogspot.com) that reveals a space that was far from empty. In fact, the space was full of grasses, trees, bushes, and, of course, trash. One might even say that it looked wild. The images highlight the fact that derelict urban spaces are commonly thought of as spaces that lack human use. Because of predetermined notions of both the Urban and of Nature, these spaces are seen as lacking both. On the one hand, a "vacant" lot is not urban because it is seemingly outside the circulation of capital and has lost its ability to generate revenue. Similarly, this "empty" space—despite being a full-fledged habitat for a variety of plants and most likely a variety of animals—is not considered natural because of its location within an urban context. Located somewhere in between our conventional understandings of Nature and the Urban, the space has become a conceptual transit point between these seeming opposites.

With these images of "urban nature" or "nature in the urban" in mind, it is interesting to note how the very name of the project framed it in language that we typically associate with the urban built environment. This is particularly interesting because so much of the lot has been dedicated to a large urban garden—a fact that also illustrates just how "full" the lot was understood to be. As of the summer of 2013, the "plaza" contains ten separate raised beds that are approximately each twenty feet by five feet (6 x 1.5 m) that are dedicated to growing a wide range of produce including artichokes, strawberries, onions, leeks, carrots, garlic, broccoli, beans, and a number of other vegetables. This abundance is complemented by almost twenty fruit trees offering plums, apples, peaches, lime, apricot, and other types of fruit. Rearticulating the notion of production, this community garden seeks value from its use rather than its potential exchange value in the marketplace. From this perspective, of course, the lot could never have been empty at all given that the reclamation project depended significantly on the presence of dirt and organic material that would allow cultivation to occur. That it was a "natural" place in the city (i.e., with biological components) meant agriculture could occur there. This privileging of the space's "naturalness" is compounded the fact the *huerto* (vegetable garden) is complemented by a horticulture project—a *jardín*—that undermines the vacant or empty quality of the space. In two locations in the parcel, the collective maintains "los árboles y plantas ya existentes en el solar, intentando alterar lo menos posible la estructura vegetal presente" ("Proyecto Original") (the trees and plants already present in the lot, trying to avoid altering the current plant life as much as possible). This xeriscaping project implicitly understood the lot as an ecological space from the beginning, a space that was not empty at all, but replete with trees and plants to be cultivated and respected.

The disjuncture between the robust dedication to horticulture and urban agriculture and the decidedly urban sounding name for the project is a reminder that this project is not an exercise in environmentalism per se, but rather conceived of as a way to "desarrollar actividades lúdicas, culturales, educativas y ambientales, que ayuden a la interacción y, por tanto, a la mejor comprensión de los vecinos con los que se comparte barrio, pero con los que generalmente no se convive" ("Proyecto Original") (develop ludic, cultural, educational, and environmental activities that help the interaction and therefore mutual understanding amongst neighbors that share a neighborhood, but with whom they rarely cohabitate). A prominent part of the space is also dedicated to events like the program *Microteatro El Terrenito* (The Little Earth Microtheater) that offered a series of short (15-minute) plays, or educational projects like those from March of 2013 entitled *Escuela pública par tod@s* (Public school for all) that offered free lectures on everything from botany to Baroque political thought. Perhaps most important is the fact that this project is organized around the principles of *autogestión* or self-management. The dedication to these princi-

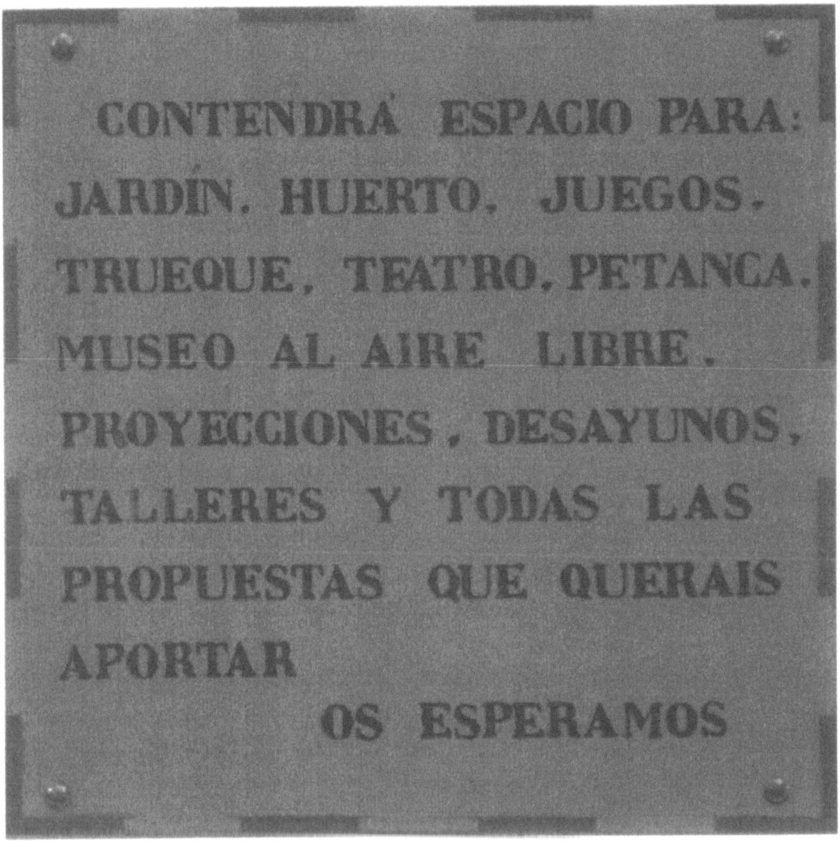

Fig. 3. Detail of sign at entrance to *Ésta es una plaza*. Photo by Matthew Feinberg

ples means that the space is organized horizontally with no leaders and that decisions are made through weekly assemblies. This nonhierarchical approach to decision making has characterized a whole range of social movements from the Civil Rights Movement in the United States to squatted social centers in Spain and other parts of Europe. More recently, general assemblies played a central role in the encampments of Spain's 15-M movement as well as those of Zuccotti Park and the other locations of the United States' Occupy movement (Martínez López Okupaciones 99 and "The Squatters Movement . . ." 871).

As Alberto Corsín Jiménez and Adolfo Estalella suggest in their research on popular assemblies in Madrid and Lavapiés, the "ultimate aim is not making decisions, but building consensus" (5). Underpinning this pursuit of consensus are other values of *autogestión* such as equality and openness. Through the concept of assembly, the space opens itself up to the rest of the city and therefore can be characterized as a border space, a term that Richard Sennett has defined as the "[zone] in a habitat where organisms become more interactive, due to the meeting of different species" ("The Public Realm").[13] Unlike "boundaries," borders make intermingling and encounters possible because of their porosity. Applying this ecological concept to human society, Sennett implies that it is through this porosity, openness, and intermingling that the cultivation of the public realm occurs. The assembly in the plaza can be considered to be one of these porous moments in the city. As a result, it is a mode of political being that highlights the fact that the cultivation of habitat for various plants found in urban gardens might not be ecological only in the conventional sense, but rather also a spatial practice for building and constructing a shared social and cultural habitat for the community through the interpersonal encounters cultivated through *autogestión* and the creation of some sense of place. Instead of merely producing capital, what was once refuse (that which was refused) now produces vegetables, culture, connection, life—indeed, it supports an ecology of urban space in which the environment is understood in biological terms as well as in social and cultural ones.

Ésta es una plaza is a name that highlights the notion of the public at the heart of a human settlement. It is in the *plaza*, the square, where citizens meet to encounter one another and begin to build a sense of place beyond the alienating urban spaces of industrial capitalism (à la George Simmel) or the consumer city of images, but rather one based on the interconnectedness of dwelling. It is not a coincidence, then, that it is in plazas like the Puerta del Sol, the Plaza Catalunya, and others around Spain where *indignados* of all walks of life came together to not only proclaim that "no somos mercancías en las manos de los banqueros" (we are not just merchandise in the hands of the bankers) but to demonstrate that new ways of being, organizing, and living might be possible. *Ésta es una plaza* is a reminder that these efforts to "rewrite urban landscapes through a commitment to social justice, self-government, and the deployment of common capacities" (Sevilla-Buitrago 49) are dispersed throughout Madrid and should not merely be looked for in its most iconic or most commercialized sites. That this type of cession of public space to community groups can be found in other parts of Madrid suggests that the economic crisis has given local citizens leverage to pressure the municipal government to acquiesce to their "right to the city." Unable to find the investment to redevelop a number of spaces, government agencies have turned them over for management to neighborhood groups and organizations.[14]

The Campo de la Cebada

Another example of this kind of grassroots initiative can be found up the hill, just on the edge of Lavapiés and the trendy neighborhood of La Latina. Called the *Campo de Cebada*, this site is particularly compelling within our framework here because like *Ésta es una plaza* it was a seemingly "empty" lot. In this case, though, the emptiness was not an urban lot filled with pioneering grasses and weeds in the process of transforming into a wild and natural place. Instead, one finds it was the urban emptiness manifested by the cold hard concrete of postindustrial abandonment.

In the same way that the iconic nineteenth-century Plaza de la Cebada marketplace built in 1872 (later demolished in the 1950s) projected Madrid's aspirations for modernity in its steel and glass façade, the recent physical transformations of this lot also make it a spatial barometer for shifts in the ecology of the urban in Madrid. After the demolition in 2009 of the public pool and *polideportivo* (sports and recreation facility) that had stood on the site since the 1960s, the expanse of concrete was like a scab left by the perpetual construction and destruction produced by the circulation of speculative and private capital in Madrid's twenty-first century urban landscape.[15] As such, it speaks to the broad shift in the late industrial period in the meaning of the 'urban' from centers of production for industrial capitalism to centers of consumption that produce capital not from manufacturing, but rather the dual economic engines of finance and real estate (and implicitly a consumer economy). In line with these evolving modes of production, the municipal government's plans originally consisted of knocking down the community's heavily-used recreation facilities to make way for a multi-level mixed-use mall. The example here is particularly striking because the municipal government—amidst hollow promises to restore the public facilities—actually planned to fund the commercial project through the sale of housing units, an approach they had used for the rehabilitation of the Mercado Barceló near the high-rent district of Chueca that had garnered some 25 million Euros for the project (Alcaide "El Ayuntamiento abrirá . . ."). The project in the Plaza de la Cebada was supposed to follow a similar tactic and leverage the symbolic capital of the neighborhood's popularity as one of the main metro stops for accessing the very popular Rastro flea market and its proximity to a trove of popular tapas bars and clubs found in La Latina.

This plan was a microcosm of Spain's economic problems rooted in speculative capital and the real estate market—a plan that fell short when the overleveraged house of cards upon which the Spanish economy had been built during the early 2000s came tumbling down in 2008, provoking the onset of the eco-

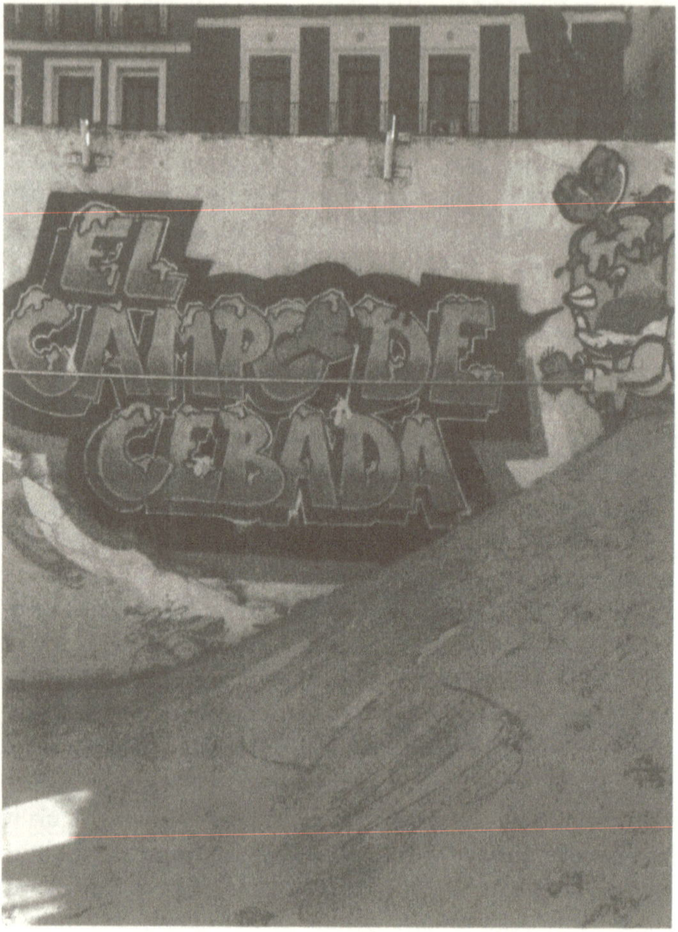

Fig. 4 Graffiti in the *Campo de Cebada* Photo by Matthew Feinberg.

nomic crisis that continues in earnest in 2014. The city's backup plan demonstrated a singular devotion to using this public space to benefit private interests since they later proposed developing a privately managed 35,000 square foot sports and recreation facility—on top of what was previously the public pool and neighborhood market (Martín "Un centro comercial gigante . . ."). Since the financial crisis occurred, the city has been unable to find anyone willing to invest in the property. The community was left with a hole in the ground—a monument to the excesses of financial capital during the speculative real estate bubble. Just as in the case of the parcel on Doctor Fourquet Street, a conven-

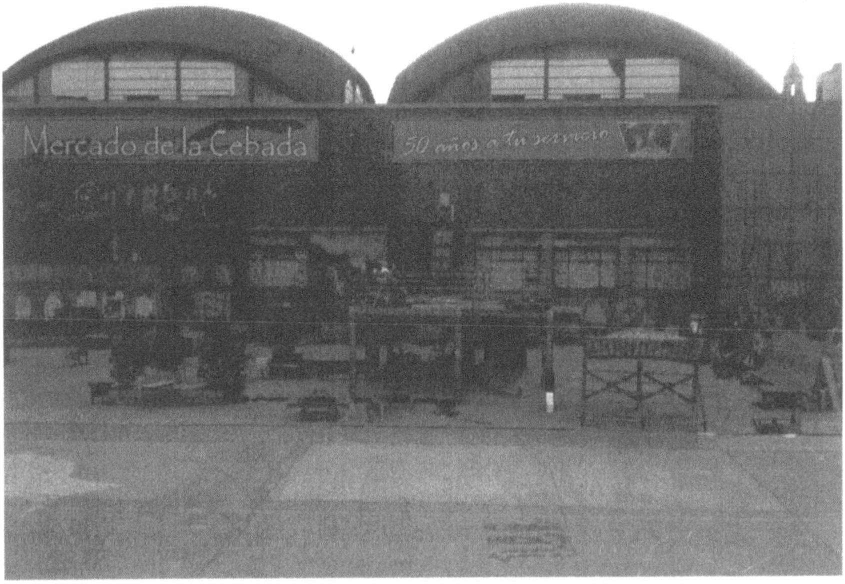

Fig. 5 *Campo de Cebada* with the *Mercado de la Cebada* in the background. Photo by Matthew Feinberg.

tional understanding of the urban would lead one to assume that the vacant lot was derelict and dead because it was unable to produce capital. But the citizens involved in the *Campo de Cebada* project saw possibility instead of vacancy. Instead of an expanse of concrete—the trash of speculative capital—they saw only a "latent garden," to use Ábalos and Herreros's term. It was a *Campo de Cebada* (barley field) where new urban possibilities could be cultivated.

The *Campo de Cebada* is a "proyecto autogestionado vecinal" ("El nacimiento de El Campo de Cebada") (self-managed neighborhood project) that began in February of 2011 after Madrid's municipal government ceded the management of the vacant lot to a coalition of urban activists made up groups like Zuloark, an architectural collective, and local and regional community associations like the Asociación de Vecinos del Centro–La Latina and the Federación Regional de Asociaciones Vecinales de Madrid (Neighborhood Association of the Centro District–La Latina and the Regional Federation of Neighborhood Associations of Madrid). These efforts began the previous year in September of 2010 when the *Campo de Cebada* was one of twenty-one sites that the Basurama architectural collective had selected during an annual event called La Noche en Blanco that filled Madrid's main streets with temporary occupations of public space intended to use culture to reinvent the relationship of citizens with their city. Interestingly, the intervention that took place in what would become the *Campo de Cebada* was organized by another collective called Exyzt and consisted of an artificial tropical island and lagoon called City Island ("Placemaking - El *Campo de Cebada*."). The effort employed a certain cognitive dissonance by creating a lush tropical respite in what had previously been a barren concrete desert nearly unbearable in the summer heat. The creation of a microclimate that made the space habitable was a collective experience for many citizens and, upon the dismantling of the project, prompted discussions that would result in the groups above taking the administrative steps necessary to have the space granted to them by the municipal authorities. Though the space remains an expanse of hard concrete, its name has reconceptualized the site. Like the overgrown lot at Doctor Fourquet (what would become *Ésta es una plaza),* the act of renaming points to a new way of understanding what was formerly "empty." At Doctor Fourquet, *Ésta es una plaza* projected the urban form of the square across a seemingly wild and "natural" space. In the *Campo,* on the other hand, the name linguistically repurposes the concrete and steel detritus of abandonment, envisioning it in almost pastoral terms.

This pastoral sensibility alludes to the agricultural trope used to describe the broader objectives of the project. The cultural and neighborhood groups that negotiated the handing over of the land did not see this postindustrial testament to financial excess as an empty stretch of concrete, but rather (as one participant describes it on the Campo's website) an opportunity for "cultivando ideas para el barrio" ("El nacimiento de El *Campo de Cebada*") (cultivating ideas for the neighborhood). This cultivation of ideas was to be carried out by making the Campo a place of encounter in order to "inventar relaciones posibles con los vecinos en el presente" ("El nacimiento de El Campo de la Cebada") (invent possible relations with our neighbors in the present). That is, the *Campo de Cebada* would be—as the agricultural allusion in its name implies—a space of *cultivation* connecting individuals in a collective project. In this sense, as the writer clarifies, these urban gardeners (in the model of Ábalos and Herrero), "No [son] agricultores políticos, [son] agricultores ciudadanos," ("El nacimiento de *El Campo de Cebada*") (are not political farmers, they are citizen farmers). The Campo was dedicated both to cultivating the urban fabric of the neighborhood and to creating, like a literal field of barley, a moving, vibrating collective made up of individuals. This agricultural topos is captured in the image found on the title banner of the Campo's web page (elcampodecebada.org) where the black and white background of the Campo and the surrounding buildings are superimposed on a deep yellow foreground, where six individuals meet in the vibrant field of barley which not only contains them, but also seems to spill out beyond the boundaries of the Campo's wall.

Of course, part of this project has resulted in a substantive urban garden, yet much of the cultivation that has taken place in the site was achieved by giving citizens a voice to express their interests. On the blog about the Campo posted by the architectural collective Ecosistema Urbano, they describe how at the outset "small chalkboards were placed at the entrance, to catch ideas passing by and communicate upcoming events" ("Placemaking–El *Campo de Cebada*"). This nineteenth-century technology was complemented by twenty-first century forms, like a web page where people could share questions, ideas and proposals. These various ideas would be discussed, debated, and scheduled at open weekly assemblies. Thus the Campo became, like *Ésta es una plaza,* a porous border that allowed for the intermingling of people and from which a range of projects ". . . began to sprout" ("Placemaking–El *Campo de Cebada*"). Some of these projects were merely assemblies that focused on specific citizen initiatives. Their purposes ranged from providing a space for the neighborhood to come together in order to discuss a resolution to the issue of the missing sports complex and the always-imminent-but-never-actualized commercial development to other kinds of "inappropriate appropriations," as they call them ("Placemaking–El *Campo de Cebada*"), such as #edumeet that continues to meet twice a week for forums on education and *Desayunos ciudadanos,* public breakfasts that happen throughout Madrid that are intended to claim public space for citizens and provide opportunities for further community discussions.

The Campo has also been used as an important place for grassroots cultural initiatives. For example, free Sunday morning concerts called *Cantamañas* have become a fixture of the Sunday morning Rastro flea market that extends for several city blocks down the hill. There has also been a festival of short films, a children's theater event, and a circus. Worth noting also is the theater group AlmaViva Teatro's performances of José Zorilla's *Don Juan Tenorio* in the Campo de Cebada in honor of All-Saints Day in 2011, 2012, and 2013. These highly dynamic performances utilize a different part of the lot for each of the play's seven acts with stagehands directing the public to periodically move from one area of the concrete theater to the next. The performances are not only an exercise in the repurposing of refuse, but the staging itself employed repurposed objects as various scenes were staged in and around the old industrial shipping container and the steel scaffolding that have become fixtures of the Campo. Since entering the *Campo de Cebada* requires descending a long sixty-foot concrete ramp, the public at these performances was surrounded by the repurposed space while they watched the theater performance incorporate the industrial refuse into the production of a classic work of Spanish theater. It is a setting that contrasted with the backdrop of residential buildings and the lights of the Teatro La Latina across the street, a juxtaposition emphasizing that the concrete field of barely is a process, in the midst of becoming. This contrast might be why César Barló, the director of the Tenorio performances, has characterized them as "intervenciónes teatrales" (Barló 11) (theatrical interventions), a description that frames the cultural activity as an interruption in the process of spatial production we call the urban and therefore suggests that grassroots cultural production plays an important role in transforming urban space.

Like hearty plants that establish themselves in even the most rocky and inhospitable terrain, culture, too, seeks out space to proliferate and come into being. In this sense the cultural repurposing of spaces of refuse in the twenty-first century city alludes to new ways of appreciating the ecology of the urban. While describing the activities of Zuloark and Basurama, Alberto Corsín Jiménez evokes an ecological understanding of the urban by describing interventions like the Campo as "finding crevices in the contemporary metropolis that allow for *building-up* collaboration" ("Three Traps Many" 12, emphasis in the original). These crevices of the city—spaces left unused and abandoned—provide rich environments for the growth and proliferation of alternative forms of cultural production and organization that reject the market as the only sphere for connection in the contemporary urban environment. It is an understanding of the urban that pushes the individualism and alienation of the capitalist city aside in favor of more collaborative relationships. As such, it echoes the openness and free exchange of the digital realm, an idea alluded to by Manuel Pascual, an architect affiliated with Zuloark who suggests that their interventions are "mucho más vinculada a la participación y el *open data* . . ." (qtd. in Bosco 72) (more connected to participation and open data). Because of these overlaps, Corsín has described the products of these collaborative efforts to repurpose the city as "open-source architectural objects" ("Three Traps" 12).[16]

Conceptualizing the Campo as an object (as opposed to a space) is an approach reinforced by the fact that the *Campo de Cebada* was awarded the 2013 Golden Nica from the prestigious digital culture organization Ars Electronica in Linz, Austria in the category of "Digital Communities." To win the award the Campo was submitted as a cohesive whole, an 'object' for review and study. For the judges of the award, the Campo was "una prueba de que la crisis también crea oportunidades . . . para el intercambio humano, la creatividad, para el encuentro de la comunidad tanto para los que vivan en Madrid como online" (qtd. in Gutiérrez) (proof that the crisis also creates opportunities . . . for human exchange, creativity, for the meeting of community as much for those who live in Madrid as those who are online). Some of these opportunities for exchange can be seen in the use of the *Campo de Cebada* as the "Campus de Cebada" (Cebada Campus) during the summer of 2013 for the Universidad Popular del Verano (People's Summer University) and the recent successful crowdfunding effort to construct a geodesic dome that will provide shade and protection from the elements for the events that take place in the Campo.

It is important to note that Corsín's term "open-source objects" refers not only to the interventions themselves, but more specifically to the repurposed objects that have been produced during workshops that have taken place at the *Campo de Cebada* and other locations around Madrid. The nature of these projects merits attention because they illustrate how the ecology of the urban is a multivalent process with connections to traditional understandings of sustainability within an identifiable cultural context. With names like "HandMade Urbanism," these efforts revolve around repurposing found materials in the city. In the case of the HandMade Urbanism workshop, an annual event that began in 2012, participants gather abandoned materials to make furniture that is later delivered to a specific community. In a certain sense, the open source nature of these projects relies on being sensitive to the cultural ecology and context of each object. This includes, as Corsín describes it, both "[sourcing] their building materials locally, [and having] local craftsmen and technicians join the project, and [documenting] as much as possible the socioeconomic context that has led to the community expressing a need for a particular piece of furniture" ("Three traps many" 13). From the detritus of the streets emerge objects that are deeply imbedded in the cultural habitat of the citizens, particularly when they are subsequently placed back into those same communities. The trash in the neighborhood, through the repurposing process, has embodied the spatial requirements, interests, tendencies, and uses of the community. From habitat it has become dwelling. This dynamic between the tangible objects that facilitated community and the spatial practices they inspired became manifest in a collaboration between Corsin and Estallela with Basurama and Zuloark during 2013 that was entitled *15Muebles* (15-M Furniture). This event not only sought to exhibit the actual materials and objects that formed the "furniture" for the encampments of 15-M and the smaller popular assemblies that proliferated in their wake, but attempted to explore the very notion of the popular assembly as "urban installations: pieces of mobile urban equipment that [take] temporary residence in the open air of plazas or streets" ("Three traps many" 14). Thus the re-purposing of objects is closely tied to the repurposing of space itself.

This network of "open-source objects" within "open-source spaces" found in grassroots urban interventions from *Ésta es una plaza* and the *Campo de Cebada* and the *Acampada Sol* and the resultant neighborhood assemblies fit into what Luis Moreno-Caballud (2012) has described as the "imaginación sostenible" (sustainable imagination). For Moreno-Caballud, these sustainable products and systems of production include the many cultural activities of 15-M and its connections with the global free-culture movement associated with Copyleft, the Creative Commons, and the movements for open-source software—new systems of production that *El País* grouped under the December 2011 headline as the "revolución cultural del procomún" (Fraguas) (cultural revolution of the commons). The two cultural projects discussed here are important because they demonstrate how the "sustainable imagination" of the 15-M movement is not merely digital but also tangible and material, and that it is occurring in the many "crevices" of the city and not only in central or high-profile plazas. To understand these connections between the commons, open-source networks of technology and culture within their urban frame suggests that what we've termed urban interventions or cultural repurposing like the *Campo de Cebada* can be understood as a sustainable mode of production, a means to utilize the resources of the urban for the benefit of many instead of for a small minority. In short, understanding the urban in terms of the commons or *procomún* allows us to imagine spaces where the digital, the cultural, the horizontal, and the participatory might converge in a kind of new urban ecology. Indeed, the cultural repurposing of space not only presents opportunities for collaborative work and alternative models for using space, but are tied to what Joana Conill et. al have called "alternative economic culture" (212), a range of practices that includes things like urban gardens and consumer cooperatives, but also hacklabs, social currency networks, free universities and others that have gained momentum because of the failures of "a disciplinary model of hardened, shrunk, financial capitalism" (Conill et al. 245). If one accepts the position of Castells, Caraça, and Cardoso that the current economic crisis was caused in large part by "deregulated global capitalism [. . .] fed by a relentless production of synthetic securities as the source of capital accumulation and capital lending" (2), then we can further understand the ecology of the urban as a dynamic between the synthetic urbanism of debt-driven capitalism and the organic production of space by citizens. In the refuse of an urban landscape transformed by economic crisis, the repurposing of the urban is one way that citizens can cultivate the "latent gardens" around them, thereby asserting their right to the city once again.

Notes

1. In September of 2013 all references to Line 2 (the red line) of Madrid's metro system began to bear the Vodafone logo and were renamed "Línea 2 Vodafone." For a range of accounts detailing the renaming of the Puerta del Sol on June 1, 2013, see "Cash-Strapped Spain," Rushton, "Próxima estación."
2. Sevilla-Buitrago's article (based on the Lefevbrian concept of the inevitability of conflict in the process of the production of social space) contends that "[c]ombining social networks with temporary encampments, the 15-M movement demonstrates that activists can devise new uses of public space and new virtual political spaces through a collective, practice-oriented production of place" (42).
3. Conill, Castells, Cardenas, and Servon's chapter "Beyond the Crisis: The Emergence of Alternative Economic Practices" from *Aftermath: The Cultures of Economic Crisis* (2012) offers compelling research about noncapitalist economic practice between the years 2008 and 2011 (210–230) in Madrid. See also the website EcoZoom! at ecozoom.mapunto.net/map for an interactive map of alternative social movements in Madrid. See Vilaseca's *Barcelonan Okupas* for a book-length study of participatory urbanism and a description of networks of squatting initiatives in Barcelona.
4. A good example of this approach can be found in *The Routledge Handbook of Urban Ecology* (2011) edited by Ian Douglas, David Goode, and Michael C. Houck where they emphasize "the relationship between the biological community and the urban environment due to the interaction between human culture and natural environment" (Douglas 3).
5. Swyngedouw's thoughts on the fetishization of nature can perhaps be understood as thinly veiled critiques of what the geographer sees as the ahistoricism of Object Theory and Actor-Network Theory. Paradoxically, they both might arrive at some of the same conclusions, but they take very different paths. For recent discussions of the implications of Actor Network Theory theory on Urban Studies, see Brenner et al., "Assemblage Urbanism."
6. Brenner's "Theses on Urbanism" is a self-reflexive manifesto that asks urban theorists to stop demarcating the urban in contrast to nonurban spaces. "[U]rban studies has conceived of the urban as a specific type of settlement space, one that is thought to be different, in some qualitative way, from the putatively nonurban spaces that surround it—from the suburb, the town and the village to the rural, the countryside and the wilderness" (88). Urban studies, he contends, turns a blind eye to an always present and undefinable "elsewhere" on which the definition of the urban depends. If we stop thinking about cities along the urban/rural divide or as discrete, distinct and

territorially bounded units, argues Brenner, we will have to pay attention to the wider urban *processes* of both concentration and extension by which modern capitalism is produced on a global, more ecologically-aware scale.

7. Iñaki Ábalos and Juan Herreros together formed the architectural firm Ábalos and Herreros in 1984. They collaborated on a series of important projects in Spain and abroad until they ended their collaboration in 2006, forming their own firms and accepting academic positions in the United States. In addition to "A New Naturalism" (2002) and *Recycling Madrid* (2000) mentioned in this essay, other coauthored publications regarding their philosophy of architecture, the urban and ways of thinking about nature include *Areas de impunidad* (ACTAR, 1997); *Natural Artificial* (EXIT LMI, 1999); *Técnica y arquitectura* (Nerea, 1992); and *Tower and Office* (MIT, 2003).
8. Pardo credits "Treintagenarios," a poem written by Spanish author Juan Bonilla, for the title of his book. "Aquí me veis, viajero / de un tiempo que se pierde en la espesura / del paso y él me da lo mismo . . . pero / nunca fue tan hermosa la basura." (*Partes de Guerra* 27) (Here you see me, traveler, from a time that gets lost in the thick of / the path's crossing and it's all the same to me . . . but / trash was never so beautiful).
9. *Urban. Nueva Serie* is a Spanish-English bilingual publication. The Spanish text is provided here: "[E]s preciso concebir otro futuro, recuperarlo como objeto de pensamiento, de debate y de diseño. Ya no se trata de la construcción de la sociedad industrial—tan sencilla de requisitos funcionales pero tan conflictiva y difícil de ejecutar—, sino de una nueva sociedad urbana llamada a desarrollar una creciente autonomía respecto a cualquier régimen económico incapaz de cumplir sus promesas de emancipación y progreso social, respecto a todo sistema que ni siquiera pueda conservar sus propios e invocados mecanismos de regulación" (5) (It's time to conceive of another future, to recover it as an object of thought, of debate and of design. It is no longer a question of building industrial society—which was so easy in terms of functional requirements but so conflictive and difficult to execute—but of conceiving a new urban society for the purpose of developing a new independence with respect to economic regimes that are incapable of fulfilling their promises of social emancipation and progress and with respect to systems that are unable to conserve their own invoked mechanisms of regulation).
10. Edward Soja's *Postmetropolis* (2000), for example, is an early attempt to draw on a variety of Social Sciences and Humanities disciplines to find out what was outdated in urban studies and urban theory, to pinpoint what it is about the contemporary city that cannot be adequately explained by theories of the urban that are rooted in nineteenth-century assumptions about urban space and capital. In his 2000 book Soja makes clear that he is neither talking exclusively about the Postmodern city, nor about New Urbanism when he says that "I have chosen the term 'postmetropolis' as a working title for what might be called the new urbanism, had not this term

been taken up by architects and designers for other and narrower purposes" (xiii). He is looking for ways, he writes, to get beyond what he terms "non-post-prefixed" theoretical terms.

11. The Casa Encendida is a project run by the Fundación Caja Madrid Especial dedicated to arts, culture, environment and education. The Casa Encendida's progressive programming is a prime example of how left-of-center politics can offer an ideological cover for capital's pursuit of profit. Caja Madrid, along with six other savings banks, were consolidated by the Spanish government as part of a bailout that was necessary in the wake of Madrid's real estate market collapse. In short, on one hand Caja Madrid funds the type of programming that inspires *Ésta es una plaza,* on the other they are one of the principal players in creating the financial morass in which Spain currently finds itself.
12. See Fraser's *Henri Lefebvre and the Spanish Urban Experience* (2011) for a book-length study analyzing how the French urban philosopher's concept of the production of social space can be used to illuminate Spanish urban culture from nineteenth-century *costumbrismo* to present-day videogames.
13. It is interesting to note that the section of Sennet's essay on the "Public Realm" relies heavily on thinking about the urban space in terms of ecology. He calls it 'context', but suggests that this is merely a term for habitat.
14. In addition to *Ésta es una plaza,* there are other important examples that have sprung up in Lavapiés and other locations in recent years. One of the higher profile examples can be found in the nearby Embajadores roundabout where the management of the iconic Tabacalera (Tobacco Factory) that was slated to be converted into a National Museum of Visual Arts by Spain's Ministry of Culture was ceded to local activists, many of whom had been involved in the *okupa* movement. It is now a neighborhood social center dedicated to artistic production and performance, gardening, community organizing and other activities. Like *Ésta es una plaza* and the *Campo de Cebada* the space is managed according to assembly-driven modes of *autogestión* or self-management. For more on the CSA La Tabacalera see Feinberg "From Cigarreras".
15. These plans were part of the Plan de Revitalización del Centro (Revitalization Plan for the Centro District) that was announced by during Mayor Alberto Ruiz-Gallardón's 2004 campaign that he referred to as "cirugia de centro" (surgery for the Centro District) (Alcaide "El Ayuntamiento abrirá . . ."). These plans fit into plans that were already underway at the time that saw intense investment by the municipal government in beautifying the city center through consumption spaces and cultural institutions. These plans were laid out in the General Urban Plan of 1997 (General de Ordenación Urbana de Madrid 1997 or PGOUM 1997). For more on the urban development in Madrid and discussion of the PGOUM 1997 see Feinberg ("Lavapiés . . ." 124–155).
16. Vilaseca's recent work (2010) on hacktivism in the TriBall neighborhood of Madrid

also highlights this potential for reconceiving the urban in terms of collaborative cultural production and open-source digital culture.

Works Cited

Ábalos, Iñaki, and Juan Herreros. *Recycling Madrid*. Barcelona: ACTAR, 2000.

______. "A New Naturalism: 7 Micromanifestos." *2G* 22 (2002): 18–26.

Alcaide, Soledad. "El Ayuntamiento abrirá un concurso de ideas para reformar la plaza de la Cebada." *El País*. Madrid Edition. 6 July 2005. Web. 15 July 2013.

Augé, Marc. *Non-Places: Introduction to an Anthropology of Supermodernity*. New York: Verso, 1995.

Baroló, César. "Introducción." *Don Juan Tenorio 2012: La creación teatral en un espacio urbano*. Ed. César Baroló. Madrid: Ediciones Antígona, 2012. 10–11.

Benjamin, Walter. "The Work of Art in the Age of Mechanical Reproduction." *Illuminations*. Ed. Hannah Arendt. New York: Schocken, 1968. 218–42.

Bonilla, Juan. "Treintagenarios." *Partes de Guerra*. Valencia: Pre-Textos, 1994.

Bosco, Roberta. "El arte electrónico premia a Madrid." *El País*.17 May 2013: 72.

Brenner, Neil. "Theses on Urbanization." *Public Culture* 25.1 69 (2013): 85–114.

Brenner, Neil, David J. Madden, and David Wachsmuth. "Assemblage Urbanism and the Challenges of Critical Urban Theory." *City: Analysis of Urban Trends, Culture, Theory, Policy, Action* 15.2 (2011): 225–40.

"Cash-Strapped Spain Will Rename Part of Madrid's Subway System 'Sol Vodafone' for Just $4 Million." *Business Insider*. 25 April 2013. Web. 5 July 2013.

Castells, Manuel. *The Urban Question. A Marxist Approach*. Trans. Alan Sheridan. London: Edward Arnold, 1972.

Castells, Manuel, João Caraça, and Gustavo Cardoso. "The Cultures of the Economic Crisis: An Introduction." *Aftermath: The Cultures of the Economic Crisis*. Eds. Manuel Castells, João Caraça, and Gustavo Cardoso. Oxford: Oxford University Press, 2012. 1–14.

Conill, Joana, Manuel Castells, Amalia Cardenas and Lisa Servon. "Beyond the Crisis: The Emergence of Alternative Economic Practices." *Aftermath: The Cultures of the Economic Crisis*. Eds. Manuel Castells, João Caraça, and Gustavo Cardoso. Oxford: Oxford University Press, 2012. 210–50.

Corsín Jiménez, Alberto. "Three Traps Many." *Culturas tecnocientíficas*. 16 May 2013. Web. 23 July 2013.

Corsín Jiménez, Alberto and Adolfo Estalella. "Assembling Neighbours: The City as Archive, Hardware, Method." *Prototyping: La crítica a paseo*. 11 July 2013. Web. 18 July 2013.

Douglas, Ian. "Introduction." *The Routledge Handbook of Urban Ecology*. Eds. Ian Douglas, David Goode, Michael C. Houck, and Rusong Wang. London and New York: Routledge, 2011. 3–6.

"Ecozoom!: El mapa general de los movimientos sociales." n.d. Web. 223 July 2013.

Feinberg, Matthew I. "From Cigarreras to Indignados: Spectacles of Scale in the CSA La Tabacalera of Lavapiés, Madrid." *Reimagining Madrid: Perspectives from Urban Cultural Studies.* Spec. Issue of *International Journal of Iberian Studies* 26.2 (2013): 21–40.

_____. "Lavapiés, Madrid as Twenty-First Century Urban Spectacle." Ph.D. Diss. University of Kentucky. 10 July 2011. Web. 20 July 2013.

Fraguas, Antonio. "La revolución cultural del procomún." *El País.* 28 Dec. 2011. Web. 9 August 2013.

Fraser, Benjamin. *Henri Lefebvre and the Spanish Urban Experience: Reading from the Mobile City.* Lewisburg: Bucknell University Press, 2011.

Gutiérrez, Bernardo. "Campo de Cebada, un premio internacional a la ciudanía en red." *20 minutos.* Codigo Abierto. 18 May 2013. Web. 7 Aug. 2013.

Harvey, David. *Rebel Cities. From the Right to the City to the Urban Revolution.* London and New York: Verso, 2012.

Lefebvre, Henri. *The Production of Space.* Trans. Donald Nicholson-Smith. Oxford: Wiley-Blackwell, 1991.

Leman Stefanovic, Ingrid. "In Search of the Natural City." *The Natural City: Re-envisioning the Built Environment.* Ed. Ingrid Leman Stefanovic and Stephen Bede Scharper. Toronto: University of Toronto Press, 2012. 11–35.

Martín, María. "Un centro comercial gigante para recuperar la plaza de la Cebada." *El País.* Madrid Edition. 10 Aug. 2010. Web. 15 July 2013.

Martínez López, Miguel. *Okupaciones de viviendas y de centros sociales: Autogestión, contracultura y conflictos urbanos.* Barcelona: Virus, 2002.

_____. "The Squatters Movement in Europe: A Durable Struggle for Social Autonomy in Urban Politics." *Antipode* 45.4 (2013): 886–887.

Moreno-Caballud, Luis. "La imaginación sostenible: culturas y crisis económica en la España actual." *Hispanic Review* 80.4 (2012): 535–55.

"El nacimiento de *El Campo de Cebada.* Septiembre 2010–Febrero 2011." *El Campo de Cebada.* 24 Feb. 2011. Web. 15 July 2013.

Pardo, José Luis. *Nunca fue tan hermosa la basura.* Madrid: Galaxia Gutenberg, 2010.

"Placemaking-*El Campo de Cebada.*" *Ecosistema Urbano.* 16 Jan. 2012. Web. 1 Aug. 2013.

"Plaza." *Diccionario de la lengua española.* Real Academia Española. n.d. Web. 17 July 2013.

"Próxima estación: Vodafone Sol." *ABC.* 31 May 2013. Web. 15 July 2013.

"Proyecto original Esta es una Plaza. Diciembre de 2008." *Ésta es una Plaza.* 30 Jan. 2010. Web. 16 July 2013.

Roch Peña, Fernando, José Miguel Fernández Güell and Álvaro Sevilla-Buitrago. "Nueva época, viejos conflictos: ciudades y teorías urbanas en la encrucijada." *Urban, Nueva Serie* 1 (2011): 3–10.

Rushton, Catherine. "Madrid Rebrands Metro in Vodafone Deal." *The Telegraph.* 24 April 2013. Web. 15 July 2013.

Seamon, David. "Seeing and Animating the City: A Phenomenological Ecology of Natural and Built Worlds." *The Natural City: Re-envisioning the Built Environment*. Ed. Ingrid Leman Stepanovic and Stephen Bede Scharper. Toronto: University of Toronto Press, 2012. 231–56.

Sennett, Richard. "The Public Realm." *Richard Sennett*. n.d. Web. 17 July 2013.

Sevilla-Buitrago, Álvaro. "'This Square is Our Home!': The Organization of Urban Space in the Spanish 15M Movement." *Progressive Planning* 189 (2011): 42–49.

Soja, Edward. *Postmetropolis*. Oxford: Blackwell, 2000.

Swyngedouw, Erik. "¡La naturaleza no existe! La sostenibilidad como síntoma de una planificación despolitizada." *Urban. Nueva Serie* 1 (2011): 41–66.

Vilaseca, Stephen Luis. *Barcelonan Okupas: Squatter Power!* Lantham, MD: Fairleigh Dickinson University Press, 2013.

_____. "The TriBall Case: 'Okupación Creativa ¡Ya!' vs. Okupa Hacktivismo." *Arizona Journal of Hispanic Cultural Studies*14 (2010): 11–30.

◆ 6

Degrowth and Ecological Economics in Twenty-First-Century Spain: Toward a Posthumanist Economy

Luis I. Prádanos

Since the global financial crisis of 2007/2008, a number of Spanish nonfiction books have adopted a cross-disciplinary perspective to critique the unsustainable social and environmental degradation caused by global capitalism. These authors, including Jorge Riechmann, Joan Martínez Alier, Amaia Orozco, Ramon Folch, Alicia Puleo, Ramón Fernández Durán, Joaquín Sempere, Óscar Carpintero, and Carlos Taibo, embrace social ecology, ecological economics, and political ecology. Such writers adopt these perspectives in order to question and redefine the innocuous European "reform environmentalism" by opposing the European Union official discourse, such as the "gospel of eco-efficiency" (Martínez Alier, *El ecologismo* 20–26). This hegemonic discourse tries to solve ecological and social problems with the same logic that causes and perpetuates them (green capitalism, environmental economics, and sustainable development).

Reform environmentalism "holds to the mainstream assumption that the natural world be seen primarily as a resource for human beings, whether economically or culturally, but it strives to defend and conserve it against overexploitation. For the most part reform environmentalists advocate measures within the given terms of capitalist industrial society" (Clark 2). The Spanish authors I introduce (and others proliferating from different Euro-Mediterranean regions) concern themselves, instead, with environmental justice, asymmetrical transnational power relations, systemic thinking, and the disproportionate social metabolism (flow of materials and energy that are required to sustain all human economic activities) of some regions. These authors propose and explore socially fair and ecologically sound holistic alternatives to the hegemonic neoliberal globalization. The neoliberal model is addicted to economic growth and, therefore, cannot be socially and ecologically sustainable. It forces all aspects of human life into the market economy by imposing a global logic based on massive privatizations of public goods and spaces, deregulation of the global financial economy (and regulation in favor of big corporations) that becomes more autonomous from real production and labor, a smaller state not focused on social or ecological wellbeing but on the global market competition, and aggressive promotion of consumerism. The postcapitalist alternatives suggested by the authors I study are based, instead, on conviviality, voluntary simplicity, slowness, and the reduction of the social metabolism of some populations in order to stay within the ecological limits of the Earth. These writers also question the tyranny of industrial time—a function of a constant production and consumption pace—and advocate for an epistemological change. For all these reasons, the social and ecological conceptions of these nonfiction writers are more likely to align with other key agents in the global movement for environmental justice (postcolonial socioecological movements such as postdevelopment or Via Campesina, critical ecofeminism, indigenous networks, and the like) than with hegemonic Euroamerican environmentalism, which they perceive as exclusively targeting the symptoms of the socioenvironmental problems and, therefore, perpetuating neocolonial and imperial global power and asymmetrical relations.

All of these authors challenge the illogical "logic" of constant economic growth in the context of a limited biosphere. By analyzing the global neoliberal economy in terms of material and energetic transformations that accelerate entropy and decrease both biological and epistemological diversity, these authors view cultural consumerism as a deadly ideology and propose the reduction of our social metabolism to allow others (both human and nonhuman) to live. I claim that, by uncovering the anthropocentric, utilitarian, and reductionist blind spots of global capitalism, these writings promote a posthumanist understanding of economy that takes into consideration the agency and interdependency of all forms of life. By re-materializing and posthumanizing the economy, these authors seem to be in line with the "material turn" that has recently occurred in environmental humanities. In this context, recent concepts such as trans-corporeality (Stacy Alaimo), vital materialism (Jane Bennett), or slow violence (Rob Nixon) are powerful in challenging the official discourses about economic growth, development, and progress, which are based on the fictitious modern separations between human and nonhuman, subject and object, observer and observed system, or economy and environment. According to systems theory all these dichotomies happen to be epistemological rather than ontological distinctions or, at the very least, their ontology is relational rather than individual. I believe that the emerging fields of material ecocriticism, ecological economics, and postcolonial environmental thought can illuminate each other and associate in fruitful ways in order to imagine and enact counter-hegemonic globalizations based on "a posthuman environmental ethics in which the flows, interchanges, and interrelations between human corporeality and the more-than-human world resist the ideological forces of disconnection" (Alaimo 142).

In the next few pages, I will lead with the way the Spanish authors I mentioned unveil some of the unscientific assumptions behind the neoclassical and neoliberal economics that currently dominate the global scenario in general, and the European Union debate in particular. I then will analyze the common ground within the alternatives proposed by these writers. Their different approaches to how to reduce society's throughput are not relevant for the purpose of this essay (that debate is currently taking place in some journals, such as *Ecological Economics, Sustainable Development*, and *Capitalism Nature Socialism*). Finally, I will briefly explore the connection between the ideas advanced by these authors in Spain and the emergence of a transnational debate in environmental humanities and posthumanism. In this regard, these authors should not be considered in the context of a national framework alone, but rather as participants in the wider critique of neoliberal globalization from a Euro-Mediterranean perspective (Prádanos). My essay confirms Ursula Heise's observation that "nonfictional genres have assumed a prominence in transnational ecocriticism that they do not have in other types of comparatist research" and this fact "has over the last few years enabled a transdisciplinary collaboration with environmentally oriented scholars" (641). However, these transdisciplinary collaboration does not need to stop at the environmental humanities level and can expand further to reach economics and other social sciences, as this essay intends to show.

Two crucial aspects are pointed out by all these Spanish authors. The first is the inseparability between current social and environmental problems, as well as the unavoidable relationship between socioenvironmental destruction, global asymmetrical distribution of wealth and power, and the degradation of a political civil society. In other words, the great acceleration of economic globalization and growth since the 1950s translates not only into a planetary environmental and energetic crisis, but also into a distribution crisis and, as a result, a crisis of democracy. The current difficulty for any meaningful communitarian participation in any relevant political decision, since all of these decisions are made by technocrats and corporations, is based on the ubiquity of the mono-logic of neoliberal political-economic ideology. The second problem is not only the ongoing globalization of an unsustainable economic system based on a biophysical impossibility (namely, constant economic growth) but, most importantly, the globally imposed anthropocentric, androcentric, reductionist, colonial, and hubristic logic behind such a system. This logic presents itself as the only legitimate way of knowing, denigrating and silencing any other epistemology. The result is that it worsens the problems that it creates by trying to address them using the same instrumental and reductionist reason that generated such problems, while preventing all possible alternatives from being envisioned and incorporated. This logic makes any possible creative solution unthinkable and inconceivable. Therefore, we are not just facing a social, economic, environmental, or political issue, but a more complex problem that encompasses all of them, namely, an epistemological vicious circle. In Alicia Puleo's words, "[a]mbos filósofos [Max Horkheimer y Theodor Adorno] afirmaron que la Ilustración se había marcado inicialmente por objetivo destruir los mitos y alimentar la imaginación con el saber pero, más tarde, la Razón instrumental se convirtió en un mito que traza nuevos límites al pensamiento" (90) (Both philosophers [Max Horkheimer and Theodor Adorno] affirm that the Enlightenment had the initial objective to destroy the myths and to feed imagination with knowledge but, later on, instrumental Reason became the new myth that imposes new limits to our thought). Today, that instrumental reason is at the service of the fundamentalism of the market economy and prevents any solution without the given terms of economic growth and debt-driven financial capital accumulation (terms that at the present are themselves at the main root of most socioenvironmental problems).

In order to avoid the thinking limitations mentioned above, all these authors deploy a systemic and trans-disciplinary approach to expose the blind spots of the hegemonic discourse. In that respect, they challenge the modern, arbitrary, and disciplinary epistemological divide that prevents creative thinking outside its obdurate reasoning; they adopt instead more appropriate, inclusive, systemic, and holistic frameworks that focus, for instance, on "socio-ecology" (Folch 91–92), "socionatural metabolism" (Sempere 164–65), "social metabolism" (Martínez Alier, *El ecologismo* 350), "ecofeminism" (Puleo), and "ecological rationality" (Riechmann, "Para una teoría" 193). All of these frameworks point out the interdependency between social, ecological, and semiotic systems. These approaches are integrated by the emerging field of ecological economics and the degrowth movement to emphasize the inseparability of the human and nonhuman spheres, in addition to culture and nature, for a proper understanding of pressing global problems.

From this vantage point, the economic system is not independent of the system of the biosphere, as perceived by neoclassical economics, nor is the environment a subsystem of the economic system, as treated by neoliberalism and its "greener" version, namely, environmental economics. The economic system is nothing but a subsystem embedded in, and dependent on, the system of the biosphere. By necessity, the activity of the economic system, to be sustainable, is restricted by the limits of the biosphere and must be compatible with the ecological laws that govern its functions. In other words, the economic system cannot ignore, for instance, the laws of thermodynamic (Carpintero 15; Riechmann, "Para una teoría" 190). In that context, an economic system based on constant economic growth, global massive urban development, and an ever-faster production and consumption of commodities is undesirable, destructive, unendurable, and unrealistic. The constant expansion of human economic activity is radically altering the global environment, so some authors are referring to this period of time as the Anthropocene (Fernández Durán, *El antropoceno* 9). This vast transformation is responsible for the ongoing collapse of many of the planet's living systems. Therefore, these authors urge us to abandon the modern social imaginary that separates the economic system and the environment. By doing so, we can start to scientifically analyze the flux of materials and energy mobilized and transformed by human economic activity (the social metabolism), and deploy that information to organize our societies in sustainable ways.

In the meantime, the global expansion of the current hegemonic economic model results in unsustainable ecological degradation. This economic model extracts raw materials and energy and transforms them into toxic waste and pollution in an ever-accelerating manner. The negative social consequences of this environmental degradation disproportionately affect the poorest people. This environmental injustice as well as the social reaction it generates has been termed "the environmentalism of the poor" (Martínez Alier) to contrast it with previous, less socially oriented, environmental movements that originated in the United States. The linear, mechanistic, and industrial time encouraged by the orthodox economic model collides with the cyclical ecological time of life and regeneration. The industrial timing of production and consumption has constantly accelerated and expanded spatially in the last few decades—thanks to neoliberal globalization. If observed outside of its own unrealistic and hubristic discourse, neoliberal globalization is an ideology of death, since it globally disrupts the ecological cycle of nutrients and depends on a constant increase of human biomass usage (Human Appropriation of Net Primary Productivity/HANPP) in detriment to other forms of life and alternative ways of human life. The more the social metabolism expands as HANPP, the less biodiversity and cultural diversity remains in the planet. In other words, the more human populations appropriate productive landscape and seascape by expanding the ecological space impacted by their socioeconomic activities (extractivism, urbanization, agroindustry, overfishing, communication and transportation infrastructures, waste disposal and pollution, etc.), the faster other species as well as nonmarket economy oriented cultures go extinct. Thus, when the global economy grows the living systems of the planet shrink and the cultural diversity decreases.

When neoclassical and neoliberal economists talk about generating wealth, they are referring to the destruction, depletion, and appropriation of the real wealth generated by ecosystems. "La contabilidad económica es por tanto falsa, porque confunde el agotamiento de recursos y el aumento de entropía con la creación de riqueza" (Martínez Alier, *El ecologismo* 350) (The economic calculation is therefore false, because it confuses the depletion of resources and the increase in entropy with the creation of wealth). The problem is that this globalizing economic system currently dominates the international political economy (WTO, IMF, World Bank, etc.) and dictates our way of life and social organization. "Es una sociedad construida sobre la economía, y una economía construida sobre la negación tenaz de la realidad" (Riechmann, "Para una teoría" 190) (It is a society constructed upon the economy, and an economy constructed upon the constant denial of reality). Thus, our mainstream economic system remains blind to the constraints of the biophysical reality. The main tenets of this system are anachronistic and obsolete, since they were constructed from the old mechanistic scientific paradigm that was dominant in the nineteenth century. It conceives the world as a machine that is made out of different articulated parts rather than as an integrated dynamic living system. "La ciencia económica tradicional es pretermodinámica, preevolutiva y preecológica" (Jacques Grinevald as cited in Riechmann 191) (The science of traditional economics is pre-thermodynamics, pre-evolutionary, and pre-ecologic). This economic system has not incorporated the scientific advances of other disciplines during the last century (Carpintero 23). It is, in sum, an ill-equipped logic to deal with the global ecological crisis that is exacerbated precisely by the globalization of that very logic.

Many of the authors highlighted in this essay point out that, to maintain the destructive system's unsustainable model, the previously described hegemonic discourse uncritically disseminates a blind faith in modern Western techno-science. Technology and science are presented as the remedy for all our present excess and abuse. Paradoxically, although this discourse presents an unwavering faith in science, it ignores or rejects what the vast majority of the scientific community confirms, as is obvious in the case of climate-change denialists. Furthermore, under the current circumstances, technological and scientific research and innovation depend on, and are guided by, corporate funding. Their main motivation is profit, not necessarily the well-being of communities and ecosystems.

Most new technologies are created under the industrial logic of fast production. They are designed to accelerate the expansion of social metabolism and to quicken the pace of the extraction of materials and use of fossil fuel energy. These technologies also tend to aggravate the ecological problem by making the scale of human biomass usage disproportionate and ecologically unsustainable. Other technological improvements, rather than targeting existing problems, generate new doubtful necessities and a greater dependency on increasing energy and material consumption. This is the case of forced mobility, created by the expansion of the car and the resulting dispersed urban model (Sempere 152). Perhaps the most important technological issue, according to many of these authors, is the role of the new information and communication technology in making the socioecological global crisis invisible. "[L]a sociedad de la imagen y la información ayuda a ocultar aún más la gravísima crisis ecológica que enfrentamos, sobre todo porque incentiva el desplazamiento de la atención de la biosfera a la infosfera (ciberespacio, realidad virtual), invisibilizando todavía más el deterioro de la Primera Piel, de la Madre Naturaleza" (Fernández Durán, *El antropoceno* 90) (The society of image and information helps to hide even more the very serious ecological crisis we face, especially because it encourages a shift in focus from the biosphere to the infosphere [cyberspace, virtual reality], making more invisible the deterioration of the First Skin, of Mother Nature). The current celebrated information society is another misleading concept, and many authors prefer to talk about society of disinformation due to over-information (Riechmann, *Un mundo vulnerable* 80). The fact is, while the new corporate mass media is disseminating and celebrating the notion of the information age, neoliberal globalization is drastically reducing the real information of the Earth (biological genetic information and cultural diversity), due to the mass extinction of species and cultures that is ongoing on a planetary scale (Fernández Durán 90–92).

In the next few pages, I will explore the alternatives suggested by these authors in order to target the issues addressed in this article. The two most crucial and urgent changes needed, according to all these thinkers, are, first, a significant reduction of the social metabolism at a global scale; and second, an epistemological change—aspects that go hand-in-hand. Since the main objective is to learn how to live better while using less material and energy, we need to rethink what it means to "live better" by redefining notions such as progress, development, and human vital needs. It is important to distinguish vital needs from superfluous or artificial needs. It is urgent to find more ecological and socially efficient ways to satisfy vital needs within each specific socionatural context. Moving faster and consuming more cannot work within the ecological limits of the biosphere. The study of human needs has to expand beyond superfluous material and energy consumption. It must incorporate emotional, psychological, social, spiritual, and environmental aspects. This will require a holistic, posthumanist, and systemic conception of health and well-being. However, this is not an unproblematic task, since vague concepts such as "quality of life" have been used repeatedly by many ideologies and groups of interest in Spain (and elsewhere) during the second half of the twentieth century to justify unsustainable neoliberal developmental projects and the environmental degradation associated to mass tourism, as pointed out by Eugenia Afinoguénova in her essay in this volume. Nevertheless, all these ideologies represented different articulations of the same Western modern anthropocentric logic and none of them proposed a posthumanist approach to the concept of "quality of life." Thus, I believe that the posthumanist approach to socioenvironmental issues supported by the authors I discuss here can dialogue better with, say, Andean and indigenous conceptions of "life" based on the relationality, interdependence, and reciprocity of the biotic community (human and nonhuman) than with the humanist EuroAmerican articulations of it that never have seriously questioned anthropocentrism nor consistently incorporated the science of ecology into their theoretical tenants._

According to Riechmann, cultural consumerism and economic growth must be abandoned and replaced by ecological rationality ("Para una teoría" 180). This new rationality will focus on the satisfaction of human needs through a sustainable (smaller) social metabolism (Sempere 228). The main goal is not to maximize, but to optimize (Sempere 223, Riechmann 180). The current economic system tends to maximize production. For instance, it produces as much food as possible in the shortest amount of time (now about one third more than it is needed to feed all humans), even though it is unsustainable long term since it produces bad quality food that threatens the health of workers and consumers while it reduces biodiversity, depletes the soil, pollutes the water and air, uses more fossil fuel calories than the food calories it produces, negatively impacts rural communities, generates food insecurity and price volatility, accelerates global warming, and so forth. An economic system based on optimization would produce, instead, as much diverse healthy food as needed within the ecological limits of the territory and without compromising the soil productive capability of the future. To achieve this new ecological rationality it is important to reject the current economic model in favor of embracing ecological economics and systemic thinking. Furthermore, thinking about the interdependency of humans and nonhumans must preclude some human societies from trying to dominate and control nature. These societies must start perceiving themselves as integrated parts of the biotic system—staying and thinking within its limitations.

In order to discontinue the ongoing global ecological and social degradation, the industrial regions of the planet must begin a socioecological transition towards a permanent and progressive reduction in their use/waste of materials and energy (Martínez Alier, *El ecologismo* 345). This change requires a cultural shift in dominant values and ways of thinking, as well as the redefinition of humans' role in the biotic community. It is relevant to understand that "todos los seres humanos son interdependientes y ecodependientes, pues el *homo economicus* competitivo e independiente es una absoluta ficción" (Fernádez Durán, *El antropoceno* 98) (all human beings are interdependent and ecodependent, since the competitive and independent *homo economicus* is a total fiction). This cultural shift calls for a socially and ecologically sustainable degrowth based on the priority of social life and relational goods (rather than individual material accumulation and consumption); creative and communitarian leisure time; the redistribution of working time; the reduction of many productive, administrative, and transportation infrastructures; the re-localization, decentralization, and simplification of economic and social life (the opposite of neoliberal globalization); and, in the individual sphere, the appreciation of voluntary simplicity, the revalorization of the ethics of care, and communitarian quality time (Taibo 52–53).

For these implementations to be possible, we need to depart from patriarchal and capitalist frames of thinking. As pointed out by ecofeminism, other economic logics are, on many occasions, organized by women in communitarian ways and function out of the market economy (Puleo 66). These changes can improve the happiness and quality of life of human communities everywhere, while enhancing the health of ecosystems. However, the vast majority of current political and economic leaders are moving in the opposite direction by focusing only on GDP growth rather than on socioenvironmental sustainability. Therefore, collective political action at a global scale is needed. Given that decisions regarding the size of the social metabolism affect the whole of humanity (both present and future), these decisions should be made with the participation of all humans (Sempere 190). The global financial crisis of 2007–2008 demonstrated once more that global public opinion is not being taken into account, and that important decisions are made paternalistically by "technocrats" and "experts" trained in the neoliberal hegemonic model. Far from helping, these technocratic decisions contribute to social and environmental degradation at a global scale since they try to solve the crisis of the model of economic growth encouraging the pursuit of more economic growth. The current disagreement in Europe about how is the best way to grow after the global financial crisis is a good example of this obduracy. Some neoliberals recommend governmental expending austerity and salary reductions to improve global competitiveness and to favor the exportation dynamic, while some neo-Keynesians say that the salaries have to grow so consumption grows as well. None of them realize that, given the socioecological situation, the first option is socially unsustainable and the second is environmentally unviable. The possibility of degrowing is not even mentioned.

Meaningful and collective participation in global political actions can only be accomplished if people are informed and are able to practice systemic and ecological thinking. Because the hegemonic system "pollutes" people's minds—what Michel Serres calls "soft pollution"—and masks the real problem, many of these authors explore how to decolonize the dominant imagery disseminated by corporate media, publicity, and neoliberal interests. These explorations take different approaches, from the insights of social psychology (Riechmann, *Un mundo vulnerable* 69) and the recommendation to avoid exposure to publicity (Taibo 71) to the deconstruction of myths related to the information and communication age (Fernández Durán, *Tercera Piel* 63–68, Folch 115–20).

The dominant corporate media along with the hegemonic political economy inertia perpetuate the pathological, hubristic, and unrealistic discourse of unlimited economic growth, constant development, techno-scientific total control and infallibility, and the separation between human activity and the ecological domain. Therefore, the decolonization of media is crucial to target the problems related to climate change (Folch 120). The hegemonic logic uncritically disseminated by the media reduces everything to a measure of monetary dimension and material accumulation. "Debemos rechazar tal simplificación de la complejidad, tal exclusión de lenguajes de valoración. Debemos aceptar, por el contrario, el pluralismo de valores inconmensurables entre sí para evitar que la ciencia económica se convierta en un instrumento del poder en la toma de decisiones" (Martínez Alier, *El ecologismo* 358) (We must refuse such simplification of complexity, such exclusion of valuation languages. We must embrace, on the contrary, the pluralism of incommensurable values in order to avoid economic science becoming the instrument of power in decision making). Thus, it is paramount to focus on complexity and to incorporate the knowledge developed by the systemic paradigm during the last century, in order to understand current problems and find effective solutions (Riechmann, "Para una teoría" 195). For a society to meaningfully participate in politics and make informed decisions, knowledge should not be only an expert's issue—since the so-called experts are biased by all the blind spots of their academic disciplines—but a human right in the public domain as well (Folch 244–45). In addition, it is crucial to listen to, and to learn from, other epistemologies and values from different traditions and experiences. Many indigenous sciences, for instance, have a lot to say about optimization and communitarian living within an appropriate social metabolism. Unfortunately, a number of these sources of diverse knowledge were (and still are) destructed, denigrated, subalternized, and silenced by the globalization of neocolonial hegemonic logic and the expansion of the global markets. For this reason, it is difficult to enact an urgent and necessary "plural landscape of knowledge forms" (Escobar 17) or "ecology of knowledge" (Santos 116) to generate creative alternatives to the globally imposed mono-logic, and its resulting mental and biophysical pollution. It is crucial to open horizons of thinking and acting beyond the andro-ethno-anthropocentric subtext of neoliberal globalization (Puleo 146).

These Spanish authors are leading the relevant transnational debate about degrowth and ecological economics. However, Spanish culture and politics have not adopted the concerns that these thinkers want to disseminate (reinforcing the biblical proverb that "nobody is a prophet in his/her own land"). This view is supported by several alarming facts: the current Spanish government (PP, conservative party) has recently cancelled some of the scarce environmental regulations in Spain for the sake of economic growth; the Spanish ecological footprint is around three times bigger than its territory; and finally, the ongoing transportation plans and strategies in Spain, according to the study of Bermejo Gómez de Segura (77–124), are completely delusional given the global energy crisis in general and the Spanish foreign energy dependency in particular. In addition, the Spanish government approved in summer 2013 an energetic reform that has been frontally rejected by most environmental organizations for not supporting renewable energies, benefitting big corporations in the sector, provoking the continuation of Spanish energetic dependency on imported fossil fuels, and setting obstacles to energy self-production. Given the energetic situation of Spain and the current context of peak oil and climate change acceleration, this reform seems to condemn Spaniards to an unsustainable economic, social, and environmental future. Other environmentally retrograde action of this government includes discontinuing the funding of the Observatorio de la Sostenibilidad en España (OSE), the main institution regarding the watch, advice, and study of Spanish sustainability. As a result the OSE closed in May 2013.

Regarding culture, until recently, few Spanish artists or cultural manifestations were ecologically committed, and ecocritical theory in Spain was almost nonexistent. Fortunately, changes have occurred in the last few years. Further, since the global financial crisis, some relevant social movements (Indignados, 15-M) have emerged and popularized notions related to degrowth and ecological economics. Urban gardening, for instance, has flourished in Spain in the last few years (in another essay in this volume, Susan Larson and Matthew I. Feinberg explore this kind of alternative communitarian initiatives that repurposed the urban spaces in Madrid). It is not by chance that many of the most interesting socioenvironmental articulations in the global North such as degrowth and the slow movement have emerged in the Euro-Mediterranean region. Some of their cultural conceptions of life based on communitarian celebrations of agrarian cycles and convivial enjoyment were constantly criticized by the more productivist Calvinism of Northern Europe. But in the current context of overproduction, unsustainable individualistic consumerist lifestyles, financial and energetic crisis, and ecological restrictions, the main weakness and sin of Southern Europe (according to an essentialist vision promoted by technocrats and developers), today seems to be an ecological virtue worthy to embrace. In any case, the European Union as a whole may be forced to lead the energetic degrowth and the shrinking of its social metabolism, as pointed out by Carles Riba Romeva: "Ahora, con las reservas de energía fósil casi agotadas y con unos recursos de la biosfera limitados en relación a su población, Europa se encuentra, a su pesar, en la punta de lanza de un cambio no tan fácil: liderar el 'decrecimiento energético' procurando mantener en lo posible los mejores valores de la propia cultura aún reconocida en todo el mundo" (272) (Now, with the fossil energy reserves almost exhausted and with limited biosphere's resources in relation to its population, Europe finds itself, involuntarily, in the spearheading of a difficult change: to lead the "energetic degrowth" trying to keep as much as possible the best own cultural values that are yet appreciated all around the world). Maybe the revival of some of the convivial and ecological aspects of the Euro-Mediterranean region (while avoiding the wealth and power distribution disparities and patriarchal traditions often associated to it), that were denigrated until very recently, can help now to create a new and sustainable cultural paradigm. By now, however, ecological rationality is only practiced in some small circles and is still far from the norm in Europe in general and Spain in particular. I believe that the hegemonic discourse that confuses progress and quality of life with economic growth and neoliberal development is still very ingrained in the mentality of many Spaniards. This is not surprising given the tenacity with which some technocrats and politicians in Spain and elsewhere have sold that discourse during the second half of the past century, as Eugenia Afinoguénova explores in her article. Perhaps, in the near future, a renovation in the field of Hispanic studies, along with the emergence of more radical Iberian and Transatlantic studies, can contribute to the global dissemination of these ideas at the cultural level by seriously adopting ecocritical and posthumanist approaches. Presently, the Spanish thinkers covered in this essay are not representative of the mainstream discourses in Spain, and contribute more to the ongoing global critique of new capitalism than to the timid ecological culture of Spain, which needs to live up to the sophistication of these authors.

In the meantime, as Martínez Alier—one of the international leaders in ecological economics and degrowth—claims in a very recent article, the potential exists for an alliance between the environmental justice organizations of the global South and the small degrowth movement in Europe. This

> alliance must be based on a common perspective against 'debt-fuelled' economic growth and the hegemonic of economic accounting in favor of a pluralism of values, the acceptance and support of bottom-up feminist neo-Malthusianism, the defense of human rights and indigenous territorial rights along with the rights of nature, the recognition of the ecological debt, and the critique of ecologically unequal exchange. ("Environmental Justice" 66)

These Spanish authors are thus more compatible with postcolonial environmental justice movements, than with the hegemonic reform environmentalism of EuroAmerican tradition, since this "reform environmentalism has become part of a system of global managerialism, closely related to institutions like the IMF or the World Bank" (Clark 120–21).

The ideas advanced by these writers can be understood as a Euro-Mediterranean contribution to the emerging transnational debate in environmental humanities and posthumanism. I claim that this debate needs to be consistently incorporated into Hispanic studies, where it has been almost nonexistent until recently. This essay—as a part of this special number of *HIOL: Hispanic Issues On Line*—is intended as a limited but necessary effort to push Hispanic studies in that direction. Many of the ideas discussed by the authors mentioned here could be the basis of fruitful dialogues with several recent ecocritical transnational debates. A brief exploration of this convergence between these authors' version of degrowth and ecological economics developed in the previous pages, and some of the most popular, recent, and interesting concepts advanced by environmental humanities, such as trans-corporeality (Alaimo), vital materialism (Bennett), and slow violence (Nixon) may be illuminating.

In her recent book *Bodily Natures: Science, Environment, and the Material Self*, Alaimo focuses on "trans-corporeality" by "attending to the material interconnections between the human body and the more-than-human world" (2). Her posthumanist and new-materialist notion of agency challenges the hegemonic conception of subjectivity and the human self (20). The idea of trans-corporeality questions all foundations of the modern philosophy that is based on individual freedom and agency and thus has political, economic, and ethical implications. The material feminism developed by Alaimo has many things in common with the vital materialism advanced by Bennett in her book *Vibrant Matter: A Political Ecology of Things*. Both books were published in 2010, but the authors never reference each other (perhaps because they are coming from different fields). Nevertheless, Bennett—heavily influenced by Latour's actor-network theory and its focus on the agency of networked assemblages of human and nonhuman actors—complains about the fact that

> theories of democracy that assume a world of active subjects and passive objects begin to appear as thin descriptions at a time when the interactions between human, viral, animal, and technological bodies are becoming more and more intense. If human culture is inextricably enmeshed with vibrant, nonhuman agencies, and if human intentionality can be agentic only if accompanied by a vast entourage of nonhumans, then it seems that the appropriate unit of analysis for democratic theory is neither the individual human nor an exclusively human collective but the (ontologically heterogeneous) 'public' coalescing around the problem. (108)

The resonance is obvious in this statement, the idea of trans-corporeality, and the main tenets of ecological economics. All of them make clear that the kind of epistemology—based on the hierarchical separation of humans and nonhumans—that currently dominates hegemonic discourses (and some non-hegemonic discourses as well) cannot account for the social and ecological violence that is perpetuated by neoliberal globalization. This "slow violence" is made invisible by "the sensation-driven technologies of our image-world" (Nixon 3). Furthermore, it is important to recognize that the same logic that exploits humans also depletes non-humans. Thus, "human rights are indissociable from environmental justice" (Nixon 265). In the last few years, more and more thinkers have concluded that social and environmental problems can neither be treated in a separate fashion, nor solved by using the same reductionist and anthropocentric logic that generated them. Environmental, social, and cultural studies need to converge in order to be analyzed and targeted through the lens of socioecological movements such as degrowth and transdisciplinary fields such as ecological economics and social ecology. As a result, postcolonial studies and ecocriticism cannot ignore each other any longer (Nixon 233, DeLoughrey and Handley 3–39).

In sum, Alaimo posthumanizes and rematerializes the conception of the human body, agency, and subjectivity; she problematizes the very notion of individual liberty that is foundational to liberalism. Bennett elaborates on the political relevance of nonhuman forces and, again, opposes the simplistic anthropocentrism of mainstream political theory. Nixon is concerned with the representational challenges of a slow violence that is made invisible by the system that infringes it: a "violence that occurs gradually and out of sight, a violence of delayed destruction that is dispersed across time and space, an attritional violence that is typically not viewed as violence at all" (2). It seems obvious to these three thinkers that the dominant logic, a logic based on the separation between human individuals and the nonhuman world, cannot track the connections among, for instance, economic activity, chemical pollution, and the proliferation of cancer cells in a child's body. A new epistemological framework is needed to understand these interconnections and interdependencies and to push for a coherent political ecology. The ecological economics advocated by the Spanish authors presented in this article intend, precisely, to posthumanize and rematerialize the economy and, therefore, their contributions are complementary to the new epistemological paradigm advanced by posthumanism and material ecocriticism. To conclude, I want to emphasize that posthumanism does not only mean that "the decentering of the human by its imbrications in technical, medical, informatics, and economic networks is increasingly impossible to ignore" but, more importantly, "that the nature of thought itself must change if it is to be posthumanist" (Wolfe xv–xvi). In this regard, the constant insistence on a change of logic advocated by the Spanish authors discussed here confirms that they are moving towards a posthumanist understanding of economy, something equally radical and vital for the future well-being of Spain, Europe, and beyond. The current context of a global crisis of imagination and creative thinking—fueled by the semiotic and biophysical pollution disseminated by the anthropocentric hegemonic paradigm—that we (the biotic community) continue to suffer can only be altered by more radical voices such as those of these Spanish authors, participating in a global critique.

Works Cited

Alaimo, Stacy. *Bodily Natures: Science, Environment, and the Material Self.* Bloomington: Indiana University Press, 2010.

Bennett, Jane. *Vibrant Matter: A Political Ecology of Things.* Durham: Duke University Press, 2010.

Bermejo Gómez de Segura, Roberto. "La política de transporte española ante el fin de la era de los combustibles fósiles." *Economía ecológica: reflexiones y perspectivas.* Eds. Álvarez Cantalapiedra, Santiago and Óscar Carpintero. Madrid: CBA, 2009. 77–124.

Carpintero, Óscar. "Introducción: la economía ecológica como enfoque abierto y transdisciplinar." *Economía ecológica: reflexiones y perspectivas.* Eds. Santiago Álvarez Cantalapiedra and Óscar Carpintero. Madrid: CBA, 2009. 11–33.

Clark, Timothy. *The Cambridge Introduction to Literature and the Environment.* Cambridge: Cambridge University Press, 2011.

DeLoughrey, Elizabeth, and George B. Handley, eds. *Postcolonial Ecologies: Literatures of the Environment.* New York: Oxford University Press, 2011.

Escobar, Arturo. "Development, Violence, and the New Imperial Order." *Development* 47.1 (2004): 15–21.

Fernández Durán, Ramón. *El antropoceno. La expansión del capitalismo global choca con la biosfera.* Barcelona: Virus editorial, 2011.

———. *Tercera Piel. Sociedad de la información y conquista del alma.* Barcelona: Virus editorial, 2010.

Folch, Ramon. *La quimera del crecimiento. La sostenibilidad en la era postindustrial.* Barcelona: RBA, 2011.

Heise, Ursula K. "Globality, Difference, and the International Turn in Ecocriticism." *MLA* 128.3 (2013): 636–43.

Martínez Alier, Joan. *El ecologismo de los pobres. Conflictos ambientales y lenguajes de valores.* 3rd Extended Edition. Barcelona: Icaria, 2009.

———. "Environmental Justice and Economic Degrowth: An Alliance between Two Movements." *Capitalism Nature Socialism* 23.1 (2012): 51–73.

Nixon, Rob. *Slow Violence and the Environmentalism of the Poor.* Cambridge, MA: Harvard University Press, 2011.

Prádanos, Luis I. "Toward a Euro-Mediterranean Socioenvironmental Perspective: The Case for a Spanish Ecocriticism." *Ecozon@: European Journal of Literature, Culture, and Environment* 4.2 (Autumn 2013): 30–48. Web. 5 Feb. 2014.

Puleo, Alicia H. *Ecofeminismo. Para otro mundo posible.* Madrid: Cátedra, 2011.

Riba Romeva, Carles. *Recursos energéticos y crisis. El fin de 200 años irrepetibles*. Barcelona: Octaedro, 2012.

Riechmann, Jorge. "Para una teoría de la racionalidad ecológica." *Economía ecológica: reflexiones y perspectivas*. Eds. Santiago Álvarez Cantalapiedra and Óscar Carpintero. Madrid: CBA, 2009. 169–213.

_____. *Un mundo vulnerable. Ensayos sobre ecología, ética y tecnociencia*. Madrid: Catarata, 2000.

Santos, Boaventura de Sousa. "A Non-Occidentalist West?: Learned Ignorance and Ecology of Knowledge." *Theory, Culture, & Society* 26.7/8 (2009): 103–25. Print.

Sempere, Joaquim. *Mejor con menos. Necesidades, explosión consumista y crisis ecológica*. Barcelona: Crítica, 2009.

Serres, Michel. *Malfeasance. Appropriation through Pollution?* Trans. Anne-Marie Feenberg-Dibon. Stanford: Stanford University Press, 2011.

Taibo, Carlos. *El decrecimiento explicado con sencillez*. Madrid: Catarata, 2011.

Wolfe, Cary. *What is Posthumanism?* Minneapolis: University of Minnesota Press, 2010.

PART III

Iberian Bio-Power: Life as a Political Matter

7

Reproductive Rights in Spain: From "Abortion Tourism" to "Reproduction Destination"

Pablo de Lora

Introduction

In the following pages I will attempt the modest task of depicting the evolution of reproductive and sexual rights in Spain since the early days of the *Transición* until our present time. In order to fully grasp the magnitude of that evolution, my departing point will be the end of the 1970s, when the phenomenon known as "abortion tourism" was fairly prevalent among Spanish women. As my arrival or contrasting-point, I will focus on a radically different sort of phenomenon: that of gay couples traveling to California in order to become parents by means of surrogate motherhood. Although there are many different dimensions intertwined in the description of that evolution, I will mainly focus on the legal and jurisprudential aspects.

But before we start, a conceptual note is in order. By "sexual and reproductive rights" I will understand the individual's immunity against the government as regards his or her decision whether to procreate. In the American legal tradition this sort of immunity is part of the so-called "right to privacy."[1] The scope of sexual and reproductive rights is of course debatable, particularly when the decision not to procreate implies the destruction of the fetus. On the other hand, the decision to be a mother or father may also imply the sacrifice of certain interests: it could entail, for instance, the controversial use of surrogate mothers, not to mention the harm of overpopulation or even the harm of merely existing, as in the case of babies who suffer from terrible diseases but are nonetheless carried to term. Although I will not entertain any of those normative aspects of the issue, these dimensions should be taking into account in order to understand the historical pathway on which I will mainly focus.

Time of Silence, Clandestine Time: 1960–1982

It is commonplace to affirm the spectacular social change experienced in contemporary Spain, and that is certainly the case as regards to women's emancipation. In the lapse of only thirty years Spanish women has gained control over their own reproduction and sexual life to an unprecedented extent: nowadays women in Spain have access to contraceptives, abortion, artificial reproduction technologies (ART) regardless of their marital status, and they can legally marry their same-sex partner and adopt children. Thirty years ago things were radically different.

In Spain, abortion, as well as contraception, has long been considered a crime. Articles 639 and 640 of the Criminal Code of 1822 stated that those who caused the termination of pregnancy with or without the woman's consent would be imprisoned for up to ten years. The pregnant woman herself could be imprisoned for up to eight years.

Later on, with the development of contraception, mere information about "the pill" was illegal and, although the legislation became more humane, abortion was punished with up to six years of imprisonment.[2] Only when the termination of pregnancy was performed as a means to hide the "dishonor" of becoming mother was the punishment lighter.[3] It is worth quoting in full the way in which this excuse was phrased in the Criminal Code of 1822: ". . . Pero si fuera soltera o viuda no corrompida y de buena fama anterior, y resultare a juicio de los jueces de hecho que el único y principal motivo de la acción fue el de encubrir su fragilidad, se le impondrán solamente uno a cinco años de reclusión" (But if she were single or a noncorrupted widow and with previous good reputation, and appeared that, under the judges' eyes, the sole motive for her action was, actually, to conceal her fragility, she will be condemned to serve only one to five years of imprisonment).

The historical exception to this longstanding situation has been Cataluña, for the brief period of the II República. The *Orden de la Presidencia de la Generalidad de Cataluña* issued on December 25, 1936, and published March 1, 1937, guaranteed the termination of pregnancy under the woman's request and criminalized those who practiced abortion while lacking medical qualifications. The aim was clearly the provision of safer abortions to Spanish women.[4] Thus, Cataluña placed itself *avant la garde* of the very few countries who already had permissible abortion laws in those days. But it didn't last long.

Whoever thinks about abortion in contemporary Spain thinks about the acclaimed novel *Tiempo de silencio* by Luis Martín-Santos, in particular the gruesome description of the procedure employed by "el Muecas," the "needle's wizard," to induce abortion in his daughter Florita.[5] Beyond fiction, the detailed descriptions given in the rulings by the Spanish Supreme Court in abortion cases are also a valuable source for grasping how abortion used to be practiced in Spain. Here are just a few from 1982:

> . . . reuniéndose el día 8 de septiembre de 1980 la muchacha y Francisca, la que cobró las 20.000 pesetas, haciendo tenderse a la joven, a que introdujo en sus órganos genitales unas agujas de hacer punto y luego, por el mismo conducto, una solución de agua y alcohol, con lo que, a las pocas horas N. expulsó el feto y tuvo que ser internada el día 12 siguiente en el Servicio de Ginecología de la Residencia Sanitaria, donde fue intervenida de una perforación de útero de la que curó a los treinta días de asistencia e impedimentos.[6]
>
> (. . . gathering on September 8, 1980, the lady and Francisca, who charged 20,000 pesetas, asking her to lay down, introducing in her genitalia the knitting needles, and then, by the same canal, a solution of water and alcohol, and as a result the fetus was expelled and then she had to be hospitalized on September 12 in the Division of Gynecology, and there she was operated of an uterus perforation that healed after thirty days of care.)
>
> . . . seguidamente volvió al bar quedando allí el amante de la embarazada, y los otros tres pasaron al domicilio de la procesada Rafaela, quien a solas y mientras los procesados Cecilia y Gustavo esperaban en otra habitación, introdujo a la embarazada con ayuda de un espéculo, una sonda en la matriz, y le dijo que al siguiente día se la quitaría, marchándose dicha mujer y sus acompañantes seguidamente. Aquella misma tarde la mujer embarazada referida expulsó el feto y la sonda, con fuertes hemorragias que duraron varios días.[7]
>
> (. . . afterwards she returned to the bar. The pregnant woman's lover remained in the bar and the other three went to the defendant's house. She acted alone, while the defendants Cecilia and Gustavo waited in another room. With the aid of a

speculum she introduced a probe in the matrix, telling her that she will remove it the next day. Afterwards they all left. That very same evening the pregnant woman expelled the fetus and the probe in the midst of a severe hemorrhage that lasted several days.)

Etelvina le mandó esperar unos minutos y regresó a su domicilio en busca de los utensilios precisos para practicar el aborto, haciéndolo en el automóvil de su marido, el también procesado Eugenio… mientras Eugenio esperaba a su esposa, en el interior de su vehículo, Ana María practicó a María Rosario la operación de introducirle por la vagina, hasta el útero un objeto punzante, y una solución necrosante de agua, jabón y ruda, que, al momento le produjo a María Rosario un gran dolor y pérdida de conocimiento del que se recuperó rápidamente y seguidamente la procesada y su esposo trasladaron en el automóvil de éste a María Rosario, hasta el domicilio de unos familiares de María Rosario, sito en Barbayeda, desde donde fue ingresada posteriormente en la residencia sanitaria de San Agustín de Avilés, donde falleció tres días después, a causa de la introducción en el útero del objeto punzante y sustancias relatadas, que le produjeron un foco séptico-toxémico, que determinó su fallecimiento.[8]

(Etelvina told her to wait for her for some minutes and returned to her house to gather the needed utensiles to terminate the pregnancy. She did it in her husband's car, the defendant Eugenio . . . while Eugenio waited for her wife, inside the vehicle Ana María introduced a poignant object through María Rosario's vagina until she reached the uterus, as well as a necrotizing solution of water, soap and rue which caused María Rosario great pain and the loss of consciousness. After her quick recovery both defendants gave her a ride to her relatives' house, in Barbayeda. Afterwards she had to be admitted at the San Agustín Hospital, in Avilés, where she passed away after three days because of the previous insertion of the poignant object and the said substances in her uterus that caused her a septic shock and subsequently her death.)

Sobre mediados del mes de febrero de 1979, la procesada Flora, soltera, nacida el 7 de marzo de 1960, de buena conducta, y sin antecedentes penales, teniendo sospecha de estar embarazada de pocos días, acudió a la casa de Leonardo, de quien le habían hablado que producía abortos, para que le resolviera su problema, y una vez se lo hubo explicado y ésta informarle de la veracidad probable de su estado, se puso en sus manos para que realizara sobre ella manipulaciones tendentes a interrumpir dicha pretendida gestación, y con tal concreta finalidad le inyectó en días distintos y por vía intramuscular varias dosis de los productos "Progynon» y "Pitucin», los cuales aplicados en dosis adecuadas pueden producir el arrastre al exterior del huevo, principalmente cuando es de pocos días, aunque no están medicamente destinados a tal finalidad, poniéndole además en la vagina unos tallos secos de origen vegetal o "palitos" los cuales producen la dilatación del cuello uterino que asimismo puede llegar a ocasionar la expulsión del feto, y con tales prácticas se llegó varios días después a la aparición en Flora de hemorragias vaginales que llevaron consigo la destrucción o expulsión del embrión ya formado, siendo por ello atendida médicamente en la Residencia de la Seguridad Social el 19 de febrero de 1979, y curando tras dicha asistencia cuatro días después.[9]

(Around the middle of February 1979, the defendant Flora, single, born on March 7, 1960, well behaved and with no criminal record, suspicious of her being pregnant, went to Leonardo's house, where she had been told abortions were performed. She asked him to help her out with her problem. Once she explained to him the situation, she relied on him to initiate the procedures that were finally going to provoke her abortion. In the days that followed he injected several doses of "Progynon" and "Pitocin," which, in adequate doses might produce the expulsion of the embryo, more likely if the pregnancy is in the early stage, although both drugs are not prescribed for such end. He also placed some dry sticks of vegetable origins or "little sticks" which prompted the dilation of the cervix, which might also have caused the egg's expulsion. After several days Flora underwent vaginal hemorrhages and the expulsion of the embryo. She had to be taken care of at the Hospital on February 19, 1979, being discharged after four days.)

Shortly after the enactment of the Spanish Constitution in 1978, the Supreme Court (*Tribunal Supremo*) ruled in a series of cases involving Spanish women who traveled abroad to terminate their pregnancies. Those cases were thereafter known as "reproductive tourism" cases.[10] More than 30,000 women (a small fraction of a total of 300,000 abortions illegally performed in Spain at that time) flew every year, mostly to London or Amsterdam, in trips that were secretly organized by semi-clandestine associations and travel agencies.[11] The number of Spanish women visiting those clinics was such that many of them hired personnel with knowledge of Spanish.

You might ask: how come the *Tribunal Supremo* had jurisdiction in those cases? The answer is that in those days Spanish judges had extraterritorial jurisdiction when the victim was a Spaniard. But how could they consider that a fetus was of Spanish nationality?[12] After all, nationality, as well as residence, marital status, and similar circumstances are attributes of "persons" and, from a legal standpoint, fetuses are not persons before birth. With an overdose of legal creativity, the judges from the *Tribunal Supremo* concluded that fetuses aborted in London or Amsterdam were Spaniards because, according to the Civil Code (article 29), we should anticipate the effects of birth if there are "favorable consequences for the fetus," and being of Spanish nationality, they argued, is certainly one of those "good consequences."[13]

"The Eleven Women from Bilbao": The 1982 Reform

After the demise of Franco's dictatorship, the hotly disputed decisions on abortion tourism as well as other famous cases such as the one known as "las once de Bilbao" (the eleven women from Bilbao) spurred the "decriminalization" of abortion by the new elected Socialist government in 1982, a legal reform that finally saw the light of day in 1985 after significant social, political, and legal turmoil. A final judgment was passed by the Constitutional Court (STC 53/1985) who, in a divided opinion, introduced minor changes in the proposed reform. According to the official polls of that time a vast majority of the Spanish citizenry was largely in tune with the legal changes that, basically, granted the possibility of abortion under three circumstances: when the woman became pregnant as a result of a sexual assault, when pregnancy compromised her psychological or physical health, and when the fetus suffered from abnormalities.[14]

The truth of the matter is that since 1985 thousands of women in Spain have been able to abort under the "health risk" heading, namely, the psychological health indication, a sort of loophole through which abortion has been extensively performed even in very late stages of pregnancy.[15] In 2007 a hidden camera investigation by some Danish television reporters showed how private clinics in Spain accepted women to terminate their late pregnancies with no oversight or assurance about the alleged psychological distress.[16] Further investigations were subsequently broadcast by the conservative channel Intereconomía in which more gynecologists were secretly filmed evincing similar leniency.[17] Admittedly this was not an unknown situation: pro-life groups of various tendencies were fiercely condemning these somewhat deceptive practices many years before the investigations were made public, but the conservative party (*Partido Popular*), which stayed in power from 1996 until 2004, systematically turned a blind eye on those protests.

2010: Abortion on Demand Lands in Spain

The legislative landscape in Spain as regards to abortion changed abruptly in 2010 when the "Ley 2/2010 de 3 de marzo de salud sexual y reproductiva y de la interrupción voluntaria del embarazo" was passed by a Parliament in which the Socialist Party was, again, hegemonic. One of the main goals of the law was to put an end to the fraud just described and also to enhance legal security for women whom, albeit sporadically, still suffered from the harassment of an eventual indictment. In order to do so, the Spanish legislation aligned with many other regulations that establish abortion on women's request within a period ranging from 10 to 18 weeks.[18] The new regulation set that period at week 14. Later abortions, up to week 22, are only allowed for medical reasons—on the part of the woman or the fetus. After week 22 the abortion is permitted if the fetus suffers from a fatal condition (anencephaly, bilateral renal agenesis, or the like) and a special hospital committee approves its termination.

Spain is below the European average in the rate of termination of pregnancies, although its rate in late-stage terminations is slightly higher. The data show that in the last two decades the trend has been to close the gap in the abortion rate between Northern and Southern European countries. It seems that whether the national legislation allows abortion on request or establishes legal indications, does not have any significant impact in the number of women who abort (Gissler 326).

Be that as it may, the *Partido Popular* has challenged the current law before the Constitutional Court and a decision is still pending. At the same time, back in office since 2011, the conservative government of Prime Minister Mariano Rajoy has repeatedly announced its intention to change the law and eliminate both the abortion on request clause and the legal indication based on fetal anomalies. In the very moment in which this contribution is being written the debate shows its resiliency as in many other countries in which consensus on the matter looks unreachable.

One of the pivotal controversies of the latest reform proposed has to do with the so-called eugenic indication. *Down España*, among many other advocacy groups for the disabled, is encouraging the Spanish Government to enforce the Convention on the Rights of Persons with Disabilities (2006).[19] Article 10 of the Convention states that "every human being has the inherent right to life" and that States "shall take all necessary measures to ensure its effective enjoyment by persons with disabilities on an equal basis with others."[20] In support of their position, full-heartedly espoused by the current Ministry of Justice Alberto Ruiz Gallardón, these groups refer to the recommendations made by the Committee on the Rights of Persons with Disabilities as regards the report submitted by Spain (Sixth Session, 19–23 September 2011).

However, in spite of the fact that the government and its social allies have insisted that changing the law to eliminate the eugenic indication is simply abiding by international law, it is worth mentioning two very relevant facts that seem to lessen the force of that legal assessment. First, the phrasing of the Committee's demand to Spain should be pointed out: "[it is recommended] that the State Party [Spain] *abolish the distinction* made in Act 2/2010 in the period allowed under law within which a pregnancy can be terminated based solely on disability."[21] Second, interestingly enough, in the United States, far from relying on the Convention to fuel their cause, some pro-life groups despise it as a pro-choice instrument and are constantly urging their representatives *not* to ratify it.[22]

So one might legitimately wonder whether *Down España*, or similar pro-disabled people advocates, are really taking a broadly anti-abortion stance on the grounds that the rights of the helpless must be defended across the board, or whether they are taking issue with only certain types of abortions for certain types of reasons, and the different term for the so-called eugenic abortion. In that sense, had the Spanish legislator established the exact same term for either type of abortion, or extended the current 14-week period to catch up the 22 weeks of the eugenic indication, there will be no quarrel. And yet, from the official public statements it seems that the conservative government is embracing the first, and more "pro-life," interpretation.

Before we move forward, let me summarize what I have been describing up to this point. Abortion was a major social problem at the beginning of our political transition toward democracy. Although there was a great pressure to decriminalize abortion, it was not until 1985, three years after the Socialist Party got into office, that a new law was passed. The new regulation satisfied almost no one: on the one hand, pro-life groups denounced the subterfuge of allowing abortion when the psychological health of the mother was at peril; on the other hand, for pro-choice groups only abortion on women's request offered legal security for women. That political move had to wait until 2010 and is currently at serious risk of becoming legal history.

In the meantime, coinciding with the political and social turmoil provoked by the abortion debate, assisted reproductive technologies have grown significantly in Spain since 1984. In the following section I will describe such evolution and the regulatory framework that has made Spain a leading destination for so-called "reproductive tourism."

1984: A Child Is Born

Yes, a very special child. The first Spanish test-tube baby, named Victoria Anna, was born at the Instituto Universitario Dexeus in Barcelona. The success has to be credited to the team lead by Dr. Pedro Barri, a prominent Spanish obstetrician. But it is also noteworthy that they acted in an absolute legal vacuum.

Indeed, the Spanish legislation on artificial reproduction (artificial insemination and in-vitro fertilization) was enacted four years later, and, as in the case of abortion, in the midst of social, legal and political controversy.[23] Beyond the irrational fears fostered by literary dystopias such as Aldous Huxley's *Brave New World*, many traditional principles were at stake. The reform, promoted by the socialist representative Dr. Marcelo Palacios, was driven by the scarcely hidden aim of, again, empowering women with regard to their own reproduction. That move, however, was not free. Women's empowerment came at the significant cost of allowing anonymous sperm donation and the subsequent sacrifice of the child's right to know his or her genetic lineage. In other countries, laws regarding the donors' anonymity is not as stringent. In Sweden, for example, a recent law grants offspring the possibility of contacting the genetic father. In other jurisdictions anonymity might be lifted only after they have reached the age of 18 (Cohen 432).

But there are other aspects of Spain's medically assisted reproduction legislation that make it a liberal country in that regard. As opposed to many other European states, in Spain since 1988 single women can have access to artificial insemination (AI). In countries such as France, Austria, or Sweden, for instance, single women are excluded from medically assisted reproduction.[24] In Italy only heterosexual couples with fertility difficulties are eligible for ART and the use of extramarital gametes is forbidden, as is embryo cryopreservation. Whereas, as we will see, the Spanish legislation is woman-centered, the Italian legislation is a redoubt of hetero-patriarchy (Navarro Michel 156, 162, 167).

Artificially assisted reproduction is accessible in Spain to married and nonmarried couples, either with their own ova and sperm or with donated gametes, and since 2006 also to lesbian couples. The nondonor husband, or male partner or female spouse or female partner, can automatically become the legal father or mother simply by signing a consent form. Interestingly enough, what is really acknowledged in that document is the fact that the father-to-be or mother-to-be is not genetically related to the offspring, not to mention the female spouse or partner if she did not even donate the ovum. Thus, biology, in this domain of artificial reproduction technology, does not seem to be as crucial as it used to be. In Spain for a large majority of the population building familiar relations by mere agreement is a bitter pill to swallow. Even postmortem AI is allowed: the widow might use the sperm of the deceased husband no later than six months after the husband's death if he previously consented to such use. Only the much-debated contract of maternal surrogacy is outlawed in Spain.

Thus, Spain's liberal legislation regarding artificial reproductive technology has made it a country of "reproduction tourism" (Pennings 2690). Popular media and academic research both in Spain and abroad have echoed this reality.[25] Spain is second in the number of clinics for medically assisted reproduction in the European Union (182) only surpassed by Italy (202).[26] But more important is the number of cycles, the rate of pregnancies and successful births (Ikemoto 282) which also contribute to Spain having become a sort of paradise for infertile couples or individuals who, in general, pursue an alternative way of reproduction; most of these couples come from Italy—mainly for legal reasons—and the United Kingdom—mainly for difficulty in access.[27] The fact that in Spain, as opposed to the United Kingdom, the clinics are allowed to "compensate" eggs' donors has surely contributed to this flow of British women to Spain in order to obtain ova (Ikemoto 297).

2006–2007: PGD and Therapeutic Cloning

But ART has also widened the possibilities of research in human embryos. It has also made feasible the early diagnoses of embryo diseases, mainly through the technique known as pre-implantation genetic diagnosis (PGD), thus empowering couples and women in their reproductive decisions. At the same time it has spurred the rage of many social sectors who claim that those techniques, to the extent that they imply the destruction of human lives, should be altogether banned. This has been the characteristic the stance adopted by the Catholic Church, and also by other religious groups and conservative politicians.[28] The successful cloning of a sheep by a team of scientists led by Ian Wilmut at the Roslin Institute in 1996 surely contributed to a state of mixed feeling in public opinion all over the world: awe for the achievement and concern for the possible ramifications of the technique. In 2001, United States President George W. Bush expressed his opposition to all forms of human cloning, announcing the restrictions of federal funds for embryonic stem cell research as well as his intention to create the President's Council on Bioethics (PCB).

Spain has not been immune to this controversy, which is ultimately a quarrel on the scope and meaning of the right to life. During the conservative government of Aznar (1996–2004), none of the necessary legal changes for the fostering of research on human embryos and therapeutic cloning were undertaken in an attempt to curtail their use. On the contrary, the act on assisted reproduction was reformed in order to limit the number of ova that could be fertilized per cycle, in an attempt to put an end to the cryopreservation of spare embryos.[29]

2008: California Dreams

The last stage of our trip leads us to California. We started this legal-historical journey describing how, at the beginning of the 1980s, the Spanish laws on abortion forced Spanish women to travel abroad in order to terminate an undesired pregnancy. Since 2008 or so we are witnessing a different instance of the broad phenomenon of "circumvention tourism": gay couples travelling to California in order to become parents by means of a surrogate mother.[30]

As I said before, there is only one aspect of our regulatory framework of the Spanish laws regarding ART that is not aligned with the most liberal regulatory schemes that one might find across the world: surrogate motherhood.

Since the 1988 Spanish "Ley de Técnicas de Reproducción Humana Asistida," gestational contracts are considered null and void, and, where there exists a split between gestational and genetic motherhood, the Spanish law grants parental rights to the woman who bears the child irrespective of the genetic link.

But for gay couples surrogacy is, obviously, the only alternative to fulfill the desire to become genetic fathers (at least for one of the members of the couple), and this is why the most affluent of those couples, who can be legally married in Spain since 2005, are taking part in the above mentioned form of circumvention tourism.[31] The landmark case took place in 2008, when the Spanish Consular Register in California refused the inscription of twins as the sons of a Spanish gay couple born from a surrogate mother who had previously relinquished motherhood and with whom they had previously arranged surrogacy according to California law. The Spanish Consular Register, however, considered that the Spanish legal ban prevailed. On appeal, the "Dirección General de los Registros y del Notariado" (DGRN) has rejected the Consular decision. As of today, those inscriptions are still challenged before the civil courts, creating great legal insecurity.[32]

In the meantime, surrogate tourism does not seem to cease, and a business of law firms counseling for gay couples eager to travel abroad for reproduction purposes is flourishing. A brief survey on some web pages and the social networks attest to this reality and to how India is becoming an alternative to the United States as a destination country.[33]

The floor is of course open for a debate in which the vindication of equality serves many different and conflicting purposes. On the one hand, as I said before, lesbian couples in Spain are considered on equal terms with heterosexual couples as regards to the use of ART: both the male and female partners of the woman who aims at becoming mother correspondingly become the father or the mother of the offspring if they consent to; that is, they are not legally required to adopt the child with whom they lack genetic links. Male gay couples, however, can only adopt, for obvious natural reasons. Could they be considered discriminated against? Discrimination is not really what is at stake when gay couples are unable to rear children, because the differential treatment is not the outcome of an institutional decision, unless, of course, one deems the mere governmental inaction to remove the hurdles posed by nature as discriminatory.

On the other hand, gay couples who cannot afford the luxury of fulfilling the California dream of becoming parents by means of surrogate mothers, also raise the ideal of equality. Their reasoning is easy and understandable. To the extent that the Spanish institutions accept the effects of those contracts, an economic discrimination is fueled. Thus, the Spanish institutions should enforce the prohibition of surrogate motherhood extraterritorially, in the exact same fashion that our *Tribunal Supremo* did in the "abortion tourism cases" in the beginning of the 1980s. But the consequence of such a policy seems a tad difficult to swallow, namely the offspring's condemnation to variable legal status as a child. According to Spanish civil law he would be considered the son of one of the members of the couple and the surrogate mother, but that would not be his condition under California law—where the gestational mother would not be considered his mother. This indeterminacy or changeable status would lead to other sacrifices of the best interests of the child, such as not being given Spanish nationality when returning to Spain, for example. The alternative seems clear: a domestic legal reform that allows surrogacy contracts.

The Spanish political and social climate does not give us much reason to think that the advent of such legal change is near. In the meantime it is interesting to question whether that scenario would be a step further in the fostering of the egalitarian ideal, or, to the contrary, the result of the pressure exerted by the globalization of services on increasingly weak national borders.

In any case, the reform to permit surrogate motherhood will entail significant costs, primarily a whole new different way of conceiving reproduction rights.

I said at the outset that throughout this essay I was going to take reproductive rights as encompassing a set of immunities and legal powers conferring individuals sovereignty on several aspects related to the decision whether or not to become parents. The resort to surrogate motherhood, both by homosexual and heterosexual couples, forces us to consider the alienable nature of that right.

Many of the "anti-commodification" criticisms of surrogacy contracts raised by the Spanish feminists (and that includes lesbian or "queer theorists" such as Beatriz Gimeno) boil down to the claim that allowing women to relinquish her reproductive sovereignty—whether to contractually waive her right to terminate her pregnancy or contractually accept the duty to abort—is a wicked form of feminine exploitation, or a disguised acceptance of children-trafficking.[34]

The very recent case of Crystal Kelley illustrates this point well. Ms. Kelley, a Connecticut resident, mother of two, entered into a surrogacy contract with a couple who offered her $22,000 to have a baby. After the bad result of an ultrasound performed at 21 weeks, Ms. Kelley was offered $10,000 if she consented to terminate her pregnancy. Apparently the surrogacy contract stated her duty to abort should the fetus suffer from severe abnormalities, which seemed to be the case. The couple's lawyer advised her that they would sue her for the recovery of the money the couple had already paid her for legal and medical fees. She refused to perform an abortion and found a couple willing to adopt.[35]

For Arthur Caplan, a renowned bioethicist, the issue is clear: "No one can contract with a woman to have an abortion. Under any circumstances. For any reason. Never. A woman controls her body and no one can make her do anything she does not want to do in terms of medical intervention with her body no matter what she has said before, signed, or promised."[36]

Crystal Kelley has not been the first woman to enter a surrogacy contract in which a clause that purports to require the surrogate to abort in certain circumstances is included. That a clause like that is null and void is hardly disputable. In general, no one can be forced to do something against his or her will, even if one contractually accepted to do so. Suppose you commit yourself to paint a house and you breach the agreement. The contractual remedy known as "specific performance" consisting of your actually being obliged to paint the house is seen with great suspicion in common-law jurisdictions because granting such remedy seems tantamount to a form of temporal slavery.[37]

Admittedly, specific performance appears to be the typical remedy when "unique goods"—goods for which courts will find difficult substitutes—are at stake.[38] The extension of specific performance to personal services is not inconceivable if it can be shown that ". . . protection of the liberty interest . . . is more important than the goals that specific performance is thought to serve."[39]

In the case of abortion, this does not seem to be the case. In surrogacy contracts the specific performance of terminating pregnancy would actually amount to physical abuse and coercion, a total infringement of a constitutional right to procreate which, as opposed to abortion, is universally recognized with marginal exceptions.[40] For that reason, as Caplan argues, even an agreed buy-out option has to be excluded. In Kelley's case the failure to abort on demand would be a material breach relieving the would-be parents from their duty to pay. However, the question still remains as to whether the surrogate mother should compensate the intended parents if she decides to perform an abortion against their wishes.[41] Whereas people generally find the waiving of the right not to give birth a difficult pill to swallow, a majority of us seem to be more confident in accepting the possibility of waiving the right to abort.[42] That terminating pregnancy is always morally suspicious, whereas not having children pertains to the catalog of our most precious liberties, might be the likely cause for that asymmetrical judgment. Or maybe is just that we are ill-equipped to confront the new realities brought by the advance of artificial reproduction technologies.

Concluding Remarks

Over the course of the previous pages I have attempted to provide a brief sketch of a winding evolution of reproduction rights in Spain. In a first stage, Spanish women have struggled for the decriminalization of abortion, a conquest that arrived partially in 1982 in the midst of massive opposition by conservative and Catholic groups. The conquest has revealed itself unstable. As in many other countries, the abortion debate seems intractable, and legislation on the issue swings with every change in the distribution of political power between conservative and leftist groups in Spain.

I have also tried to show that abortion does not exhaust the wide range of controversies linked to reproduction rights, neither in Spain nor elsewhere. We have witnessed disputes regarding the use of artificial reproduction technologies, and, more recently, the debate on surrogate motherhood fostered by the aim of homosexual couples to become fathers. Their involvement in circumvention tourism, either to California or India, to fulfill their desires is exerting great pressure to modify Spain's current legislation and permit surrogate motherhood in the name of "equality." In a fairly unexpected twist, this (provisionally) last stage of the evolution of women's reproduction rights reintroduces, albeit in a much more complex and paradoxical light, the recurrent themes of women's exploitation and the nature of women's rights to reproduction.

Notes

1. The landmark decision is *Griswold v. Connecticut*, 381 U.S. 479 (1965). Article 16 of the Universal Declaration of Human Rights states that: "Men and women of full age, without any limitation due to race, nationality, or religion, have the right to marry and to found a family."
2. Article 413 of the Criminal Code (1973).
3. Article 414 (1973).
4. See Hernández Rodriguez, *El aborto en España*.
5. It is worth quoting the passage: "La muchacha, en lugar de en la posición arriba indicada más favorable para provocar la expulsión del contenido uterino, yacía de lado en el jergón y con el cuerpo engatillado. Sus gritos dotados de sentido habían ido haciéndose más débiles conforme aumentaba la pérdida de líquidos vitales a lo largo de las horas transcurridas desde que la operación iniciada por el mago de la aguja tuvo su insatisfactorio comienzo. Este mago debía de haber equivocado la trayectoria del instrumento punzante, o tal vez la punta del mismo, a causa de su excesivo uso, había perdido la eficacia tantas veces demostrada. Era también posible que su excesiva juventud diera, tanto a los tejidos propios como a sus productos, una consistencia o una elasticidad diferentes de las acostumbradas. O bien que la contracción de la matriz, otras veces suficiente para el desembarace de las atribuladas hembras, esta vez sólo sirviera para dilatar las venas perdedoras de sangre y para hacerla sentir los rítmicos dolores que sus espaciados gritos indicaban." (*Time of Silence* 107)

   (Instead of taking the upright position favorable to the expulsion of the uterine content, the girl was lying on her side on the mattress, her body doubled like a cat's. Her cries had gradually weakened in force with the increasing loss of vital fluid and the time lapse since the start of the operation performed by the magician with the needle. The magician must have mistaken the trajectory of his pointed instrument, or perhaps the point had with frequent use loss the efficiency it had so often displayed before. It was also possible that the extensive youth of the patient was the reason for an unusual consistency and elasticity of the tissues and their products. Or could the contraction of the matrix, sufficient in other cases for the delivery of pregnant women, in this case have serve only to dilate the blood-letting veins, producing the rhythmic pains that her cries indicated.)

6. STS 605/1982, 18 de diciembre de 1982.
7. STS 1293/1982, 7 de julio de 1982.
8. STS 1181/1982, 1 de julio de 1982.
9. STS 424/1982, 6 de mayo de 1982.
10. Quite recently, in the midst of the controversy over a new reform in abortion legislation, *El País* published the story of a woman named "Cinta" who travelled to Amsterdam along with other women in one of those "abortion trips." See "A Londres o a la abortera de barrio," *El País*, 5 May 2013. Web. 9 May 2013. The first edition of *El País Semanal*, published in 1976, was also devoted to the phenomenon of abortion tourism; "Abortar en Londres" (3 Oct. 1976).
11. According to official data, the number of abortions in Spain in 2011 was 118,359. Almost 90 percent of them were performed in private clinics, and 10 percent were provided to foreigners. See "Interrupción voluntaria del embarazo. Datos definitivos correspondientes al año 2011", Ministerio de Sanidad, Servicios Sociales e Igualdad.
12. One might also wonder whether the extraterritoriality of criminal law is itself justified; for a superb and recent discussion see Alejandro Chehtman, "The Extraterritorial Scope of the Right to Punish."
13. This legal provision is a vestige of Roman law and it has been incorporated into many legal systems. What Roman lawyers had in mind is that the fetus should be able to inherit before acquiring personhood.
14. The survery is number 1456 of the Centro de Investigaciones Sociológicas conducted in 1985.
15. This is the reason why Spain, along with Hong Kong, New Zealand, Israel, South Korea, and Ethiopia has been considered a country with liberal abortion laws; see Gilda Sedgh et al., "Induced abortion: incidence and trends worldwide from 1995 to 2008."
16. See "Noticia sobre abortos en establecimientos de Carlos Morín," *YouTube*. YouTube. Web. 27 Jun. 2014.
17. See "Abortion in Spain/Asi se aborta en España." *YouTube*. YouTube. Web. 27 Jun. 2014.
18. Currently the vast majority of European States (19) have this kind of legislation. See M. Gissler et. al., "Terminations of pregnancy in the European Union."
19. See "No cabe un trato desigual con el aborto por razón de discapacidad." *La Razón*, 15 May 2013. Web. 27 Jun 2014; see also "Down syndrome association: Spain's eugenic abortion law violates human rights of disabled." LifeSiteNews, 12 Feb. 2012. Web. 27 Jun. 2014.
20. See "Convention on the Rights of Persons with Disabilities." United Nations. 2006. Web. 27 Jun. 2014.
21. The full text can be found on the web site for the United Nations Office of the High Commissioner for Human Rights.

22. See "Hidden Abortion Agenda in UN Convention on Disability Rights." *LifeNews*. LifeNews, 20 July 2012. Web. 27 Jun. 2014. See also "House Members Letter Highlights Problems with UN Disability Treaty." *Turtle Bay and Beyond*. 1 Aug. 2012. Web. 27 Jun. 2014..
23. See Ley 35/1988 de Técnicas de Reproducción Humana Asistida.
24. See ESHRE Central Office, "Comparative Analysis of Medically Assisted Reproduction in the EU: Regulation and Technologies (SANCO/2008/C6/051). Final Report," p. 24.
25. "España concentra hasta el 40% del turismo reproductivo." *ABC*. 31 Jul. 31 2012. Web. 22 May 2013. See also "Passport, tickets, suncream, sperm . . . ", *The Observer*, 15 Jan. 2006. Web. 22 May 2013.
26. "Comparative Analysis of Medically Assisted Reproduction in the EU: Regulation and Technologies (SANCO/2008/C6/051)," p. 46. The data corresponds to 2008.
27. See Itizar Alkorta Idiakez, "Donación de óvulos"; F. Shenfield et. al., "Cross border reproductive care in six European countries," pp. 1363–64; "Comparative Analysis of Medically Assisted Reproduction in the EU: Regulation and Technologies (SANCO/2008/C6/051)," pp. 56, 68, 71; and Guido Pennings and W. Van Hoof, "Extraterritorial Laws for Cross-Border Reproductive Care: The Issue of Legal Diversity," p. 189.
28. See the Instruction *Donum Vitae* issued by the Congregation for the Doctrine of Faith in 1987.
29. See Ley 45/2003, de 21 de noviembre.
30. The term has been coined by Glenn Cohen; see "Circumvention Tourism".
31. One of the most famous gay couples involved in the procedure admitted having paid 16.000 euros to the Californian surrogate mother; see "La pareja gay de Valencia pagó 16.000 a la madre de alquiler de sus dos hijos," *Las Provincias*, 3 Sep. 2012. Web. 27 Jun. 2014.
32. Spanish popular media has widely echoed the story: see "La pareja gay con el hijo de un vientre de alquiler se siente discriminada," *Levante EMV*, 27 Jun. 2009. Web. 27 Jun. 2014. See also "Una pareja gay no puede registrar hijos de una 'madre de alquiler,'" *El País*, 2 Dec. 2011. Web. 27 Jun. 2014; and "El juez anula el registro de los hijos de una pareja gay nacidos de un vientre de alquiler," *El Mundo*, 12 Jan. 2011. Web. 27 Jun. 2014
33. See for instance the Center for Surrogate Parenting or the Spanish law firm NQ Abogados or Dema Abogados.
34. Besides her activism in the lesbian movement in Spain, Gimeno is a well-known feminist theorist. In her blog and elsewhere she has robustly opposed surrogate motherhood; see "Vientres de alquiler," BeatrizGimeno. Web. 13 Jun. 2011. Web. 27 Jun. 2014.
35. The story has been amply covered; see for instance "Surrogate mother fled across country to give birth and save baby after parents wanted child aborted when ultrasound revealed series of disabilities," *Daily Mail*, 4 Mar. 2013. Web. 27 Jun. 2014;

and "Surrogate mom offered $10K to abort baby after parents discover she would be born with disabilities," *New York Daily News*, 5 Mar. 2013. Web. 27 Jun. 2014.

36. See "Surrogacy Case Reveals Moral Holes, Bioethicist Says," *NBC News*, 6 Mar. 2013. Web. 27 Jun. 2014.
37. The classic piece on specific performance in the Anglo-American literature is Alan Schwartz, "The Case for Specific Performance." The remedy in case you refuse to do what you committed yourself in doing is, typically, of pecuniary nature.
38. See Anthony Kronman, "Paternalism and the Law of Contracts," p. 784, and Alan Schwartz, "The Case for Specific Performance," p. 280.
39. Schwartz, "The Case for Specific Performance," p. 274.
40. The Chinese single-child policy might be the paradigmatic one.
41. During the last U.S. presidential campaign, the public came to know that the son of the Republican candidate (Tagg Romney) and his wife had signed a surrogacy contract in which it was granted to the surrogate mother the liberty to terminate the pregnancy at will. This strikingly conflicted with the pro-life stance of Mitt Romney.
42. For a good and thorough discussion see Glenn Cohen, "The Constitution and the Rights Not to Procreate."

Works Cited

Alkorta Idiakez, Itizar. "Donación de óvulos." *El País*, 28 Mar. 2006. Web. 13 Nov. 2013.

Chehtman, Alejandro. "The Extraterritorial Scope of the Right to Punish." *Law & Philosophy* 29.2 (2010): 127–57.

Cohen, G. "Response: Rethinking Sperm-Donor Anonymity: Of Changed Selves, Nonidentity, and One-Night Stands." *The Georgetown Law Journal* 100 (2012): 431–47.

_____. "Circumvention Tourism." *Cornell Law Review* 97 (2012): 1309–98.

_____. "The Constitution and the Rights Not to Procreate." *Stanford Law Review* 60 (2008): 1135.

Gissler, M., et. al. "Terminations of Pregnancy in the European Union." *BJOG International Journal of Obstetrics and Gynecology* 119 (2012): 324–32.

Hernández Rodríguez, Gerardo. *El aborto en España. Análisis de un proceso socio-político.* Madrid: Universidad Pontifica de Comillas, 1992.

Ikemoto, Lisa C. "Reproductive Tourism: Equality Concerns in the Global Market for Fertility Services." *Law and Inequality Journal* 27.2 (2009): 277–309.

Kronman, A. "Paternalism and the Law of Contracts." *The Yale Law Journal* 92.5 (1983): 763–98.

Martín-Santos, Luis. *Tiempo de silencio.* 1962. Barcelona: Seix Barral, 1999.

_____. *Time of Silence.* Trans. George Leeson. New York: Columbia University Press, 1964.

Navarro Michel, Mónica. "Reproducción asistida, modelos de familia y derechos reproductivos." *Sobre Bioética y Género*. Eds. María Casado and Albert Royes. Cizur Menor (Navarra): Thomson-Civitas, 2012.

Pennings, Guido. "Legal Harmonization and Reproductive Tourism in Europe." *Human Reproduction* 19.12 (2004): 2689–94.

Pennings, Guido and W. Van Hoof. "Extraterritorial Laws for Cross-Border Reproductive Care: The Issue of Legal Diversity." *European Journal of Health Law* (2012): 187–200.

Schwartz, A. "The Case for Specific Performance." *The Yale Law Journal* 89 (1979): 271–306.

Sedgh, Gilda, et. al. "Induced Abortion: Incidence and Trends Worldwide from 1995 to 2008." *The Lancet* 379 (2012): 625–32.

Shenfield, F., et. al. "Cross Border Reproductive Care in Six European Countries." *Human Reproduction* 25.6 (2010): 1361–68.

8

Mar adentro and the Question of Freedom

Paul Begin

On February 20, 1995, Ramón Sampedro appeared on Spanish television in an interview with Mercedes Milá, providing him the opportunity to explain himself to a wider audience and gather public support for his challenge to the Spanish government for the right to an assisted suicide. He would appear on television at other times as well, including the recording of his own suicide. The broadcast images of Sampedro and his arguments for the right to die on his terms brought discussion of the very taboo subject of euthanasia to a fevered pitch in Spain. The major presses (*El País*, *La Vanguardia*, *ABC*, etc.) were publishing articles almost daily. Public intellectuals (Rosa Montero, Fernando Savater, Juan Manuel de Prada, to name only a few) contributed opinion pieces to the debate.[1] *El País*, for its part, kept an online space devoted exclusively to the social issue. The controversy has endured to this day, to the extent that legislation was just recently put in place to offer palliative care while, significantly, the debate continues over the legality of euthanasia and assisted suicide.[2] In 2004, Alejandro Amenábar entered the fray with his film *Mar adentro*, a biographical portrait of Ramón Sampedro and his quest for death. The film was nothing short of a *cause célèbre*, sweeping the Goya Awards and earning an Oscar for Best Foreign Picture. In conjunction with other prominent cases of euthanasia (Vicente Humberto, Madeleine Z, Inmaculada Echevarría, Jorge León), *Mar adentro* played a key role in furthering the public debate over one's right to die on his or her own terms.

Regardless of Amenábar's attempt to distance himself from the controversy, the timing alone made the film a touchstone in the ongoing public discourse about what constitutes life and the role of the government in that life.[3] It is furthermore taken to be obvious that *Mar adentro* is decidedly pro-euthanasia (Bradshaw, French, Holden, Smith). One possible hiccup to this argument may be the disjunction between the protagonist's desire to end his life and the inspiration that he provides for others through his fight to die (Holden, Wilson).[4] Also of contrast is the sheer beauty of the film (considered by many in Spain to be a filmic masterpiece) and its bittersweet denouement. For Paul Julian Smith, what stands out is the film's televisual subtext "that underwrites *Mar adentro*'s cinematic virtuosity and helped facilitate the film's massive popularity in Spain" (112). While a few critics have looked at the film through a disability studies lens (Marr, Minich), most of the critical attention that *Mar adentro* has received revolves around the film's ostensible support of euthanasia, its seemingly facile discourse ("an extraordinarily shallow euthanasia weepie" according to one critic), the virtuoso cinematography, and of course Javier Bardem's subtle performance as Ramón Sampedro (Bradshaw).[5] Within this general reception two central issues have been largely ignored: (1) the fact that the film is about assisted suicide proper, as distinct from euthanasia (a general term meant to signify "good death"), and (2) the corresponding moral argument made *within the film* for assisted suicide.

The difference between assisted suicide and euthanasia broadly is key. Ramón Sampedro falls out of categories that justify the implementation of euthanasia but he does fulfill the categories of having been unwavering in his petition and can claim a form of long-term suffering. He would end his life on his own if he could. The prohibition is therefore a sort of discrimination against those with disabilities. This leads to the second issue: *Mar adentro* is indeed a film about one man's quest to end his own life, but it is this quest for the legal right to end his that leads a larger question about the very definition of individual liberty within modern democratic societies. While many might expect the real life debate in Spain to be largely a battle between religious views versus secular views of life, the real issue seems to be one of autonomy and freedom. As María José Guerra notes, "it was possible to detect a public reaction of a different kind, in support of the conviction that the right to die is an issue that has to do with the whole recognition of the value of personal autonomy" (429). It corresponds that the underlying tension in *Mar adentro* is the question of freedom: How is it possible that a free society based on Enlightenment principles can justify not allowing a person to end their own life on their own terms? One helpful way to clarify the issues related to this paradox is by turning to Isaiah Berlin's distinction between "negative" freedom, freedom from obstruction or intervention, and "positive" freedom, which is the capacity for self-rule or autonomy. What I propose is that *Mar adentro*, through its filmic discourse (dialogue, mise-en-scène, montage), highlights this important distinction and implicitly participates in a larger movement to redraw the frontier between individual liberty and the authority that theoretically protects such liberties. Ultimately, *Mar adentro* functions to humanize an ongoing struggle *within* society between rationalist ideals from the Enlightenment and developing values that are informed by pluralism and empiricism.

The first point to examine is how Amenábar develops a moral case that works to justify assisted suicide and thus bring the viewer along to *see* this particular conclusion. In order to understand *Mar adentro* qua moral argument it is helpful to observe what Ramón is free to do. To begin, the film uses a cinematic trick that Amenábar has employed in previous films, such as *The Others* (2001) and *Abre los ojos* (1997), by which the spectator does not know if the on-screen images and sounds are part of a dream or the diegetic reality. A dark screen appears and we hear a voiceover instructing us to become increasingly calm ("Tranquilo. Estás más y más tranquilo"). We are told to imagine a screen. A screen within a screen appears and expands to reveal an image of a white sand beach that is accompanied by a soundtrack of waves gently hitting the shore. It turns out to be an insight into the mind's eye of Ramón Sampedro as he meditates. It is not until the moment of inner peace is broken that we realize that our experience is meant to give us a glimpse into Ramón's inner experience. Ramón's eyes open and we see Gené (Clara Segura), Ramón's legal advocate, who is reading to him from a book called *Potencia tu fuerza mental* (*Building Mental Muscle* by David Gamon and Allan D. Bragdon). As if looking through Sampedro's eyes, the camera pans his room, revealing books and personal effects. As Gené leaves she puts on music for him, Mozart. If one is able to suspend belief, for a brief moment we experience the world in a way similar to how Ramón might. This is one of multiple instances in which the spectator is forced into the enclosure of Ramon's living space and offered insight into the protagonist's mental state.

The next sequence is his first visit from Julia (Belén Rueda), the lawyer who will be bringing his case to the Spanish courts. In establishing his case she asks about his daily routine. He lists eating breakfast, listening to the radio (mostly debates), and writing on his "computer," a wooden contraption of Ramón's own design by which he can write using a pencil attached to a back-scratcher. He notes that he enjoys inventing things. The opening sequence establishes in clear terms that, in spite of his severely limiting disability, the protagonist is free to enjoy the pleasures of eating, reading, and listening as he wishes. Ramón is likewise able to exercise his mental faculties at a high level, reasoning as he listens to debates, inventing things and practicing meditation. In other words, despite his disability, Ramón is actually "free to" do virtually whatever he desires with the help of others. He is, in almost every way, able to do as he pleases *and is able* within the confines of his home.

Once Ramón's "freedom to" is established, the film wastes no time in posing the critical question as to why Ramón wants to die. He has two main arguments: (1) a lack of dignity in his condition and (2) notions of private property, namely that his body is his sole possession and that he should be able to determine whether or not he should live without any interference from the state. Asked why he will not accept a wheelchair, Sampedro replies that this would be like accepting the crumbs of what was once his freedom ("migajas de lo que era mi libertad"). The key image is a close up of Ramón's hand next to Julia's. While foreshadowing their relationship, the image of these two hands, one able-bodied the other disabled, is also a metonymy for the disparity between his desire and his capacity. [Fig. 1]

The contrast between his life before the accident and his current physical state, the constant need for help in all of life's daily needs, the mental anguish (which is on display throughout the film) all lead Ramón to the decision that his life is undignified, that he is the author of his own values, and thus he should put an end to his suffering through assisted suicide. The problem, if one

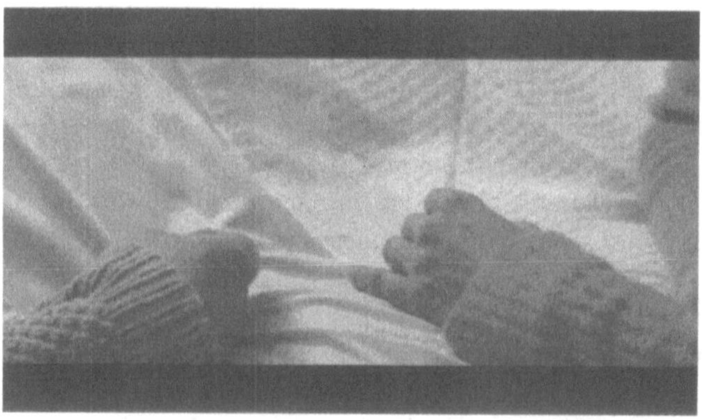

Fig. 1. The insurmountable distance between desire and ability.

can be so obvious, is that Spanish law prohibits assisted suicide as part of a larger social contract. The social contract is established to protect human liberty and it does so by protecting its citizens against death to the extent that this is possible. Death, clearly, would mean an end to freedom. The case of Ramón, therefore, is a paradox: Ramón is simultaneously free to commit suicide and also constrained from doing so in order to guard his freedom. Can both ideals exist at the same time?

Ramón clearly shares in John Stuart Mill's view of freedom: "that of pursuing our own good in our own way, so long as we do not attempt to deprive others of theirs [. . .] Each is the proper guardian of his own health, whether bodily, or mental and spiritual" (17). The argument for self-determination—pursuing our own good in our own way—is not only articulated at certain intervals by characters in the film, it is asserted in cinematographic terms. Individual liberty, in fact, is *the cinematographic* argument of *Mar adentro*. Amenábar presents the argument for autonomous self-rule by focusing almost exclusively on the subjective experience of Ramón and only touching on how it might impact those closest to him as well as the local community and the Spanish community at large. We have already noted how Ramón is confined to his room. Of the few forays into the external world, the most psychologically informing scenes are the dream sequences and flashbacks, functions of memory and consciousness that only contribute to the sense that Ramón is trapped, physically, emotionally, and even politically. Whereas Ramón is reticent to open up and share his emotional state with others, the film viewer is privileged insight into his past, his self-image in an ablest state and, more importantly, his present longings for love and companionship, a desire that is linked to an ability to use his body. In terms of mise-en-scène and sound, the film is overtly private and intimate. This lends an empathetic quality to the film viewing experience that effectively steers the right to die debate toward one of individual liberty. Moreover, this introverted perspective has another, more political function, that is, it keeps the question of assisted suicide and individual liberty confined to one exceptional case without taking into account larger social implications, for if there is one place where I should be free to act as I wish, it should be in the private sphere of my home as long as my actions do not impinge on the freedom of others.

By and large, *Mar adentro*'s emphasis on individual liberty circumvents many of the counter arguments against assisted suicide, in particular arguments about morality, social impact, and the so-called slippery slope. In fact, Julie Minich observes that the film offers only an abridged version of the debate (26). This may be true of much of the dialogue, however, inscribed within the cinematography are some counterarguments that merit critical attention because of the way in which they help us to understand the nuances of freedom. The opening sequence in which Ramón lays out his case is followed immediately by a subtle yet powerful argument against this view of human freedom as total freedom from institutions. Cutting from the initial meeting between Julia and Ramón, the lawyer then interviews his brother, José (Celso Bugallo), who states: "Las ideas son libres pero lo que él pide no está bien. Queremos lo mejor para él [. . .] No es racional, como él dice." (Ideas are free but what he is asking for is not good. We want what is best for him. [. . .] This is not rational, as he says). It is notable that the mise-en-scène used for José's arguments mimics the techniques of a television interview. The character of José is one of the trio of voices against assisted suicide within the film, which includes a priest (José María Pou) and Rosa. The latter, as we know, eventually comes to embrace Sampedro's point of view and assists him in taking his own life in the name of love. Like the priest, José is depicted rather schematically as a rustic and conservative Galician. Julie Minich correctly observes how, "the film treats Galician as a language of stifling social conservatism" (28). Within this context, José's moral position is likely to be viewed as a product of ignorance. (Ramón in fact uses that very word—*ignorancia*—to describe his brother later on in the film.) José's argument touches on two fundamental ideas when considering the ethical implications of assisted suicide/euthanasia: (1) the notion of imposing what is best for someone over and above their own wishes (i.e. paternalism) and (2) the corresponding idea that no rational person would choose to end his or her own life.

Both considerations merit some expansion. Aristotle, to use just one foundational example, argued that it is only rational to do justice/good to oneself and, conversely, that it is irrational to do harm intentionally to oneself. Because life is a good in itself, and death puts an end to that good, it would be irrational to put an end to one's life. The conclusion is that the person who commits suicide, with suicide being the most extreme form of self-harm, is by definition, acting contrary to his or her own true wishes (Aristotle 97). Put another way, he or she would be acting contrary to reason and against what would naturally be their best interest (to live). In his recent work on disabilities and Spanish cinema, Matthew Marr argues that an important subtext that has gone undetected in *Mar adentro* is Ramón's long-term depressive condition. Marr notes the presence of prescription tranquilizers and notes, "Ramón's unyielding suicidal drive . . . *is in itself* a persuasive indicator of his character's struggle with an undiagnosed psychiatric illness" (102). Specifically, Marr asserts that Ramón likely suffers from hypomania or manic depression. Once the symptoms are elaborated, we can easily begin to identify this as a highly plausible interpretation and, importantly, that the protagonist may have suffered from a manic-depressive state since *before* the accident that left him a quadriplegic. If Ramón does in fact suffer from a form of clinical depression, and Marr makes a very strong case that he does, then this fact would indeed have bearing on moral and legal arguments for assisted suicide. It would explain why Ramón expels so much effort on being rational. His plan to appear before the Spanish court is to demonstrate that he is lucid and in complete control of his mental faculties as he argues for self-termination. Although it is likely not meant to be taken as a philosophical rebuttal, José's comment about doing what is in Ramón's best interest does point to a larger political question of self-mastery, which constitutes the positive aspect of freedom.

In thinking through concepts of political freedom in *Mar adentro*, another component would be the religious argument against assisted suicide and specifically the debate between Ramón and the priest, Padre Francisco (José María Pou). The film was released only six years after Ramón Sampedro terminated his life and effectively re-energized what had grown into a polarizing debate about the right to die as well as Spain's relative constitutional worth as compared to other European nations. Not surprisingly, the Roman Catholic Church has played and continues to play a prominent role in this debate on life and certainly informs the conservative party's strong opposition to assisted suicide and the historical Ramón Sampedro's case specifically. The filmic confrontation between Padre Francisco and Ramón is clearly designed to embody these two opposing worldviews. The problem is that the religious argument is dismissed out of hand as superstitious and oppressive. Even if we accept some of Padre Francisco's statements at an intellectual level, he is treated diegetically as a caricature of religious fundamentalism, insulting the entire Sampedro family by suggesting that Ramón does not want to live because he is not sufficiently loved. Metaphysical arguments about the nature of human existence are dismissed by omission. There is no debate about the arguments for or against the existence of God or the genesis of man, ideas that, historically, have been critical in defining the concept of human dignity and therefore considering questions of self-immolation, abortion, and capital punishment. In the Judeo-Christian account, human dignity is a product of identification with God, whereas with Kant it is a product of personal autonomy (Debes 45). Views toward suicide tend to flow out of these two camps, perhaps all too dogmatically. In *Mar adentro* the two sides are already drawn. Before Padre Francisco's acolyte can finish what is depicted as a couched statement about eternity and living as God's creations within that context, he is cut off by Ramón. In fact, the climax of the argument between the two quadriplegics is directly related to the nature of freedom in political, social, and philosophical terms. After Ramón is chided by the priest for his "bourgeois sense of private property," he points out that he is entitled to freedom of thought and notes the hypocrisy of the Right to reject assisted suicide while accepting the death penalty. Ramón also points out that the Spanish public by and large approves of assisted suicide, a fact that is supported outside the narrative world of the film.[6] He claims to represent a majority viewpoint. If this is true, and it seems to be the case, the most heated part of their exchange is worth keeping in mind:

> Padre Francisco: Una libertad que elimina la vida no es libertad.
> (A freedom that eliminates life is not freedom.)
>
> Ramón: Y una vida que elimina la libertad tampoco es vida.
> (And a life that eliminates freedom is not a life.)

It seems that both statements are true. But how can these two rights or freedoms coexist? To have someone assist Ramón in dying would be, on his own view, an end of consciousness and result in a loss of freedom. To force him to live, also by his account, would be to deny him his freedom to self-determination. What prevails is a continuation of the debate as to what constitutes liberty in the political sense.

What *Mar adentro* brings to the fore is a debate about freedom and personal autonomy in a liberal, progressive state such as Spain. Ramón is free on one level but constrained on another. It is a paradox that is most readily understood through Isaiah Berlin's classic formulation of two concepts of freedom, in which he delineates between "negative" liberty and "positive" liberty. Negative liberty, on Berlin's view, is freedom from constraint so as to do or be however one chooses. It may be thought of as a lack of interference in order to pursue our own good in our own way. We could think of this as a the freedom to purchase property, to access 500+ cable programs, to eat organic salad or choose fast food every evening. One may be limited by circumstances such as financial resources, education, or ability, but I am technically free to choose how to live in the private sphere without any restrictions. Meanwhile, positive liberty is "the wish on the part of the individual to be his own master" (179). If negative liberty answers the question of what one is free to do, positive liberty answers the question of who is in control of one's life. I may, for example, be free to choose to eat salad every night but if I am driven by an internal compulsion to eat fast food and I succumb to that desire each and every night then I am not my own master, I am a slave to my desire, and therefore lack freedom in the positive sense, that is, the ability steer my life in a positive direction. This is where the social contract becomes important. Laws are normally fine with letting my desire compel me to each cheeseburgers every day. They do, however, curb my desire insofar as it impinges on the freedom of others. I am not free to kill or maim or slander or rob. I consent to this agreement because I would not want someone to do these things to me either. In other words, I consent because it is in my best interest. Berlin writes: "A rational (or free) state would be a state governed by such laws as all rational men would freely accept; that is such to say, such laws as they would themselves have enacted had they been asked what, as rational beings they demanded" (191).

Thus freedom in the positive sense occurs when that which we would choose reasonably coincides with our ability to do so: "A rule cannot enslave me if I would impose it upon myself" (Berlin 190). It is a view that is prevalent to some degree in the writing of many prominent Enlightenment thinkers, such as Locke:

> So that however it may be mistaken, the end of law is not to abolish or restrain, but to preserve and enlarge freedom. For in all the states of created beings, capable of laws, where there is no law there is no freedom. For liberty is to be free from restraint and violence from others, which cannot be where there is no law: but freedom is not, as we are told, liberty for every man to do what he lists (for who could be free when every other man's humour might domineer over him?), but a liberty to dispose, and order as he lists, his person, actions, possessions, and his whole property, within the allowance of those laws under which he is, and therein not to be subject to the arbitrary will of another, but freely follow his own. (323–33)

This seems perfectly plausible if we consider murder or theft. Separate and apart from metaphysical claims, these acts must be prohibited or curbed because they would deny the freedom of another. But the case of assisted suicide, of helping another to end their life, is another story. What if this patient suffers from bipolar disorder or is depressed from a romantic event gone awry? Jennifer Michael Hecht has recently argued against suicide (not assisted suicide) in part on the basis that the vast majority of people who attempt suicide and fail never try again, the implication being that motivators for suicide are all too often temporary (12). Hecht also signals the problem of social scripting; if Ramón's case indicates that life as a quadriplegic is not worth living, how might this impact other physically disabled beings.[7] Also, to take a view of the body as a possession opens the door to exploitation by beings whose economic situation may lead them to sell organs, work in prostitution, or even sell oneself for food.[8] And if Ramón is given permission to end his life because it is degrading, could inmates serving life-sentences not claim the same thing and pursue the same fate? One can reasonably understand why a social contract that aims to protect citizens and look out for their best interests may have an imperative to prohibit assisted suicide (since it cannot reasonably prevent suicide). If it is the citizens who dictate laws, and those laws allow for them to live in accordance with their will, then the laws are not an imposition nor do they enslave—they protect liberty. Even Mill wrote that to sell oneself into slavery would be to violate one's own freedom and therefore should be prevented, "It is no freedom to be allowed to alienate his freedom," and by extension (as he does not mention it specifically) so too would the act of suicide, or any form of self-destruction (95).

But what happens when my ultimate values differ from those of the prevailing order, or when my definition of freedom as absolute personal autonomy does not accord with inherited beliefs and laws? The laws that are put in place to protect individual liberty are done so under certain assumptions about ultimate values and the highest aims of man, assumptions that were developed by and large by men of the eighteenth century. Enlightenment thinkers such as Locke, Rousseau, Jefferson, and Mill operated under the idea that reason would lead us to discover immutable universal truths about the nature of man. Based on natural law and derived from reason, these truths could be applied broadly, across cultures and used to form the basis of governments. Chief among these truths is the claim that one of man's highest values, what Rousseau would call absolute values, is freedom. Jefferson's may be the most recognizable example of this thinking:

> We hold these truths to be self-evident, that all men are created equal, that they are endowed by their Creator with certain unalienable Rights, that among these are Life, Liberty and the pursuit of Happiness. That to secure these rights, Governments are instituted among Men, deriving their just powers from the consent of the governed.

The Enlightenment notion of liberty as a universal right and, by extension, a universal value is one of the key precepts in modern democratic societies such as Spain. While freedom has many forms, the dominant view (Hobbes, Kant, Locke, Mill, among others) is that *being alive* is a pre-condition to freedom. It follows that governments, in preserving the ultimate value of freedom, are

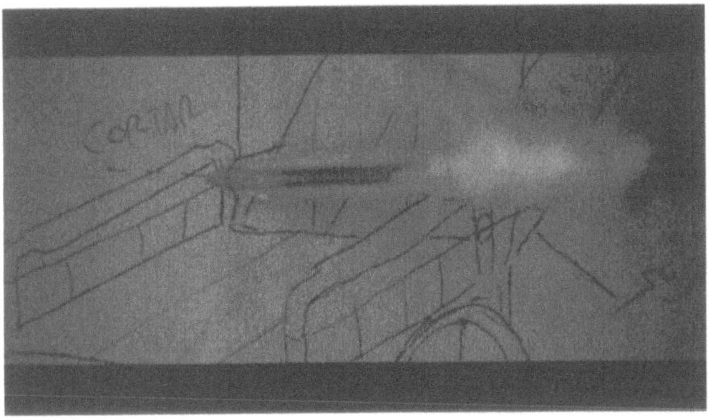

Fig. 2. Ramón's drawings, combined with his practice of poetry, rhetoric, and philosophy establish him as a veritable Renaissance man.

expected to preserve the lives of its citizens as the critical piece of preserving that freedom. In outlawing assisted suicide the idea is that if a citizen cannot recognize life as an ultimate value synonymous with freedom, the government will do it for you. But who is to say that this definition of freedom is the only correct one?

Who is to say that reason, driven more by empiricism than notions of natural law, cannot lead one (in this case Ramón Sampedro) to identify with freedom as the right to die? What is at stake, ultimately, is the very definition of freedom *and* the sacred cow of Enlightenment thought—reason. *Mar adentro*'s discourse on the nature of rational thought and the question of freedom is brought into bold relief during the sequence in which Ramón decides to make a physical appearance in court. Three visual components illustrate this point. The first is the sequence through which Ramón engages his family in the construction of a wheelchair that he will use for the trip. [Fig. 2]

During this sappy, feel-good section of the film, Amenábar offers several close-ups of the diagram that Ramón made of a modified wheelchair. The images of these drawings are of very little narrative importance but their function as part of a larger argument is to provide further empirical evidence of Ramón's rational capacity. He can argue, appreciate music, write poetry, and engineer a wheelchair—a veritable Renaissance man. It is safe to speculate that, two centuries ago, a similar level of intelligence might not consider suicide as an option, but values have changed because assumptions about human dignity have

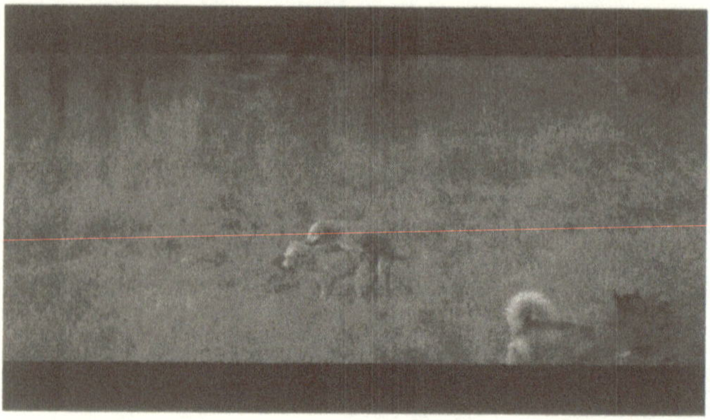

Fig. 3. Like Segisumndo in *La vida es sueño*, Ramón might argue that dogs have more freedom than he, even though he is a rational being protected by democratic law.

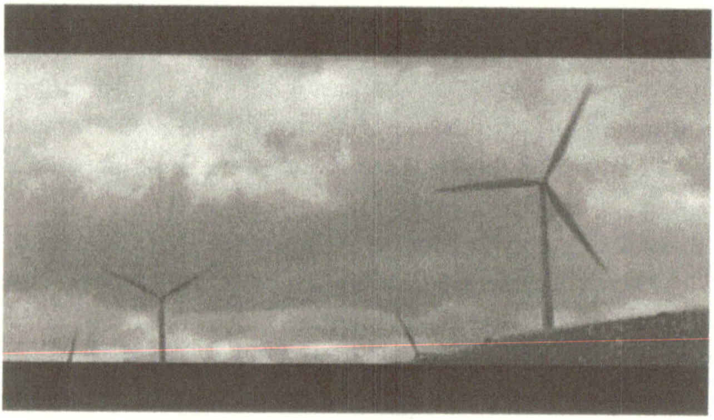

Fig. 4. Wind turbines provide an updated version of the classic symbol of a quixotic quest to transcend the constraints of reality.

changed. Second, en route to the court case, the car drives through the verdant Galician countryside such that the spectator is proffered a series of symbolic images: a mother scolding her child for running away, a woman walking home, a teenage couple sneaking off to the forest holding hands, two dogs copulating, a cyclist. [Fig. 3] [Fig. 4]

The images are not gratuitous. They form a visual list of precisely those physical activities that Ramón cannot do. As the ride comes to an end, a windmill slowly emerges—that perennial Spanish symbol of unrealistic ideas. The activities of biking, walking, running, and sex are but a chimera for Ramón,

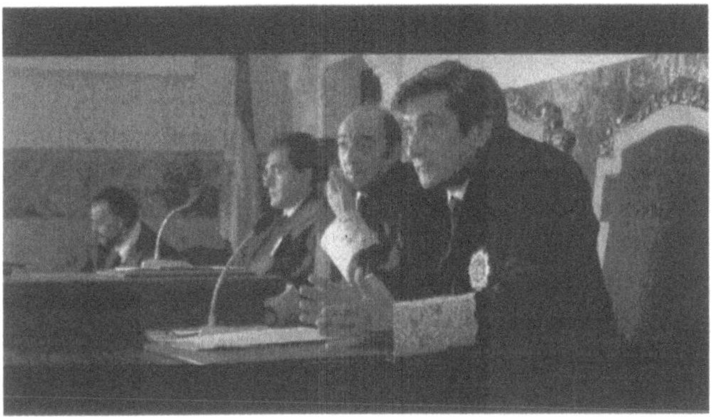

Fig. 5. The sartorial appearance of the Spanish judges functions as an index of the disjunction between current law and popular opinion.

never to be achieved. The world of physical activity is celebrated but also simultaneously off limits, a microcosm of Ramón's condition. Here we are expected to recognize along with Ramón, that to be cut off from physical world is tantamount to a death sentence. He is, in his own words, "a head attached to a corpse" (qtd. in Davison). The point seems to be, as Ramón observes the natural world and its inhabitants, that physical activity is necessary to feel complete and inform our judgments about what constitutes life. It is view that challenges the idea of human value and dignity as rooted in notions of natural law and Judeo-Christian metaphysics both.

The third portion of this filmic sequence is the court hearing itself. Ramón's inability to participate in the physical activities is, to him, degrading. The result is a loss of dignity. It is argued that the state has a duty to protect human dignity. In conferring on Ramón's condition, Marc (Francesc Garrido), Ramón's lawyer, requests that *reason* prevail, not metaphysical beliefs about the sacredness of life. The judges, meanwhile, stand as the embodiment of Enlightenment *reason* and natural law. Through mise-en-scène, their laced robes and ornate velvet chairs all point to a political system based on centuries-old ideas about human nature and ultimate values. [Fig. 5]

The pivotal moment in terms of philosophical differences takes place when the lawyer requests that Ramón be able to plead his case in person. The response given is curt: this is not the place ("No ha lugar"). Court procedure will not allow Ramón to speak. Here the film is true to life: Ramón Sampedro's case was denied due to a geographic technicality—he should have started his petition in Galicia, not Barcelona. However, on a symbolic level, what the court sequence highlights is a fundamental divide between two forms of logic, one that is steeped in ideas about natural law and natural rights versus another that views life as closely tied to physical capacity and total personal autonomy. On the latter view, the body is not sacred. It is a biological fact and a possession, like a house or a car. This is the crux of the matter: that one definition of liberty, rooted in an Enlightenment worldview and hinging on speculative reasoning and metaphysics, precludes another, based on an empirical view of the body as a socio-biological fact. Put succinctly, if man does not possess a soul, is not "endowed by a Creator," but is a socio-biological being, then he is not bound by any teleological constraints. Life is just a period in which the organism lives and breathes; it has no purpose to fulfill. On this view, if that organism is in pain and no longer wants to continue in its existence then why not assist it ending that painful existence? The film as a whole, as Minich notes, "has negative implications for the political inclusion of people with disabilities" (30). But the problem of political inclusion is not simply an inability to incorporate minority identities; there is a larger issue at stake that has to do with opposing views of the ends of man and their corresponding ideas of justice and liberty on the whole. Consequently, there exists a tension in defining liberty in societies such as Spain's, containing as it does a citizenry that is pluralist, with some members holding a sacred view of man and others holding an empirical view of existence.

Berlin writes that, "in a world where a man seeking happiness or justice or freedom (in whatever sense) can do little, because he finds too many avenues of action blocked to him, the temptation to withdraw into himself may become irresistible" (185–86). This is precisely what took place with Ramón Sampedro, both in real life, and as filmic representation. Enlightenment definitions of "liberty as life" do not leave space for Ramón's specific case. Without the political freedom to be the author of his own values and then carry them out, Ramón withdrew into himself. The title, *Mar adentro*, alludes to this fact. The irony is that when reason before authority failed to provide an adequate solution, Ramón sought a rather anti-Enlightenment, even romantic solution to his dilemma: he petitioned assisted suicide from his friend Rosa on the basis of love. He convinced Rosa that if she loved him she would respect his autonomy and that she would help him die. Ultimately it was love that allowed Ramón to pursue freedom as he saw fit and end his life.

The idea that helping someone to commit suicide may be an act of love is not limited to *Mar adentro*. Released just seven months after Amenábar's film, *Tapas* (José Corbacho and Juan Cruz 2005) also screens assisted suicide. While *Tapas* is of a lower production quality and less renown, it does gloss several important social questions, such as Chinese migration to Spain, health care and aging, and assisted suicide. Five stories intertwine in the tapas bar of a small neighborhood in Barcelona. One of the stories is that of an elderly couple, Don Mariano (Alberto de Mendoza) and Doña Conchi (María Galiana). Don Mariano is dying of cancer. Doña Conchi is selling drugs out of the local bar in order to pay for them to go to another country to receive what they believe will be better medical care. Sensing that his death is inevitable, Mariano reveals that he wants to die on his own terms. Conchi, who has access to the necessary drugs but who is initially against the idea (though not vehemently), eventually agrees. They share a fine dinner and Mariano provides gifts for both of them. It is more of a celebration than a somber event. Unlike *Mar adentro*, the 2005 film does not engage with the political issue of freedom at all. Questions of morality and legality are not deterrents either. The question of assisted suicide is treated sentimentality: if the person who you love is sick and wants to end his life, then you assist them, out of love. This is the message that one may extract from *Tapas*.

While both films treat the issue of assisted suicide at a crucial moment in Spanish society, they cannot be lumped together. While *Tapas* is interested in exploring in brief the day-to-day realities of contemporary Spanish life, it is ultimately sentimental. *Mar adentro*, on the other hand, is a filmic argument in favor of total personal autonomy. It is somewhat surprising that the movie does not end with Ramón's final monologue and suicide. We are instead reminded that there were originally two people aiming to commit suicide: Ramón and his lawyer, Julia. The latter suffers from CADASIL (Cerebral utosomal dominant arteriopathy with subcortical infarcts and leukoencephalopathy), an inherited condition that causes strokes and other impairments. The National Institute of Neurological Disorders and Stroke (United States) describes the symptoms as follows: "CADASIL is characterized by migraine headaches and multiple strokes progressing to dementia. Other symptoms include cognitive deterioration, seizures, vision problems, and psychiatric problems such as severe depression and changes in behavior and personality. Individuals may also be at higher risk of heart attack." Julia, knowing what the future likely would hold, had committed to providing assistance to Ramón before taking her own life. For reasons that are not made explicit in the film, Julia backs out, leaving Ramón in despair. After Ramón's death, the film cuts to a scene in which Gené (Clara Segura), Ramón's close contact from the Death with Dignity group, brings a letter to Julia and comments about how he goes on living through letters even though he is physically dead. Julia, however, is now the inverse of Ramón – a functioning body attached to a nonfunctional mind. The point seems to be, as Gené's teary reaction indicates, that Julia would be better off dead. Her condition does not constitute life. Life, so it seems, is frolicking on the beach with your family, which Gené proceeds to do immediately after her visit with Julia. This sequence functions as an epilogue that reinforces the overall tension between freedom and functionality, both physical and mental. The implication is undeniable: if one is mentally incapacitated, they would be better off dead, just as Ramón argued that his situation is not "life" because of his physical condition. At minimum, the suggestion that Julia and Ramón both are better off dead is denigrating to someone with the same or similar disabilities, as Minich has pointed out. Either way, what is certain is that the film has a proposition to make about what constitutes life, as Sampedro put it, "Quiero vivir completo, o entero o nada" (qtd. in Prades) (I want to live fully, wholly, or not at all).

On the question of the nature of freedom in liberal democratic Spain, *Mar adentro* clearly puts forth a verdict in favor of personal autonomy with regard to the freedom to end one's life on one's own terms, a verdict that goes against current law. One may justifiably side with our Enlightenment forefathers, the likes of Jefferson, Mill, Locke, Rousseau, those who believed in natural law and its universal truths regarding the aims of life, especially the view that liberty is a universally cherished value, and that in order to be truly free one must be alive. It follows that in order to protect liberty governments must protect its citizens, even from themselves. Here one may even signal a confluence of Enlightenment values and Catholic values. But the film enters the public debate by viewing the abstract issue through the specific case of Ramón Sampedro. This drama humanizes the question and invokes empathy, perhaps expanding an already wide gulf between public opinion and official law. In the end, it tries to resolve the paradox of assisted suicide by appealing to spectator empathy and begging the question: would you want the same right if you were in Ramón's position? This is precisely the reason for which Ramón stares directly into the camera while he offers his final monologue. By breaking the cinematic taboo of making direct eye contact with the spectator the film elicits a response.[9]

As the filmic discourse suggests, we are in the midst of a major shift in how we define human life and values. Pluralism (itself a fruit of Enlightenment freedom of thought) and empiricism have both played a major role in this development. If natural law previously lead thinkers to forbid men from doing that which was destructive to their own lives because of claims to what Rousseau called "absolute values," current beliefs about what constitutes the good life increasingly seem to rest on more empirical accounts of human flourishing, with the body as a primary site of identity and value. In this way we can see a link between Ramón's claim to his body as personal property and issues such as the debate over stem-cell research, modern eugenics, and the expanding field of neuroscience. Stem-cell research aims to improve human life through biological re-creation. Modern scientific forms of eugenics, that is, the active choosing of traits through DNA sorting, is the manipulation of physiology as a means of trying to secure a better life, usually for one's children. Personality disorders are increasingly explained by looking at neural pathways instead of childhood experience. Indeed, this steady but palpable change in empirical understanding of what it means to be human, as well as the critique of values based on metaphysical claims, seems to have upended the Enlightenment ideal of what defines human flourishing. If humans are increasingly looked at as socio-biological beings, not metaphysical ones endowed with certain traits, then the body logically is a major site of exploration and should hold vast importance in how laws are conceived. In a strange way, the body is simultaneously sacred, not as a receptacle for the soul, but as an end in itself and a prized possession, as a brief scan of supermarket magazines make clear. Ramón refers to it in the film as his "greatest good," and if the body is indeed of paramount importance then a hopelessly nonfunctional body would be felt as a form of injustice. This would certainly impact the way in which we consider assisted suicide in Ramón's case. One might even argue that assisted suicide, on the basis of biological functioning and long-term suffering, is not much different from the euthanizing of pets. There is, in fact, a scene in *Mar adentro* in which Ramón's brother, Juan, becomes uncomfortable with the way in which his son, Javí, is participating in the discussions and progress of his uncle's case. Juan pulls Javí aside and reminds his son that if Ramón wins in the courts he will be given an injection and die, "like a dog." Here one can detect a sort of symbiosis in the social movement to allow humans more control over their own deaths at the same time as there is a push to protect animals from cruel treatment and even death, such as with the ongoing debate over bullfighting in Spain. As many human groups increasingly take an evolutionary view of themselves, they also see a need to preserve the lives of the species that share the same DNA.

What is freedom? And what is the government's role in preserving that freedom? To what extent might the government infringe upon one's freedom in order to preserve the best interests of the commonwealth? (How would "best interests" even be defined?) There are, at times, no easy answers. What is certain at this juncture is that *Mar adentro* illustrates how our concepts of liberty are not as stable many of our Enlightenment forefathers believed. Through its narrative discourse and dialogue the film effectively enters the public debate in Spain over assisted suicide by putting forth a filmic argument that defines liberty as freedom from obstruction with regard to one's own body. This marks a subtle but rather important shift in understanding political freedom as aiming to protect *life* as a prerequisite for sharing in the sovereignty of the state. The Spanish government has yet to adopt policies that conform to the now (reportedly) more popular view that is in favor of assisted suicide, one that is reflected in films such as *Mar adentro* and *Tapas*. There is, to be sure, concern with giving increased license to the practice of ending a life because it does not meet a physiological standard of well-being; it might have further implications, such as with long term health care. For now, at least, the shift in ideas of freedom is not official.

Notes

1. Savater and Montero express support for tightly controlled euthanasia while de Prada expresses a more sacred view of life. See: Savater, Fernando. "Lo moral y lo legal." *El País*. 17 Feb. 1998. Web. 21 Jan. 2014; De Prada, Juan Manuel. "Eutanasia." *ABC*. 27 Sept. 2003. Web. 21 Jan. 2014; Montero, Rosa. "La vida." *El País*. 10 Feb. 1998. Web. 21 Jan. 2014.
2. See Ana Alfageme. "El Gobierno anuncia una ley de muerte digna y cuidados paliativos." *El País*. 20 Nov. 2010. Web. 21 Jan. 2014.
3. See "Amenábar considera insensate animar a morir y obligar a vivir" *El País* 2 Feb. 2005. Web. 21 Jan. 2014.
4. Similar comments may be found in reader posts on sites such as nytimes.com and imdb.com.
5. Smith, while not taking away from Amenábar's cinematographic flare, has drawn attention to the film's relationship with television formats. See pp. 105–121.
6. "Various opinion polls since 1995 have shown that a majority of Spaniards favor decriminalizing assisted suicide." See "A Suicide Tape in Spain" and Guerra, p. 432.
7. It is for precisely this reason that Minich critiques *Mar adentro*: "Indeed, *Mar adentro* is a representation that, despite containing an apparently positive portrayal of a person with a disability, has negative implications for the political inclusion of people with disabilities" (30).
8. In 2001 in Rothenburn, Germany, Bernd-Jurgen Brandes responded to an internet ad requesting someone to be willingly killed and eaten. The man who placed the ad was Armin Miewes, a 42-year old computer expert. He had always had the fantasy of killing, frying, and eating another human. No monetary compensation was offered. An agreement was eventually reached. Meiwis killed Brandes, cut up the body, and stored it in bags in his freezer. He had cooked (with garlic) and eaten approximately 42 pounds of his victim by the time he was arrested. Meiwes could not be convicted of murder because Brandes consented to be killed (and eaten). He could only be convicted of assisted suicide, which under German law carries a maximum penalty of five years in prison. Eventually he was imprisoned for life for manslaughter. See,

for example, Luke Harding, "Victim of cannibal agreed to be eaten," *Guardian*. 3 Dec. 2003. Web. 10 Jan. 2014.

9. See P. Begin, "When Victim Meets Voyeur: An Aesthetic of Confrontation in Hispanic Social Issue Cinema" *Hispanic Research Journal* 9.3 (June 2008): 261–75.

Works Cited

Anon. "A Suicide Tape in Spain Inflames the Issue in Spain." *New York Times*. 9 March 1998. Web. 27 Dec. 2013.

Aristotle. *Nichomachian Ethics*. Trans. Joe Sachs. Newburyport, MA: Focus, 2002.

Berlin, Isaiah. *Liberty*. Oxford: Oxford University Press, 2008.

Bradshaw, Peter. Review of *Mar adentro* ("The Sea Inside"). *The Guardian*. 11 Feb. 2005. Web. 14 April 2014.

"CADASIL." *National Institute of Neurological Disorders and Stroke*. Web. 21 Jan. 2014.

Davison, Phil. "Why death would be a victory for Ramon: In Xuno de Sieira Phil Davison meets a man who wants to die, but is incapable of suicide. Spanish law decrees that no one can grant his wish." *The Independent*. 3 August 1994. Web. 21 Jan. 2014.

Debes, Remy. "Dignity's Gauntlet." *Philosophical Perspectives* 23 (2009): 45–78.

French, Phillip. "The Last Judgment" *The Guardian*. 13 Feb. 2005. Web. 21 Oct. 2013.

Guerra, María José. "Euthanasia in Spain: The Public Debate after Ramón Sampedro's Case." *Bioethics* 13.5 (1999): 426–32.

Hecht, Jennifer Michael. "Stopping Suicide" *The Chronicle Review* Dec. 6 2013: B10–B13.

Holden, Stephen. "A Lifetime Battle for the Right to Die." Review of *Mar adentro* ("The Sea Inside"). *New York Times*. 17 Dec. 2004. Web. 21 Oct. 2013.

Jefferson, Thomas. "Declaration of Independence." *National Archives*. 10 Jan. 2014.

Locke, John. *Two Treatises of Government*. London: Cambridge University Press, 1970.

Mar adentro. Dir. Alejandro Amenábar. Perf. Javier Bardem and Belén Rueda. Sogepaq, 2004. Film.

Marr, Matthew. *The Politics of Age and Disability in Contemporary Spanish Film*. London: Routledge, 2013.

Mill, John Stuart. *On Liberty*. Oxford: Oxford University Press, 1998.

Minich, Julie A. "Life on Wheels: Disability, Democracy, and Political Inclusion in *Live Flesh* and *The Sea Inside*." *Journal of Literary & Cultural Disability Studies* 4.1 (2010): 17–32.

Prades, Joaquina. "La verdad sobre el 'caso Ramón Sampredro.'" *El País*. 1 Feb. 1998. Web. 21 January 2014.

Smith, Paul Julian. *Spanish Screen Fiction: Between Cinema and Television*. Liverpool: Liverpool University Press, 2009.

Tapas. Dir. José Corbacho and Juan Cruz. Perf. Ángel de Andrés López, María Galiana, Elvira Mínguez and Rubén Ochandiano. Filmax, 2005. Film.

◆ 9

Still Different? Biotechnology, Politics, and Culture in Spain

Sainath Suryanarayanan
and Katarzyna Olga Beilin

In 2011, WikiLeaks cables revealed that the Spanish government had asked the United States government to press Brussels not to enact additional laws regulating biotechnology (William). At the dawn of the third millennium, Spain has distinguished itself among the European Union (EU) and major EU nations in being the most enthusiastic supporter of all sorts of biotechnologies, even as the EU continues to adopt an ambivalent, precautionary stance in light of the risks posed by these emerging technologies of life (e.g., Todt and Luján, Ferretti and Pavone). Proponents of Spain's pro-biotechnology policies situate biotechnology research and development at the center of an emerging high-tech economy, which is denominated as "knowledge-based," and portrayed as unshackled from Spain's retrogressive, anti-Enlightenment past (e.g., *Nature Biotechnology*, Bertero). Critics assert that Spain has implemented biotechnology policies in a hasty manner that renders Spanish society beholden to the interests of powerful multinational corporations and governments, and that this may be viewed as a continuation of Spaniards' historically subservient orientation toward those in power. In this sense, Jorge Riechmann states that "Spain is still different" (*Qué son* 23) and in February of 2014 Eshter Vivas publishes in *El público* an article on GMOs based on the same idea and with a title similar to our essay's: "Transgénicos: 'Spain is different.'"[1] While certain attitudes toward biotechnology in Spain can be, perhaps, interpreted in this way, we show that they result from a complex set of debates and historical experiences, where hegemonic and alternative discourses switch places, and cultural and political changes transform the meaning of enthusiasms and cautiousness towards science. Science and technology appeared as an unwanted good in various discourses connected to progress and civilization since the nineteenth century, and especially in Unamuno's vision, but they were also viewed as desirable, albeit lacking in national quality, by Regenerationists and others who wanted the Europeanization of Spain. In contrast to common belief, emerging historical works show that science and technology under Franco occupied a prominent space in the national economy (e.g., Camprubí, Néstor and Roqué), although political propaganda did not connect it to Spanish national culture. Thus, interpreting the embrace of science and technology in democracy as opposed to its supposed rejection under Franco may be not completely correct. Biotechnology is, however, a particular kind of technoscience, one that is set to transform essential biological qualities of life, and due to this fact the risks are arguably much higher.[2] Additionally in the twenty-first-century economy, biotechnology develops hand in hand with the financial interests of multinationals, which follow a neoliberal vision of the privatization of all resources, including living organisms, with the purpose of maximizing profit.

Biotechnologies are the means to transform and harness "life itself" in heretofore unprecedented ways that involve transfer of genetic and genomic material between cells, often across species. They are playing an increasingly central role in the social, political, and economic trajectories of societies in the twenty-first century. Their power is based on the sheer breadth of applications, from biomedical therapies and enhancements to the "gene revolution" and lab-grown meat in agriculture to novel industrial applications such as biopolymers and biofuels, and engineered nonhuman ecologies. Biotechnological processes and products are not only technoscientific, but also fundamentally cultural, political, and social. In having the capacity to manipulate the genetic bases of life and generate novel life forms, they trouble people's deeply held categories, values, and beliefs about life, and even what it means to be human. Biotechnology can already produce new hybrid forms of life containing the genetic material of different species, including humans. It entails the distribution of certain benefits and risks, which invariably leads to political questions about who will benefit, and who will bear the burdens and thus, in Riechmann's words, they become a "nexus of socio-political relations" (*Qué son*). Biotechnologies are social in the sense that their development is occurring hand-in-hand with interconnected processes of neoliberalization and globalization, which are transforming the industrialized societies of the nineteenth and twentieth century into today's "knowledge economies" (Jasanoff, Moore et al.). This shift is concomitant with the emergence of a "neoliberal bioeconomy," where biological materials and living organisms are not simply extractable resources but have become the very foundation upon which markets operate (Pavone). Biotechnologies have become the vanguard of the business of life and key generators of what Kaushik Sunder Rajan calls biocapital (Sunder Rajan). Despite the leveling tendencies of neoliberalizing international and state policies, multinational corporate actors, international researcher networks, nongovernmental organizations, and globally circulating capital, bioeconomies have manifested in a heterogeneous manner in different nation-states and regions, primarily due to differences in the relations between state, science, industry and civil society which are culturally particular and historically influenced (Jasanoff).

While every nation-state offers a potentially unique lens into the political and cultural dynamics of biotechnology, Spain's case is particularly compelling. In contrast to other major EU states such as the United Kingdom, Germany, and France, Spain has consistently ranked amongst the most enthusiastic toward commercializing biotechnologies across the spectrum of "red" (biomedical), "white" (industrial), and especially "green" (agro-food) applications, where the contrast between Spain and the rest of Europe is the greatest. Close to two decades since the first arrival of corporate-owned genetically modified (GM) corn from across the Atlantic Ocean to European shores, citizens and governments of the major EU continue to show deep ambivalence toward the commercialization of GM crops and foods. By contrast, Spain was for a long time the only EU land on which GM crops could be planted, and is today the EU nation with the largest acreage of GM corn, and the third-largest producer of corn in the EU. In 2012, GM corn occupied seventy-six thousand hectares constituting 30 percent of all Spanish corn (Spanish Bioindustry Association [ASEBIO]).

While EU and national-level surveys suggest that Spanish citizens show higher levels of optimism toward issues of science and technology, including toward novel biotechnologies compared to citizens of other EU countries (e.g., European Commission [EC]), the Spanish governments' openness to implementing various biotechnologies has exceeded its citizens' enthusiasm. Furthermore, while Spaniards specify the need for regulation by strict laws as a condition in indicating their approval of emerging biotechnologies such as synthetic biology, as an important difference with some other European countries, their vision of these regulations is based primarily on the advice of scientific experts, as opposed to the arguably more democratic approach of deliberations reflecting public or ethicists' participation (EC).

The question of why Spain is different with regards to its orientation toward novel biotechnologies has invited contrasting explanatory discourses. The most prevalent range of explanations frame Spain as an "emerging" knowledge economy, whose investment in the science and technology as a means of economic growth suffered in the "dark age" of Franco's dictatorship, and which only picked up since integration into the EU (e.g., *Nature Biotechnology*, Bertero). This discourse of Europeanization, promulgated by political authorities, scientists, promoters of the biotechnology sector, and Spanish media, situates the development of science and technology, and biotechnology in particular, at the very heart of the social and economic progress of the democratic Spanish nation.[3] Humanists, artists, and philosophers such as Américo Castro, Federico García Lorca, Miguel de Unamuno, and many others, defined Spanish culture as opposed to the preoccupation of the rest of Europe with materialist science and technology, and as specializing in matters of "spirit, nurtured through the contemplation of death." Thus the Spanish government's enthusiastic commercialization of various biotechnologies is implicitly seen as a Europeanizing process of the Spanish nation that is most welcome. In other words, Spain's emerging bioeconomy contributes to a cultural construction of a nation that no longer wants to be Europe's "other."

These widely held perspectives are countered by more critical ones, propagated by nongovernmental organizations (NGOs), civil society organizations (CSOs), and academics such as Jorge Riechmann, Carlos Amorín, Salvador López Arnal, Francisco Garrido, and others who, although they do not dismiss biotechnology altogether, call into question its trajectory of development as being premised on neoliberal models of incessant growth that exacerbate existing levels of social inequality and environmental degradation. In this alternative frame, Spanish citizens' deference to political and scientific authorities regarding issues of science and technology, and in particular biotechnology, can be seen as a historical repetition of their subservience to those in power. In these counter-narratives, the politics of biotechnology suggests a Spain whose political culture remains as complicit with the agendas of powerful interests as under prior authoritative regimes in Spain's modern history—that is, Spain is still different from Europe. This vision needs to be however problematized through analyses of discourses from Spain's past when authority and science were located at the opposite poles, when hegemonic discourses in culture defined Spanishness as incapable and unwilling to pursue the rational regime of "proof" as in Unamuno's "Sobre la europeización" (on Europeization) from 1906.

Here, we examine contesting narratives about the development of biotechnology in the context of debates on Spanish national culture. We begin by delineating key sociopolitical characteristics of Spain's developing bioeconomy. We then review debates in the Spanish media and among Spanish intellectuals regarding the commercialization of biotechnologies in Spain, identifying distinctions and overlaps among a spectrum of critical voices. Next, we highlight certain crucial phases in the historical formation of modern Spain, which featured contrasting attitudes towards modernity and science, and which, as we show, may have indirectly shaped contemporary discourses on biotechnology.

The Emerging Spanish Bioeconomy

Spain's biotechnology firms are clustered predominantly in the Madrid area and Catalonia, with the rest distributed across Andalusia, Valencia, Galicia, and other parts of the country. This emerging biotechnology "sector" features broadly two kinds of firms: newer small- to medium-sized firms of less than 50 employees, whose core activity is biotechnology, and which are based on American-style start-ups by entrepreneurial scientists; and older industrial firms with 250 employees or less, whose core is based not on biotechnology, but who have nevertheless diversified and entered into the biotechnology business (Díaz et al.). With biotechnology marked by successive Spanish governments as an area of strategic action in the Spanish National Plan for R&D (e.g., see *Nature Biotechnology*), and with support from the EU, Spanish biotech firms are seeking to "to dominate the largest possible number of sectors" in Spain (Díaz et al 37) and increase their presence in a globalizing marketplace. Toward this, biotechnology firms are mobilizing new social relations with universities, public research organizations, investment banking firms and large multinational corporations.

Three key aspects of this evolving web of relationships that shape emerging neoliberal bioeconomies are worth noting in relation to Spain. First is the rise of a new model of public research and development—"the ICREA model"—that seeks to turn Spanish universities and research institutions into ""economic engines." "Established first by the semi-autonomous government of Catalonia in 2001, the Institució Catalana de Recerca i Estudis Avançats (ICREA) as a virtual institute is a highly regarded attempt to improve Catalonia's "innovation performance" by attracting internationally renowned scientists and scholars outside the Catalan education system to conduct their research work in Catalonia (Technopolis 2011). Through ICREA, the Catalan government sought to circumvent perceived institutional, cultural, and economic barriers to recruiting the world's most talented researchers and jumpstarting the Catalan knowledge economy.

In the "traditional" Spanish and Catalan systems of research and higher education, recruitment and employment of researchers in universities and research centers occurs on the basis of "national civil service terms and conditions," which is "endogamic" (Technopolis). Existing academic staff tends to recruit junior academics with whom they have close ties, who in turn end up with long-term positions in one place. Connectedly, higher education institutions are unable to offer competitive salary and research support packages due to rules for national civil service salaries. ICREA has enabled public universities and Catalan research institutes to offer to its recruits "research-only" "permanent" positions with "competitive salaries" and the opportunity to "lead a new research group," 72 percent of whose focus is on science and technology and the rest on the social sciences and humanities (Technopolis). Significantly, the ICREA model of internationalizing Catalan research and development is tightly linked to the spread of an entrepreneurial culture of innovation that seeks to establish stronger connections between "upstream" technoscientific research and "downstream" potential market applications. It actively supports researchers to "patent their inventions," collaborate with industries and private sector enterprises, and generate "spin-offs" (Technopolis). As of 2011, only six standalone Catalan research centers accounted for 63 percent of all ICREA recruits, including the Center for Genomic Regulation (CRG), an international biomedical research institute created in 2000 by the Spanish and Catalan governments in Barcelona. This culture of academic innovation based on public-private collaborations and the commercialization of university research is set to spread to other parts of Spain with efforts such as IKERBASQUE in the Basque region and another in the Madrid region that seek to replicate the ICREA model. ICREA research centers such as the CRG are key primary partners of biotechnology firms looking to develop new products.

It is noteworthy that during a period marked by drastic downturn in key indicators of social and economic wellbeing, Spain's biotechnology sector has shown "highly positive results" in terms of financial and material growth (ASEBIO 2013). This is of course not to suggest that all is well with Spain's biotechnology industry. As the evolving biotechnology sector strives to become central to Spain's society and economy, across-the-board cuts in governmental funding for scientific research create new challenges and opportunities (Vela). The intensifying austerity stands to create the conditions for further rounds of commercialization of Spain's university research environments. It increases the likelihood that Spain's small biotechnology firms—faced with high costs entailed in procuring state-of-the-art equipment, skilled personnel and resources, and in developing a marketable product—will be relying on foreign investors for survival in a highly competitive marketplace. This sets up a situation wherein the R&D programs and associated profits of Spain's biotechnology firms are steered not toward the needs of Spain's civil society, but toward the concerns of far-off venture capital firms and shareholders. Furthermore, the ongoing Spanish economic "meltdown" fuels the conditions for multiple mergers of existing small biotechnology firms and strengthens the biotechnology arms of a few multinational corporations that possess greater reserves of capital and privileged access to powerful actors of national and international governance.

A second important characteristic of Spanish bioeconomy has been the rise of scientized regulation. In emerging neoliberal knowledge economies policymakers adhere to the recommendations of highly specialized scientific advisory bodies, which tend to exclude social and political criteria (Moore et al.). Indeed, scientized regulations have helped push through the commercialization of biotechnologies without much public debate in Spain (Todt, Todt and Luján), even though the regulatory "red tape" entailed in obtaining commercial permits for any particular technoscientific product with ecological and social implications is a perennial source of teeth-gnashing for industry, including in the biotechnology sector (see Díaz et al.). Biotechnology came to have a privileged place in the Spanish science policy of the 1980s, in part due to the "ambitions of successive governments" after the transition to democracy, and also due to "a small but influential group" of biochemists and molecular biologists of international acclaim in the area of biomedicine, who had "close connections with political and scientific authorities" (Luján et al. 182).

The challenge of governing biotechnology emerged in full force after Spain's admission to the European Commission (EC) with the EC's first directives (in 1990) regarding risks related to the containment and release of genetically modified organisms (GMOs) into the environment. In principle, the administrative authority for addressing the EC's directives in Spain lay with the Directorate General of Environmental Policy in the Ministry of Public Works, Transport and Environment. However, in practice, this ministry gave scant weight to environmental concerns; instead, a provisionally formed body called the National Biosafety Commission (CNB), comprising of policymakers, and laboratory-based biochemists, molecular biologists, agronomists, and chemists (but without ecologists), not only took an advisory role but also acted as the "Competent Authority" on GMO release by becoming the "*de facto* policymaker" on the Spanish government's decision to commercialize GMOs in 1996 (Luján et al.; Todt and Luján).

In the aftermath of the EC's 1990 directives on GMOs, the governments of the United Kingdom, Germany, and the EU, instituted a variety of systematic initiatives to enhance public participation of citizens and civil society organizations in debates about GMOs (Jasanoff), although these participatory endeavors in Germany were seen by NGOs and CSOs as being purely legitimation exercises on behalf of those in political power, where invited civil society groups could discuss the risks and benefits of GMOs but "could not call into question those technologies in themselves" (Ferretti and Pavone 291). In Spain, unlike Germany, the United Kingdom, and the EU, it was not until after GMO crops and foods became commercialized that a public debate erupted (Todt and Luján). Apart from the lack of serious consideration of public perspectives in the Spanish government's scientized regulation of biotechnology, there are relatively weak "political opportunity structures" of public participation (Kitschelt, Todt), which, may be considered as an historical legacy of the authoritarian regimes that governed Spain before the transition to democracy.

A third important factor in the unfolding scenario of biotechnology in Spain is a "countervailing" one (Moore et al.), that is, the rise of "new social movements," NGOs and CSOs that are engaging with issues of science and technology. However, for a variety of reasons these movements have not been able to meaningfully influence the development of events in Spain in their struggles for reforms in realms of environmental, ethical, social, and personal health. Since Spain's turn toward democracy a number of environmental organizations have appeared and disappeared inconsequentially. Joaquín Fernández claims there have been between seven hundred and one thousand such small organizations. In his words, "el fervor ecologista" (73) (the environmentalist fervor) was such that a party, PEE (Partido Ecologista Español), was created in 1977, but it failed politically, receiving only four hundred votes in the elections of 1978. A great number of protests, especially anti-nuclear manifestations that passed through Spain in those years can be contrasted to the lack of positive political participation by Spain's publics. This may have been due to a lack of institutions of public participation. When GM foods and crops were commercialized in Spain in 1996, the situation was not radically different. In addition to the lack of institutional infrastructure, the inherited distrust towards political authorities made NGOs and CSOs suspicious that formal participation might end up legitimizing the political authorities' policy decisions as it did in Germany (Todt, Ferretti and Pavone). Excluded from formal discussions about biotechnologies, Spanish NGOs and CSOs have taken a confrontational route in devising public media campaigns that directly address citizens about the risks entailed in various biotechnologies, and calling into question the neoliberal system of technoscientific development on which novel biotechnologies have to date been premised (Ferretti and Pavone). Our analysis of the debate that has occurred in Spanish media since GM products began to be commercialized suggests that, in contrast to institutions of public participation, media culture has been significantly democratized.

Debates on Biotechnology in Spain

Although multiple analyses suggest that the Spanish government's commercialization of GM crops and foods in the 1990s occurred without much sustained public involvement (Todt and Luján), unlike the case of other major European countries (Jasanoff), beginning with the twenty-first century, Spanish journalists, NGOs, civil society groups, and intellectuals were involved in vigorous public debates in media about the intertwined development of biotechnology and society and their contesting visions of a just, democratic, and progressive Spain (e.g., Todt and Luján, Ferretti and Pavone). Based on our analysis of the *El País* archive, it is possible to state that an incipient debate started at the end of the 1990s with a considerable number of articles appearing in 1999. Although a comparative study of different media debates exceeds the scope of this article, it seems significant to note that the debate on GMOs was much more detailed, intense, and polarized in *La Vanguardia*. Between 1999 and 2014, *La Vanguardia* published approximately three times more articles that featured GMOs (over nine hundred) than *El País* (with only two hundred eighty two). Just in 1999, the Catalan newspaper published one hundred forty-two press articles on the subject as opposed to less than twenty in *El País*. Although in both newspapers diverse opinions were represented, *La Vanguardia* tended to highlight dangers rather than benefits while *El País* maintained overall the tone of an "objective" information campaign. Below we highlight some exemplary publications on the subject.

On May 28, 1999, a brief editorial note in *El País* announced that Austria had forbidden transgenic (GM) corn because of emerging research that it kills beneficial insects such as monarch butterflies. On October 29, 1999, an article entitled "El cultivo de los transgénicos crece este año un 44%, según Monsanto" (Plantations of GMOs grow this year about 44% according to Monsanto) informed the public that among the countries which had opened to GMOs were Portugal, Ukraine, and Romania, curiously without mentioning that Spain had done it earlier. The article, however, admitted that debates on the health effects of GMOs were intensifying. On December 15 of the same year, another brief editorial informed Spaniards that a United States-based environmental group, FED (Foundation for Ecologic Development), led by anti-biotechnology activist Jeremy Rifkin in the United States, was claiming that Monsanto had violated anti-monopoly laws. The article concluded that the controversy about GMOs cannot be reduced to a simple difference between the United States and Europe since there is also a growing debate on the other side of the Atlantic. Perhaps the most interesting publication from 1999 in *El País* is an article by Javier del Pino, which informs readers that the multinational agribiotechnology firm Monsanto had promised not to commercialize seeds, called "Terminator technology" by ecological activists, which become infertile after one year forcing growers to buy new seed from the producer at the end of the season.

It was only around 2001 that the wider debate on GMOs began in *El País* (In *La Vanguardia* the most intense debate was in 1999). This year, among the many new words incorporated by the new dictionary of Academia Real Española, such as "zapear" (to zap), "guay" (cool), or "talibán," (Taliban) was the adjective "transgénico" (*El País,* 11 Oct. 2001) (transgenic). Jorge Riechmann's book *Qué son los transgénicos?*, whose purpose was to inform and warn about the risks posed by GMOs, was first published. *El País's* rhetoric remained "balanced," sometimes optimistic and at other times reflecting fear and worry. A great example of the optimism in relation to the desire for modernization can be found in an editorial focused on the newly elaborated list of the most developed countries in the world, with Spain listed as the twentieth, and where human development was connected simultaneously to technoscientific progress and the eradication of poverty. As the article suggested, quoting the United Nations Development Program (UNDP) from 2001, technological development and the eradication of poverty can be simultaneously achieved through genetic modification of crops. In this context, the leading role of Spain is provided by the subtitle of the article: "España se perfila como un líder potencial en cuanto a innovación y adelanto tecnológico, según el informe del PNUD hecho público hoy" (*El País* 2001) (Spain is emerging as a potential leader in terms of innovation and technological progress, according to the UNDP report published today). This publication projects the Spanish government's openness to GMOs as being motivated in part by an ambition to have Spain among the world leaders in human development, according to UNDP's criteria. Interestingly, an interview with Hugh Grant, vice president of Monsanto, by Javier Sampedro from the same year points to a possibly different understanding of the relationship between the concept of human development and openness to biotechnology. In the introduction to the interview, the journalist uses dark metaphors, stating that "El nombre de Monsanto, el gran productor estadounidense de semillas transgénicas, resuena en muchos oídos europeos como un sombrío poder tecnológico que amenaza con pervertir la pureza bucólica de nuestros campos" (the name of Monsanto, the great producer of GM grain based in the U.S., sounds in European ears as a dark technological power which threatens to subvert the bucolic beauty of our fields). During the interview Sampedro leads Grant to admit that his company is not in fact concerned about remedying world poverty but rather about increasing its own profit and that the European demands for transparency forced the company to reluctantly embrace huge changes in its *modus operandi*. Grant insists that it is "a myth" that Monsanto plans to commercialize "Terminator" grains, not because they lack the desire to do so but rather because they did not manage to purchase the firm Delta & Pine that originally patented this technology. The vice president of Monsanto characterizes Europeans' distrust towards GMOs as a misplaced fear caused by the coincidence of mad cow disease, and as an exaggerated preoccupation with what they put on their plates. By the end of the conversation, the Monsanto spokesperson was praising Spain for having a more "scientific" attitude towards GMOs than the rest of Europe and being, next to United Kingdom, the most important partner of Monsanto.

Around 2004 *El País* began to publish articles on successful biomedical experiments involving genetic modifications. One of them features the research of Manuel Serrano, who after postdoctoral practice at the Cold Spring Harbor Laboratory in New York, returned to Madrid to start his own research group at the National Center of Biotechnology and produced mice resistant to cancer. This researcher, who seems to have fulfilled the early dream of Pedro, the main protagonist of Luis Martín Santos' *Tiempo de silencio* (1962) (Time of silence), announced his hopes that through similar genetic mutations humans at high risk of getting cancer could be protected from it in the future. By contrast, around 2004, genetic engineering in agriculture began to be questioned by consumers who were rejecting GMO products in the markets of the most developed countries, including Spain. In May 2004, *El País* published an article entitled, meaningfully, "Monsanto renuncia a comercializar el trigo transgénico por el rechazo del mercado" (Monsanto gives up commercialization of GMO wheat due to the rejection by the market) and subtitled: "Greenpeace celebra la medida como una victoria para el medio ambiente" (Greenpeace celebrates this as a victory for the environment). The difference between the mostly enthusiastic rhetoric employed in articles focused on biotechnology for medical purposes and a mix of enthusiasm and concern when agricultural biotechnology is at stake has been a characteristic of *El País* publications since 2004, with the debate becoming much more intense during recent years. Out of the two hundred and eighty-two articles on the subject, two hundred have been published since 2011.

Both *El Pais* and *La Vanguardia* have featured diverse informational and polemical pieces ranging from narratives of outright rejection to heavily regulated precautionary frameworks to emancipatory narratives of hope, enlightenment and progress. The sheer heterogeneity and number of press articles and publications debating biotechnology suggest a fairly robust and open political culture. Among various voices, we highlight below some key discourses of contention about the politics of Spanish biotechnology in relation to public and environmental health, and social inequality. While echoing enthusiasms and concerns of voices in other places, these discourses also illustrate particular aspects of the political culture in Spain.

The Limits Critique

Intellectuals such as Riechmann, Lopez Arnal, Amorín, and Garrido are troubled not only by the risks posed by biotechnologies for public and environmental health, but also, and most importantly, by the conjoined development of biotechnologies and neoliberalizing, globalizing capital in emerging bioeconomies. The spread of GMOs entails unknown risks for the environment where accumulating toxins can produce long-term synergies with other ambient environmental factors, whose effects are hard to predict (Riechmann). Huge monocultures of genetically modified crops loaded with insecticides, herbicides, and petrochemical fertilizers may be destroying pollinators and other beneficial organisms. Riechmann quotes scientists Quist and Chapela, whose experiments showed that artificially inserted genetic elements are finding their way into remote populations of plants that were originally non-GMO, revealing the limits of technical claims about containing the runaway flow of transgenic DNA. Because growers become reliant on the agri-biotechnology industry as they purchase GM seeds and associated chemical and mechanical technologies needed to fully express the GM traits in their crop, the food chain becomes more and more controlled by a handful of multinational corporations. In 2011, according to Riechmann, only ten multinationals controlled close to 70 percent of the world grain market.[4] GMOs constitute, then, not only a risk to environmental and human health, but also a biopolitical risk of the privatization of "life itself," providing legal bases for the ownership of natural resources that are increasingly administered according to corporate financial interests rather than public and environmental interests.

In all of these fields, the concept of "limits" is crucial. While the productivist projects of biocapital aim at transcending the limits of nature, critics argue that we need to transform our culture to respect natural limits and within those limits take care of existing human and animal life. This discourse, again reminiscent of *Tiempo de silencio*, where Pedro decides that it is the culture and not the genes that needs to change, has been taken up by the political groups to the left of PSOE. In 2013, Izquierda Unida admonished PSOE for their vote against forbidding GM crops in Andalucía (*El* País, 26 Sept. 2013). Likewise EQUO, a green/left party founded by Juan Uralde in 2010, argues for a self-sufficient sustainable economy without GMOs, but with new renewable energies. Here, however, certain complexities arise because some of the renewable energies, notably, are designed to be produced with the help of genetic engineering.

For example, as Alicia Rivera writes in *El País*, transgenic bacteria have recently been patented as a way to produce bio-fuel from marine algae. One of the protagonists of Belen Gópegui's novel *El padre de Blancanieves* faces an ethical dilemma created by a similar transgenic technology. Should he take the job in the multinational which researches a truly promising biofuel that could do so much good, but is administered by a company which benefits from and contributes to a deeply unjust world order? A similar dilemma is faced by environmentalists, because while "green capitalism" seems to be incapable of limiting the capitalistic drive to grow that ends up destroying life-sustaining ecologies, it seems equally impossible to overcome capitalism given the existing political order.

The concept of limits is also meaningful in the context of biomedical research. As biotechnologies blur the boundaries between disease-therapeutics and health-enhancement and generate novel life-forms, they create the conditions for overcoming the limits imposed by our ecological (bio-physical) and evolutionary conditions. For example, the biomedical research on anti-ageing interventions, which uses stem-cell therapies and regenerative medicine, potentially reframes the very process of aging as a "disease." At the same time, new biotechnological therapies entail enormous financial stakes and interests inseparable from capitalistic processes, which for Riechmann are characterized by a fundamental drive to escape all biophysical and biological limits imposed on "the human condition." His opponents, however, argue that human civilization can be viewed as a plight to overcome the limits of the body and that the higher life expectancy is one of its results that nobody deplores.

Riechmann criticizes the Spanish government's enthusiastic adoption of GM crops and foods in deference to pressure from the United States government and multinational agrichemical and seed corporations, which was in contrast to most of the other EU nations. He argues that the Spanish government neither defended consumers nor citizens nor nature, but rather the large agri-biotechnology corporations, and that too in a manner that lacked transparency and public participation. Riechmann calls for limited and strongly controlled use of GMOs based on judicial, ethical, political, and economical frameworks defined by the principle of precaution. The principle of precaution specifies that if a substance suggests risks to human health or environment, in the absence of decisive scientific evidence, it should be considered as harmful and avoided. In the case of biotechnology development in Spain, this principle has been applied by the government rather ambiguously, after, not prior to, the commercialization of GM foods and crops, and with the implicit aim of preempting social conflict (Todt and Luján).

Biotechnological Emancipation

While the intellectuals critical of bioeconomy affected the public opinion through discussions in press, public conferences, university seminars, and alternative media, some of the most famous public intellectuals, such as, for example, Jesús Mosterín, systematically wrote and spoke of biotechnology enthusiastically, insisting in its scientific basis. Mosterín has consistently expressed his trust for a wide range of technoscientific solutions and has refused to criticize the intensifying alliance between institutions of science and transnational biotechnology firms. In an article entitled "Quién teme a los transgénicos?" (Who is afraid of GMOs?), he argues that humans have been genetically engineering organisms for ten thousands of years, crossing different species of fruits and vegetables and "nature" has done the same; hence there is nothing new about this procedure. He states that, according to scientific research, agri-biotechnology has not had any negative effects on human or animal health, while it has significantly increased agricultural production making it possible to feed more people. In an article entitled "El hombre transgénico" (Transgenic man) Mosterín expresses enthusiasm towards the possibility of genetic modification of humans, justifying his lack of fear in a similar way to the case of agricultural products. He claims that human beings are the most genetically modified organisms on earth, being at the culminating point of million years of biological and cultural evolution.

The hope in the possibilities offered by genetic engineering of human bodies has been consistently expressed in Spanish media as well as in articles written by Spanish researchers in international journals. In 2013 *El País* article "Curar células no es diseñar bebés," (Healing cells is not designing babies), Javier Sampedro, who had appeared critical towards GM grains produced by Monsanto, praised British achievements in eliminating harmful genes from fertilized embryos, insisting that it has nothing to do with "designing babies" and that those who cannot see a difference are better off saying nothing. The fact is, however, that no clear criteria are set to distinguish between therapy and enhancement. The very existence of genetic engineering technologies makes deliberations on all its possible uses necessary.

In the debates on genetic modification of embryos, scientists differ on the criteria that would condition future developments. In a feature article of the *Journal of Medical Ethics*, Paula Casal, leading Spanish bioethicist, considers children's lives as a common good of humanity rather than as a private property of the parents-to-be.[5] Rather than enhancing an individual's life, she argues for an enhancement of humanity toward bringing up those lives which are "barely worth living" (724). She believes that this goal could be achieved through the reduction of sexual dimorphism (a phenotypic difference between males and females of the same species) in favor of female traits in all future embryos, because women are genetically less predisposed toward violence than men (this should not be confused with the elimination of males). Casal explains that thanks to higher oxytocin levels in females "the caring impulse spreads to anything that resembles their offspring, making all creatures with disproportionally large heads, hesitant steps, and incompetent vocalisations seem adorable" (723). She quotes examples of a female bonobo, Kuni, who cared for a small starling, of a female chimp, Washoe, who risked her life to save a drowning youngster she hardly knew, as well as the famous case of female elephants that saved baby rhinos from drowning. She points to research showing that while oxytocin is responsible for caring, testosterone is connected to crime. The recent discovery of the "monogamy gene" RS3 334, present in some males but not all, making those who have it into more faithful partners, better fathers, and less frequent perpetrators of crime, leads Casal to suppose that it would be desirable that all males have it. Thinking of the world's future, Casal imagines "Equalia, an imaginary peaceful, social and gender egalitarian society, populated by equally sized and long lived coparenting couples and Dimorphia with its more violent, non-parenting, harem-keepers towering over the females and dying before them" (727). She concludes that there are "good reasons to move closer to a society like Equalia, which lacks various regrettable features typically linked to polygamy, such as inequality, sexual violence, and homicide" (729). Condoning genetic enhancement for reasons of justice, equality and environmental improvement, Casal's discourse is diametrically opposite to that of Riechmann's, even though both philosophers share political values, environmental concerns, feminism and even the belief in the need for animal rights. It is perhaps mainly their divergent attitudes toward science that are responsible for the differences in their views on enhancement. Attitudes towards science have been shaped by Spanish history in a number of complex and often contradictory ways and both Riechmann's and Casal's approaches can be derived from discourses of the past in opposition to the *status quo.*

Spanish National and Political Culture and Technoscience

At the end of the bitter war of 1808 between Spain and the invading armies of France led by Napoleon Bonaparte, victorious Spaniards famously cried out, "Vivan las caenas!" (Long live the chains!), making clear their desire to be free from the French and their associated enlightenment model of liberty. In 1814 this was mainly due to the French invasion, but tensions between the French ways of the Spanish aristocracy and Spanish people had surfaced by the mid- to late eighteenth century, when the royal government intervened in the oligarchical takeover of farmland by ecclesiastical and private interests. Under the French-influenced monarchy, while the emerging bourgeois of merchants and their descendants in Barcelona and Madrid had reasons to be satisfied with the spread of capitalism and the enlightened policies of the monarchy, the great majority of working classes were descending into deeper squalor; and along with the rural peasantry, their loyalties remained firmly with the Church, because they felt that progress was not for them. In 1808, the invading French brought with them more promises of modernization, liberty, and emancipation that they claimed would rid Spain of its woes. But Spaniards, supported by the English, defeated the French and demanded of them to leave Spain. Spain's path toward modern nationhood was thus constituted by Spaniards' rejection of an Enlightenment model adopted by key European nations. The rationale for this rejection can be viewed, however, in mutually opposed ways: as being externally imposed, not liberating at all and deepening people's poverty or, alternatively, as being motivated by the adherence to the authoritarian models of the power.

A second key historical moment appears at the end of the nineteenth century in the form of the Regenerationist movement (regeneracionismo), which brought modern Spain's evolving identity into a more direct dialogue with European Enlightenment and technoscientific modernity. Increasing discontent since the 1870s and the 1898 military defeat, followed by the loss of its last remaining imperial possessions—Cuba, Puerto Rico, and the Philippines—to the United States in the "War of Independence," sparked a heterogeneous and loosely knit movement, regenerationism, among influential intellectuals, who perceived Spain's decline to be the outcome of backwardness in science and technology. Regenerationist intellectuals were deeply influenced by the ideas of German philosopher Karl Krause. His philosophy, in presenting a harmonious reconciliation of rationality and faith, and science and religion, arguably facilitated a bridge between Catholic Spain and a secularizing Europe. Regenerationists such as Joaquín Costa sought to redefine the national "spirit" of Spain and to propose social and economic reforms that were in alignment with European enlightenment ideals of liberalism and technoscientific modernization. Their ideas of a Europeanist Spain gained influence in the early years of the twentieth-century, when political authorities implemented policies and invested in infrastructure to promote the development of science education in schools (Ruiz-Castell), and to raise the level of scientific research. Another contemporaneous set of powerful intellectuals, who came to be known as the "Generation of 1898," were also influenced by Krausism and Regenerationism at the outset, but eventually came to emphasize tradition over universalism, Hispanicization (i.e., "Spain is different") over Europeanization, and literature, art, and philosophy over science and technology, made famous by Unamuno's assertion. "¡Que inventen ellos!" (Let others invent!). While important aspects of the Generation of 1898 thought became appropriated by Falangist ideology, and also idolized by a spate of Hispanist scholarship, contemporary Spain's headlong rush to kick-start its bioeconomy can be connected to the Europhiliac yearnings of nineteenth-century Regenerationists and elements of Spanish society to become a part of a modernizing Europe.

On the other hand, however, and to make the cultural debates with history more complex, it is impossible not to notice certain parallelisms between Unamuno's rejection of science—as a blocking of access to "wisdom" in favor of religion and other forms of spirituality—and Riechmann's belief that even the most cutting-edge science will not forge human happiness, which is to be found in human relations and harmonious coexistence with nature and art. Similarly, the insistence on life within limits and the opposition to genetic transformations of "nature" in order to accommodate human growth should remind us of Unamuno's disdain of material goods. But the similarities end here. In his lecture from 1906, "Sobre la europeización" (On Europeization), published in *La España Moderna,* Unamuno affirms: "Hay dos cosas de que se habla muy a menudo, y son la ciencia y la vida. Y una y otra, debo confesarlo, me son antipáticas" (Recently people talk a lot about life and about science. I have to confess that I do not like either). Instead, he claims that the object of wisdom is death and to be wise it is necessary to prefer death to life. He rejoices in the fact that Spaniards are not part of European modernity, as some foreigners claim, because as he suggests, modernity is not a culture of death, which is so important for Spaniards. He literally says that it is not possible to "morir bien" (die wisely) outside of Spanish culture, that is, having abandoned it for modernity. This leads Unamuno to certain expressions of cultural superiority as he classifies modern Europeans as "desgraciados" (disgraced) or even "hombre[s] que no son hombre[s]" (humans that are not humans) because they attempt to find wisdom in life and in science which is not possible.[6] Furthermore, Unamuno aestheticizes poverty and is reluctant to consider the need for politics to alleviate it. To the contrary, life and not death is the greatest value of the environmental philosophy of Riechmann, activist writings by Vivas and their collaborators. Those in the environmentalist left worry that the alliance neoliberal economy and science will harm life, increase poverty, and intensify the dependence of poor people on powerful multinationals. Their philosophy is egalitarian and political. Although critics of bioeconomy believe that it is necessary that the rich limit their material consumption, it is for the sake of sufficiency for the poor and for future generations. In this way, while Riechmann-like views can be connected to the discourses which rejected the violence of Enlightenment (Long live the chains!), which criticized the imperfection of scientific "proof," and the incapacity of science to produce "spiritual" wisdom (Unamuno), these views need to be also connected to the emancipating political and vanguard ethical rhetoric; equal distribution of resources, protection of health, and consideration of the rights to life of all vulnerable creatures. Similarly, Paula Casal's vision has heterogeneous discursive roots. While her enthusiasm for enhancement can be traced to pro-Enlightenment discourses and Regenerationism, her fantasy of "Equalia" achieved through the reduction of sexual dimorphism reminds us of anarchists' visions of the ideal society.

While the (in)famous slogan "Spain is different" was part of a governmental campaign under Fraga in the 1960s to promote Spain as an attractive tourist spot, symbolized by exotic flamenco dancers, sexy bullfighters, and sultry beaches, it nevertheless captured a series of acts and attitudes since the eighteenth century, which, according to historian Américo Castro, could be distinguished by their enduring intimacy with death, where the principle theme of "man as a naked and absolute reality" overshadowed prevalent European discourses of material progress (Castro). Franco's regime capitalized on these discourses, propagated by Falangist intellectuals, who once read Unamuno and who took as their point of departure developments prevalent in the rest of Western Europe, to project Spain as the Oriental Other that western tourists could experience in their own proverbial backyards without having to give up their European tastes. The implications of this slogan were much broader than just for Spain's tourism industry; it became an enduring strand in the construction of "Spanishness" itself. An overemphasis on the spiritual essence of Spanish civilization by leading humanists and Spanish political authorities in the latter half of the twentieth century helped to maintain Spain's difference from the technologically obsessed but spiritually inferior "West". In creating a discursive split between spiritually inclined literature, humanities, and art on the one side and materially inclined science and technology on the other, the "Spain is different" image has conversely fueled linear narratives of the progressive transformation of Spain from its "dark age" of regressive science and technology under Franco to today's enlightened and EU-integrated knowledge economy (*Nature Biotechnology*; Bertero).

But emerging historical studies about scientific and technological research during the fascist dictatorship suggest a more complex relationship between the development of science, technology and the nation of Spain than the widely promulgated linear narrative. Spain under Franco charted a path of national, technoscientific, and cultural development that emphasized autarky and difference. Scientists who did not fit in the regime's autarkic projects were exiled, expelled, and in this sense, intellectual development of science and technology indeed was suppressed (e.g., Santesmases and Munoz). At the same time, however, the political and economic constraints imposed on scientific institutions by Franco's regime also presented an opportunity for other sciences and scientists to become active participants in engineering the various nation-building projects of the fascist dictatorship (e.g., Camprubí ; Nèstor and Rocquè). Camprubí traces the coevolution of a "vertically integrated system" (499) of rice production and distribution with the development of new techniques of artificial rice breeding and genetics by Spanish agricultural scientists between 1939 and 1952. Underlining the importance of scientific and technological development for Franco's autarkic ambitions, Camprubí shows how "agricultural scientists were active agents within the corporatist political economy, promoting their plans for the nation and for rice improvement at the same time" (500). Similarly, Néstor and Rocqué dispel the myths that Franco's regime did not care about science and technology and was even opposed to it, in their analysis of "the rise of modern physics during Francoism (1939–1975)." They argue that early Francoist policymakers utilized an array of repressive measures (war, exiles, purges) to shape the development of physics "around military-related and applied fields such as optics and the material and nuclear sciences" and to incorporate physics into the political and religious ideologies of the regime, "changes to which physicists actively contributed" (207). Prominent physicists such as Carlos Sanchez del Río and José Maria Otero criticized the atheist separation of Science and Religion and articulated their visions of a "scientific humanism," which "will only be achieved through the integration of science in the Christian scheme of the world" (Néstor and Rocqué 222). In other words, a variety of projects of scientific and technological research related to agricultural genetics, physics, etc. during Francoism developed in tune with the national project of gaining economic self-sufficiency, i.e., autarky. Even though that project of autarky failed, the ensuing science and technology projects were highly productive, with the research of Spanish physicists and biochemists gaining recognition at both national and international levels. Significantly, scientists themselves were active agents in the process of autarkic state-building that did not eschew international exchanges completely. The newly emerging strands of historical research thus suggest that neither was Francoist Spain backward in scientific research and technological development, nor were key institutions of science and technology aloof from the political developments occurring under Franco. In other words, technoscientific research institutions and scientists were key cobuilders in Franco's nation-building project of autarky, just as the biotechnology sector and its scientists today are key collaborators of powerful political authorities in the intertwined development of Spain as a neoliberal bioeconomy. In conclusion, the opposition between Francoism and science, which motivated the enthusiasm for science in post-Franco Spain as a way of overcoming the oppressive ways of the Francoist era, is not justified. It has existed more in cultural discourses than in economic reality.

The democratic transition saw Spain becoming "the most Europhiliac" of EU member states (López and Rodríguez). Spain's "almost completely uncritical" "Europeanism" (López and Rodríguez 12) extended to its development of technoscience and, in particular, to biotechnology. The transition to democracy in the 1970s oversaw the creation of "new state agencies for research based on the recommendations of the OECD [Organization for Economic Cooperation and Development], and the growth of a university system" (Nèstor and Rocquè 233), which led to the dismantling of the research councils that had been created under Franco and a curtailment of the Catholic Church's influence in the state's decisions. The scientific research that developed in the newly democratic Spain was more outwardly oriented; publishing in international journals, international collaboration, and the elevation of basic research over applied research became the norms (Néstor and Rocqué). That said, research and development did not feature prominently in the democratic Spanish state's macroeconomic policy, which, although tailored to adapt to the emerging global economy, remained dependent on Franco's legacy of mass tourism, property development, and construction (López and Rodríguez). By the time Spain had integrated with the EU in 1986, the government had constructed the National Center for Biotechnology and earmarked biotechnology development in its science policy with the aim of becoming a knowledge-based economy on par with the most advanced nations of the EU. The "difference" that the biotechnology sector and the Spanish government sought to highlight in marketing campaigns that promoted the expansion of the Spanish bioeconomy was no longer with Europe, but with Spain's anti-Enlightenment past (e.g., *Nature Biotechnology*). Indeed, by the 1990s and 2000s (before the recession of 2008), amidst flourishing construction and housing sectors, Spain's booming economy was seen by EU's commentators to have bettered the rigidities of "old Europe" (López and Rodríguez). Spain was again "different" from Europe, this time not in a retrograde and reactionary fashion, but as the leading economic edge of the EU. This leadership, however, had been achieved at the cost of putting life at a risk, whose seriousness is an object of intense debates, as the material progress that had been attained may turn out to be damaging in many ways.

Conclusion

It is in the context of Spain's shifting positioning vis-à-vis Europe that we can locate its different orientation toward all sorts of biotechnologies. The nation of Spain, "born again" into the European fold of technoscientific modernization, is perhaps akin to a convert, who ends up exceeding the norms and expectations of the community that the convert seeks to be a part of. Seen in another way, ever since its democratic transition, Spain has been in a hurry to become one of the leading knowledge economies of "the West," and this is arguably why the Spanish government commercialized biotechnologies without much of a sustained public debate, and has subsequently made an aggressive push to develop Spain's bioeconomy, again without seriously taking into heed the critiques and alternatives posited by NGOs and CSOs. This has ironically occurred at a time when the EU and other major western European nations have been engaged in much hand-wringing about the very same biotechnologies. Notwithstanding the preeminence of United Kingdom, Germany and France in the EU's bioeconomy, their reluctance toward agri-biotechnologies, synthetic biology and animal cloning, all of which are relatively well accepted in Spain, has provoked enthusiastic commentators to aver that Spain is forging ahead even as the rest of Europe is being left behind.

Observing Spain's implementation of agri-biotechnologies, Riechmann presciently wondered if in fact Spain is identical to what it has always been, submissive to powerful authorities, and not challenging them sufficiently. Building on Riechmann, we have argued in this paper that the political and institutional legacies of successive authoritarian regimes shaped the development of a political culture that is subservient to the agendas and interests of those with political power, contributing to the subsequent lack of public debate and participation surrounding the commercialization of biotechnologies. In other words, Spain is no longer different in the way this difference was defined in oppositional terms to the rest of Western Europe. However, Spain may still be different, in a more fundamental sense of a political culture that is subservient to regimes of power. The regimes of power have shifted, they are no longer connected to authoritarian monarchy or dictatorship allied with Catholic Church but rather to globalized multinational interests. The authority is constructed by the expertise of scientists codesigning a bioeconomy and the scientized regulations that they create. In attempting to erase its marks of otherness in relation to "the West," Spain may have ironically ended up repeating the same pattern of relationships between political authorities and civil society.

But, there is also a different possible way of interpreting Spain's recently achieved leadership in biotechnology, namely as a result of yet one more introspection during the years of the Transition which has led it to exceed EU norms by measuring itself to its anti-Enlightenment past and not worrying as much about what other countries were doing.[7] In contrast to the project of the Generation of '98, during the Transition and democracy Spain's self-interrogation ends up being materially progressive, but it happens at a time when it becomes clear that further material progress is damaging in many ways, when in other words, it is not really progressive anymore. This realization, which can surprisingly be also traced to the Generation of '98 is now taking the form of environmental and political contestations embraced by new social movements and the new political left.

Relations between Spain's civil society, political structures of power, and scientific research institutions may be entering a new phase with the bursting of Spain's economic bubble. Since 2008, Spain has been caught in a deepening economic recession, leading it to register the highest levels of economic inequality and rates of unemployment amongst all the twenty-eight EU nation-states.

Madrid's Puerta del Sol became the epicenter for the rise of the *Indignados*, a new social movement echoing the "Arab Spring" and Occupy movements in the Middle East and North America respectively. The *Indignados* call into question the neoliberal model of democracy that is being instituted in Spain, "because the parties in power do not look out for the collective good, but for the good of the rich. Because they understand growth as the growth of businessmen's profits, not the growth of social justice, redistribution, public services, access to housing, and other necessities" (Beatriz García, quoted in López and Rodríguez). The *Indignados*' broad critique resonates with critical narratives of the mechanisms and institutions upon which the development of Spain's biotechnology sector has been premised (e.g., Riechmann). In the words of López Arnal, "La cuestión política de fondo es nada más, y nada menos, que preguntarse quién controlará la biodiversidad, los recursos genéticos, las fuerzas de la vida, y en beneficio de quién, es decir, el combate entre quienes están a favor y quienes estamos en contra de la creciente privatización de la vida y de los procesos vitales" (The political question is to ask who and for whose benefits will control the biodiversity, the genetic resources, the forces of life. The battle is between those who are in favor and those of us who are against the growing privatization of life and vital processes).

Notes

1. At that time our essay had just been submitted to HIOL, but we considered this coincidence of visions with the famous journalist as a sign that we were on the right track.
2. Technoscience is a concept used in science and technology studies. It points to mutual codependence of science and technology as they historically emerge and as they function in present day society.
3. "Indeed, the Spanish Parliament has unanimously voted 2014 to be the Biotechnology Year in Spain."
4. See also the website of GMwatch for data.
5. This vision emerges from work by Robert Sparrow that deliberates on parents' right to choose their future baby's gender.
6. "¡Desgraciados países esos países europeos modernos en que no se vive pensando más que en la vida!" (Disgraced are the modern European countries where people live without thinking of death.)
7. We owe this reflection to William Viestenz.

Works Cited

Amorín, Carlos. *Las semillas de la muerte. Basura tóxica y subdesarrollo: el caso de Delta & Pine*. Madrid: Catarata, 2000.

Bertero, Michela. "Science Attracts Spain." *EMBO Reports* 10.8 (2009): 817–19.

"Biotechnology in Spain." *Nature Biotechnology* 24.9 (2006): S3–S13.

Camprubí, Lino. "One Nation: Rice Genetics and the Corporate State in Early Francoist Spain (1939–1952)." *Historical Studies in the Natural Sciences* 40.4 (2010): 499–531.

Carlos Amorín, *Las semillas de la muerte. Basura tóxica y subdesarrollo: el caso de Delta & Pine*. Madrid: Catarata, 2000.

Casal, Paula. "Reform, not destroy: reply to McMahan, Sparrow, and Temkin." *Journal of Medical Ethics* 39.12 (December 2013): 719–41.

Castro, Américo. *The Meaning of Spanish Civilization*. Ed. José Rubia Barcia. Berkeley: University of California Press, 1977.

Díaz, V., E. Muñoz, J. Espinosa de los Monteros, and Jacqueline Senker. "The socio-economic landscape of biotechnology in Spain. A comparative study using the innovation system concept." *Journal of Biotechnology* 98 (2002): 25–40.

"El cultivo de transgénicos crece este año un 44%, según Monsanto." *El País* 29 Oct. 1999, Web. 10 Dec. 2012.

"El nuevo diccionario de la Real Academia incluirá terminos como 'guay,' 'zapear,' o 'talibán' " *El País* 11 Oct. 2001. Web. 11 Dec. 2012.

European Commission. "Europeans and Biotechnology in 2010: Winds of Change?" Luxembourg: Publications Office of the European Union, 2010.

Fernández, Joaquín. *Ecologismo español*. Madrid: Alianza Editorial, 1999.

Ferretti, Maria Paola, and Vincenzo Pavone. "What Do Civil Society Organisations Expect From Participation in Science? Lessons From Germany and Spain on the Issue of GMOs." *Science and Public Policy* 36.4 (2009): 287–99.

García Lorca, Federico. "Teoría y juego del duende." 1932. *Biblioteca Virtual Universal*. Web. 12 Oct. 2009.

Garrido, Francisco Javier. "Biotecnología S.A. Una aproximación sociológica." *Política y Sociedad*, 39. 3 (2002): 641–59. Web. 22 Dec. 2012.

Gópegui, Belén. *El padre de Blancanieves*. Barcelona: Anagrama, 2009.

Grant, Hugh. Interview with Javier Sampedro. "La información sobre biotecnología debe ser transparente." *El País* 13 Oct. 2001.Web. 19 Dec. 2012.

"Grupos agrarios y ecologistas demandan a Monsanto por las semillas transgénicas" *El País* 15 Dec, 1999. Web. 23 Jan. 2013.

"IU reprocha a los socialistas que voten contra el pacto de Gobierno: El PSOE rechaza prohibir los ensayos con transgénicos en Andalucía, como el pedía su socio." *El País* 13 Sept. 2013. Web. 2 Jan. 2013.

Jasanoff, Sheila. *Designs on Nature: Science and Democracy in Europe and the United States*. Princeton: Princeton University Press, 2005.

Kitschelt, Herbert P. "Political Opportunity Structures and Political Protest: Ant-Nuclear Movements in Four Democracies." *British Journal of Political Science* 16 (1986): 57–85.

"La 'fuga de cerebros' hace perder miles de millones de dólares a los países en desarrollo." *El País* 10 July 2001. Web. 30 Dec. 2012.

López Arnal, Salvador. "Conocimientos científicos y decisiones políticas." *Aporrea* 17 Dec, 2011. Web. 23 Dec. 2012.

"La moratoria contra los transgénicos abre una nueva pugna comercial entre la UE y EE.UU." *La Vanguardia* 26 June 1999. Web. 13 Nov. 2013.

Lucas, Ángeles. "Equo presenta una economía sostenible y de autosuficiencia." *El País* 8 Mar. 2012. Web. 23 Nov. 2013.

Luján, Josè Luis, Orlando Mirabal, Daniel Borrillo, María J. Santesmases, and Emilio Muñoz. "Spain: Transposing EC Biotechnology Directives Through Negotiation." *Science and Public Policy* 23.3 (1996): 181–84.

Martín Santos, Luis. *Tiempo de silencio.* Barcelona: Seix Barral, 1999.

"Monsanto renuncia a comercializar trigo transgénico por el rechazo del mercado." *El País* 24 May 2004. Web. 22 Dec. 2013.

Moore, Kelly, Daniel Lee Kleinman, David Hess, and Scott Frickel. "Science and Neoliberal Globalization: A Political Sociological Approach." *Theory and Society* 40 (2011): 505–32.

Néstor, Herran, and Xavier Roqué. "An Autarkic Science: Physics, Culture, and Power in Franco's Spain." *Historical Studies in the Natural Sciences* 43.2 (2013): 202–35.

Pavone, Vincenzo. "Ciencia, neoliberalismo y bioeconomía." *Revista Iberoamericana de Ciencia, Tecnología y Sociedad* 7.20 (2012): 145–61.

Pino, Javier del. "La multinacional Monsanto renuncia a comercializar semillas transgénicas 'Terminator.'" *El País* Oct 1999. Web. 20 Dec. 2012.

Riechmann, Jorge. *Qué Son Los Alimentos Transgénicos?* 2002. Barcelona: RBA, 2011.

_____. *Transgénicos el haz y el* envés. Madrid: Catarata, 2004.

_____. "Hacia una agroética: Consideraciones sobre la ética ecológica y actividad agropecuaria." *Cuidar la T(t)ierra. Políticas agrarias y alimentarias sostenibles para entrar en el siglo XXI.* Barcelona: Icaria, 2003.

_____. *Interdependientes y ecodependientes.* Barcelona: Proteus, 2012.

Rivera, Alicia. "Bacterias transgénicas para producir biocombustible de algas marinas." *El País* 19 Jan. 2012. Web. 13 Feb. 2013.

Rodríguez, Emmanuel. "The Spanish Model." *New Left Review* 69 (2011). Web. 31 Jan. 2014.

Ruiz-Castell, Pedro. "Scientific Instruments for Education in Early Twentieth-Century Spain." *Annals of Science* 65.4 (2008): 519–27.

Sampedro, Javier. "Curar células no es diseñar bebés." *El País*17 June 2013. Web. 15 Oct. 2013.

Santesmases, María Jesus, and Muñoz, Emilio. "Scientific Organizations in Spain (1950–

1970): Social Isolation and International Legitimation of Biochemists and Molecular Biologists on the Periphery." *Social Studies of Science* 27.2 (1997): 187–219.

Spanish Bioindustry Association (ASEBIO). "Report: 2012 News and Trends from the Spanish Biotech Sector." June 2013. Web. 15 Jan. 2014.

Sunder Rajan, Kaushik. *Biocapital: The Constitution of Postgenomic Life*. Durham: Duke University Press, 2006.

Technopolis. "Evaluation of ICREA 2001–2011." Mar. 2011. Web. 24 Jan. 2014.

Todt, Oliver. "Social Decision Making on Technology and the Environment in Spain." *Technology in Society* 21 (1999): 201–16.

Todt, Oliver, and José Luis Luján. "Spain: Commercialization Drives Public Debate and Precaution." *Journal of Risk Research* 3.3 (2000): 237–45.

Unamuno, Miguel de. "Sobre la europeización." *La España Moderna*. Madrid, 1906.. Web. 12 Dec. 2012.

Vela, Carmen. "Turn Spain's Budget Crisis Into an Opportunity." *Nature* 486.7401 (2012): 7.

William, Nigel. "EU Challenged on GM Crops." *Current Biology* 21.2 (2011): R53–R54.

PART IV

Reassembling the Archive through the Concept of Life

10

Iberian Cultural Studies beyond the Human: Exploring the Life History of Marcos Rodríguez Pantoja in Spanish Anthropology and Popular Film

Daniel Ares López

6. "Marcos" portava, a l'hora de produir-se l'abandó, un bagatge cultural que li possibilità, juntament con la iniciació del vell cabrer, la supervivència i la integració a aquell medi. 7. Tot fa suposar que la vinculació de "Marcos" amb la naturalesa era tan profunda que arribà a sentir-se un element més de d'aquesta realitat. 8. "Marcos" aprèn el llenguatge—a partir de les formes lingüístiques que domina—del lloc on habita: Els renous amb els quals ell se identifica els diversos animals, el cant del ocells y el crit del cérvol; els renous amb el quals ell creu que s'entenia amb la guineu, amb els llops o amb la serp. 9. Però sap que existeix una clara diferència entre ell i els animals. Ha vist que pot servir-se de les mans i que li és possible pensar. És conscient d'aquesta superioritat i creu que també ho són els animals.

—Gabriel Janer Manila[1]

(6. Marcos had, when he was abandoned, a cultural experience that allow him, along with the initiation of the old goatherd, his surviving and integration in that environment. 7. We can deduce that the connection of Marcos with nature was so deep that he got to feel about himself as one more element of that reality. 8. Marcos learnt the language—starting from the linguistic forms he already knew—of the place he lived in: the sounds with which he identifies the diverse animals, the birds' songs and the yell of deer; the sounds with which he believed he communicated with the fox, with the wolf, or with the serpent. 9. However he knew there was a clear difference between him and the other animals. He saw that he could use the hands and that it was possible for him to think. He was conscious of this superiority and he thought the animals also were.)

> I will call it "anthropodenial": a blindness to the humanlike characteristics of other animals, or the animal-like characteristics of ourselves.
>
> —Frans de Waal[2]

> To knot companion and species together in encounter, in regard and respect, is to enter the world of becoming with, where who and what are is precisely what is at stake.
>
> —Donna Haraway[3]

> Yo lloraba y se tiraban a mí dando saltos y me cogían los brazos con la boca hasta que yo reía; luego, me señalaban el camino hasta la cueva de ellos, la lobera, y desde allí yo ya sabía irme.
>
> —Marcos Rodríguez Pantoja[4]

> (I cried and they jumped up at me and grabbed my arms in their mouths till I laughed; then they showed me the way to their cave, the den, and from there I knew the way.)[5]

Introduction

Entrelobos (Among Wolves) is a fiction feature film of Spanish production directed by the recognized documentary filmmaker Gerardo Olivares that was released in Spain in 2010 to significant box-office success.[6] As it is customary in the contemporary film industry, the movie's promotional campaign included an official web page, illustrated features in the national press, as well as a number of interviews with the director and cast members, either published in newspapers and magazines or broadcast through other media. By means of this promotional campaign, *Entrelobos* was presented to the public, firstly, as a fiction film based on the extraordinary real life story and personal experience of Marcos Rodríguez Pantoja during his childhood and teenage years. When he was seven or eight years old, Rodríguez Pantoja (b. 1946) was obliged to leave his poverty-stricken rural home in order to live as a goatherd in a lush and remote valley of the Northern Andalucía range of Sierra Morena. Even though for a very brief period an old goatherd lived with him, this person soon disappeared, leaving the child almost totally isolated from human contact until he was discovered and captured by the Civil Guard twelve years later (in 1965). In the mid-1970s the cultural and educational anthropologist Gabriel Janer taped and transcribed Rodríguez Pantoja's own narration of his life and published it a few years later as part of an anthropological study written in Catalan on the "case" of Marcos as a "feral child" (Problemàtica).

Even though Janer's book was also published in English some years later (*Wild Child*), Marcos's case as a feral child did not have a wide public resonance at that time in Spain, and when Olivares decided to make a film about it, "the wild child of Sierra Morena" was a largely unknown or forgotten issue. The idea that *Entrelobos* was recuperating the real life experience of Marcos for the public was highlighted by Olivares in a number of promotional interviews. This idea is also conveyed in the director's autobiographical account (published in a Spanish newspaper as part of a special feature on the film) of his one-year search for, and investigation about, Rodríguez Pantoja, which included conversations with Janer (Olivares, "Como Conocí"). The photographs and footage—including the final sequence of the film—that show the now aged Marcos participating in the movie also transmit the notion that the film narrates a real biographical experience ("El Mundo. Especiales"). There are two other aspects of *Entrelobos* emphasized not only in promotional texts and interviews but also in most of the critical reviews on the film published in the Spanish press and abroad. One is the fact that Olivares's movie is an adventure-genre film aimed at popular or family audiences (Belinchón; Ocaña; "Días de cine"; Holland). The other is the aesthetic significance and high technical quality of the representation of wild fauna achieved by means of an independent film crew specialized in filming wildlife (Alonso; Sánchez; "Días de cine"; "El Mundo. Especiales"). For Olivares, his film "integrates nature and fiction [so that] is not possible to understand [one without] the other" in order to tell "Marcos' story of love for nature" (qtd. in Belinchón; my translation).[7]

In this essay, I study the film *Entrelobos* and Janer Manila's anthropological study (on which the film's script is freely based) in ways that situate my work between the tradition of cultural studies (anthropocentrically concerned with critical views of cultures and societies) and the developing field of animal studies (concerned with non-anthropocentric critical perspectives on the human and the nonhuman). I want to analyze these cultural texts in order to explore two overlapping relations that run through some of the disciplinary boundaries that have separated the study of the human *cultures* from the study of the nonhuman *nature* and the environment of the Iberian Peninsula. The first is the relation between personal memories and life histories of the Iberian rural working-class during Franco's fascist regime and the cultural representation/appropriation of these memories and narratives by contemporary Spanish film industry and media, both as cinematic spectacle and as part of a "commodified" popular historical memory of the Francoist period. The second is the relation between contemporary Iberian wild *nature* and human *culture*. This relationship comprises both the material, embodied historical encounters between human and nonhuman animal species (especially those species considered wild or nondomesticated) in the Iberian Peninsula and the symbolic representations of those encounters and cohabitation in scientific, intellectual and artistic discourse. Since Olivares's film constructs its cinematic narrative by means of an intertextual transformation/recodification of Rodríguez Pantoja's personal memory of social oppression and everyday cohabitation with wild animals—largely mediated by anthropological discourse—*Entrelobos* and Janer's work constitute exceptional texts to examine contemporary representations and imaginaries of human-nonhuman animal relations vis-à-vis social injustice within the Iberian cultural and artistic field.

I argue that the representation of the "wild" nonhuman animal in Olivares's film simultaneously goes beyond and falls within its traditional anthropocentric function as a symbol or metaphor of the human/social. On one hand, we can find in *Entrelobos* a profuse deployment of diverse genre-film visual and narrative conventions. On the other hand, the film is still valuable and innovative within the Iberian cultural field from a non-anthropocentric critical perspective because of Olivares's efforts to translate into cinematic sounds and images Rodríguez Pantoja's non-anthropocentric oral life history of everyday social-like relations with nonhuman animals. With the assistance of recent groundbreaking theoretical work on human-nonhuman animal relations such as Donna Haraway's, I argue that *Entrelobos* concurrently reproduces and disrupts local configurations of deep-rooted anthropocentric modern worldviews and humanist discourses historically erected in the West as an ideological Great Divide between the human and everything else.

The script of *Entrelobos* draws intertextually from Rodríguez Pantoja's rich memorialistic narrative that was recorded and analyzed by Janer as a "feral child case": a new case to be placed and understood within the body of knowledge configured by the psychological, pedagogical or anthropological studies of children who grew up in isolation from humans published between the eighteenth and the twentieth centuries (Benzaquén 3–11). Working within the anthropocentric episteme of twentieth-century Western thought and the scientific paradigm prevalent in the cultural and educational anthropology of 1970s Spain, Janer argues that Marcos's account of everyday relations with domestic and wild animals is the product of a psychological fantasy generated as a mechanism of survival in a situation of absolute isolation. However, a non-anthropocentric reading of Rodríguez Pantoja's own narration suggests that the embodied complexity and ethical significance of his relationships with animals cannot be intellectually accounted for by means of sociocultural theories based on rigid traditional anthropocentric dichotomies such as nature/culture, human/animal or wild/domesticated.

As we will see, some episodes and sequences of *Entrelobos* point to a non-anthropocentric artistic reinterpretation of the scientific archive (both in the conventional and Foucauldian sense of this term) of Marcos as a feral child. In some parts of the film Marcos's historical embodied relations with wild animals are represented on the screen (and performed in front of the camera by human and nonhuman actors) as a social subject-forming process that takes place across biological difference and that cannot be understood as either belonging to the "natural" or the "cultural" domain. Donna Haraway has described interspecific bio- and socio-cultural processes of this type as the reciprocal and ongoing "becoming with" of "companion species" in what she calls "naturecultures." In these processes, according to Haraway, "the partners do not precede the meeting [and] species of all kinds . . . are consequent on a subject- and object-shaping dance of encounters" (4). In other words, the ways historically situated sentient beings develop their subjectivity and are themselves experienced as objects emerge from those multiple encounters among different species. These encounters, for Haraway, are the most visible and ethically meaningful in "mundane" everyday companionate relations between human and nonhuman animals.

While *Entrelobos* shows to be a worthy effort to represent artistically embodied human-nonhuman animal companionate relations and reflects contemporary conservationist sensibilities to Iberian wild fauna, the fact that the film recodifies Rodríguez Pantoja's life history through a full set of genre-film narrative conventions mitigates the non-anthropocentric critical potential of the film. Furthermore, it makes highly problematic its function as a cultural representation of subaltern memory. As we will see, on one hand, the use of adventure-film genre codes, narrative structures, and imaginaries constructs Marcos's experience as a new local version of popular film feral-child heroes such as Tarzan or Mowgli. On the other hand, the recodification of his personal subaltern memory is tailored to fit a "commodified" historical memory of the Francoist period. Furthermore, by merging the cinematic rhetoric of adventure film and wildlife documentary, Olivares's film represents the wild animal, and particularly the Iberian wolf, as a thrilling spectacle to look at and as an uncritically anthropomorphic projection of human moral values.

An Isolated "Feral Child," or Multi-Species Companionship?

In the mid-seventies the Balearic cultural and educational anthropologist Janer Manila met Marcos Rodríguez in the city of Palma de Mallorca where the Andalusian young man was working and decided to do research on him. As the title of his book suggests, Janer taped, transcribed, and analyzed Rodríguez Pantoja's oral account of his life experiences as a case study in a pedagogical and anthropological investigation on "wild" or "feral" children. In this way, the information he collected about Marcos's life, as well as his own scientific analyses and interpretations, became part of an *archive*, and constitutes an *archive* itself, both in the conventional and Foucauldian use of this term. For Foucault, "in the density of discursive practices [there are] systems that establish statements as events (with their own conditions and domain of appearance) and things (with their own possibility and field of use)" (128). These systems of statements (whether events or things) is what Foucault calls the "archive." In this sense, for Foucault, the archive is neither "the sum of all the texts that a culture has kept . . . as documents," nor "the institutions [which make] possible to record and preserve" those statements, but the system of their functioning and what defines their enunciability from the outset. In Foucault's definition, the archive is "the law of what can be said" or "the general system of the formation and transformation of statements" (129–30). Janer's scientific research on Marcos is an archival practice in the conventional sense because it implies a recollection, classification, preservation, and analysis of Marcos Rodríguez's words and biographical information. Furthermore, what he says about Marcos is the result of certain implicit archival laws and procedures that determined its enunciability and functioning as scientific knowledge. The formation of a scientific archive on "Marcos's case" required that Rodríguez Pantoja be conceived as a "case study." This case study entails the consideration of a human subject as an object of anthropological scientific inquiry and as a challenging "problem" of social and cultural deviance or abnormality that demands a (unspecified) pedagogical intervention. Within the scientific archive Rodríguez Pantoja is not conceived, thus, as a reliable human interlocutor with a historically situated perspective and a valuable voice of his own to dialogue with or to learn something from.

Since Janer conceives Marcos's as a "feral child case," what he says about his life and behavior is meaningful as long as he can establish comparisons and connections with other documented cases of "feral children" that shaped the history and body of knowledge of this particular "field" since the eighteenth century. Furthermore, the idea of "feral child" implies a series of assumptions about *abnormal* or *deviant* relations to both nature and society, as well as the notion of an absolutely exceptional or *extraordinary* lived experience that would always remain as an insurmountable obstacle to the "wild" person's adaptation to civilized life. According to Janer, "feral children have for long fascinated sociologists, psychologists, and educators simply because their special plight raises so many . . . questions about social alienation in general. Can a child raised in exceptional circumstances attain the same degree of social adaptation as a normal child?" (*Wild Child* 162). Janer affirms that for "most authors" the answer is negative and his own study demonstrates that "the conditioning at a key age by isolation from human beings and interaction with animals will never quite be erased" (*Wild Child* 162).

Janer interpreted Marcos' account of his relations with wild animals as the consequence of the harsh socio-economic conditions of the Spanish Post-Civil War—which led to the abandonment and isolation of Marcos—and as the product of the child's imagination. For Janer, this imagination, as well as his earlier experiences in society, would help him to overcome his lonely condition and the necessity of socialization:

> [Marcos] had to learn to be entirely self-sufficient . . . in the wilderness. He had to interpret the wild, adapt his thoughts to it, organize that strange world and turn it into an operable one . . . susceptible of being rationalized . . . He had to face untamed surroundings and integrate himself into their ways, which he came to believe—and even now believes—he was part of: the lives of wolves and foxes, of eagles and deer, of partridges and snakes. These were the forces that formed his personality until the day of his capture. (*Wild Child* 8)

According to Janer, Marcos's tale constitutes "a fascinating and powerful picture of a human being in extraordinary circumstances" (*Wild Child* 21). Only those "extraordinary circumstances" can explain, in Janer's view, the fact that Marcos, in his "lost and solitary life," got to believe that "he [was] not alone, that he liv[ed] with those animals, and [that] he attempt[ed] to communicate with them [and] to understand them, to establish a relationship of affection to overcome his solitude" (*Wild Child* 152). Furthermore, for Janer, the fact that Marcos describes and narrates his own experience as a "wild child" in a very positive light (as a period of relative happiness, material well-being, and carefree life within an animal community) can only be explained as a consequence of the "trauma" of his capture (conceived as a common feature in feral children cases) and the subsequent difficulties "to become fully one with civilization" (*Wild Child* 162) which led him to nostalgically idealize the past (*Wild Child* 153). According to the Balearic anthropologist, "only if Marcos became capable of establishing in society the kind of emotional balance he believes to have witnessed among the animals, would he find an end to the emotional turmoil in which he presently [in the seventies, when the investigation was made] appears to live" (*Wild Child* 163). Janer also recognizes, however, that Marcos's life and perspectives as a young man at the time when he was interviewed were conditioned by his social situation as part of the service-industry proletariat in the city of Majorca.

In the first edition of his book, Janer traces the historical and cultural roots and development of the separation of human culture from *nature* since the ancient world to the twentieth century. His literature review includes classics of structuralist anthropology (such as Lévi-Strauss's studies on myths) as well as brief readings of well-known works from the Western literary and philosophical canon (Problemàtica 13–80). Janer's statements on Marcos's lived experience are implicitly based on that modern nature/culture dualism which, according to Bruno Latour, dominated Western thought and scientific practice since the early-modern Scientific Revolution (Latour 33). Other related implicit dichotomies in Janer's study are those that oppose human to animal, rational to irrational, wild to civilized, and feral to social. Furthermore, these dichotomies seem to construct a rather homogeneous idea of what *normal* "society," "humanity," and "civilization" is. For Donna Haraway, following Bruno Latour, this type of modern dualist thought forms the ideological Great Divides that have separated the human from everything that is nonhuman (15). These Great Divides govern the statements of the scientific archive, determine the "law of what can be said" to count as anthropological knowledge, and also explain Janer's views on human-nonhuman animal interactions. These views, along with the ideological Great Divides that underpin them, can be problematized from non-anthropocentrical critical perspectives not only in the light of contemporary critical theory on human-nonhuman relations such as Haraway's, but also from the perspective of cognitivist and evolutionary developments in the scientific study of animal minds and behavior.[8]

In this essay, I am proposing the possibility of a non-anthropocentric rereading of Rodríguez Pantoja's recorded memorialistic narrative about his life. This reinterpretation requires extracting Rodríguez Pantoja's oral narrative from the scientific archive into which it was originally conceived as a "feral-child case" and read it as a *non-anthropocentric oral life history*. What makes Rodríguez Pantoja's life history non-anthropocentric is the fact that the social or historical actors and agents present in the narrative belong to both human and nonhuman animal species. The voice of the oral life storyteller obviously is human, but many of the values, beliefs, perspectives and experiences expressed (sometimes anthropomorphically) throughout his narrative are not human-centered. In Rodríguez Pantoja's oral life history human and nonhuman animals are conceived and represented as agents and actors within a naturecultural "social field" that acknowledges the existence of companionate and affective relations across species boundaries.

The idea of affective companionate relations with animals is expressed throughout Rodríguez Pantoja's oral life history recorded by Janer. In his narrative, Rodríguez Pantoja describes everyday bodily, physical interactions and communicative exchanges with different wild and domestic animals such as goats, foxes, mice, eagles, wolves, and snakes. In different moments throughout his life history, Marcos describes how he kissed, pet, fed, played with, and slept with nonhuman animals. He tells how he was able to communicate with animals, how he helped and sometimes got help from them, and even how the occasional "misbehavior" of some of his wild animal companions caused him some domestic troubles in the cave where he lived. Moreover, Marcos describes the many practical things he learnt by observing other animals' behavior, but also the sense of superiority he noticed because of the things he could do (such as thinking or using his hands) in comparison to other species. The development of companionate relations, however, was neither always easy nor possible with all animal species. For example, Rodríguez Pantoja recalls how he had to punish physically some individual wolves in order to teach them to respect his goats, and how a relationship with the wild boar was impossible because "it's the only animal that makes no friends with any other living creature [and] it's deadly dangerous" (*Wild Child* 34).

Rodríguez Pantoja's words describe what Haraway calls "ongoing companionate relations" between human and nonhuman animal species. According to Haraway, in these types of interspecific encounters and "naturecultural" processes the form and meaning of "subject" and "object" are not pre-established but develop through the historical encounters and relationships among multiple human and nonhuman animal species. For Haraway, this historical and ongoing "dance of encounters" among human and nonhuman animal species is defined as a coshaping "becoming with" rather than a predictable relationship among pre-established biological entities.

The non-anthropocentric paradigm suggested by Haraway's thought opens up new possibilities for a non-anthropocentric interpretation of Marcos's rich memorialistic account of his companionate relationships with a number of wild and domesticated nonhuman animals that were part of South Spain's mountain ecosystems in the 1950s and 1960s. Janer seems right in pointing out the role of human anthropomorphic imagination in Rodríguez Pantoja's interpretation of his everyday relations with nonhuman animals. However, Marcos's rich narrative about his everyday life with animals also shows many other human and nonhuman cognitive, behavioral, and material processes. Among many others, Rodríguez Pantoja's oral life history bears testimony of his genuine curiosity and reflection about animal life; his feelings of empathy and respect for animals; and his willingness and ability to take care of animals or play with them, as well as to develop nonspecieist relationships of mutual acceptance, trust and even affect based on basic nonexploitative rules of cohabitation.

I think that anyone that has lived with a companion animal for many years, has been actively engaged in the animal's wellbeing, and has developed an affective relation of companionship with him or her knows well what embodied everyday human-nonhuman animal relations may get to be. In the case of Marcos, these relationships shaped his non-anthropocentrical, "abnormal" sense of self and otherness that manifests in his way to reflect on and talk about animals as if they were his neighbors, friends or relatives. Furthermore, *through* his everyday interactions with animals, the young Rodríguez Pantoja seems to have developed abilities and attitudes that allow him to affect and to be affected by nonhuman animals in multiple ways. These interactions include hunting animals, feeding them, caring for them, petting them, touching and being touched by them, responding to them, developing relations of mutual trust, setting up basic rules of cohabitation, carefully observing and reflecting on their habits, learning practical lessons from these habits, or exchanging basic information with the animals through vocal, corporeal, and tactile expression.

As Haraway argues, these kinds of actions and exchanges are something perfectly "mundane and ordinary" (as so many everyday human relations with both "domestic" and "domesticated" wild animals demonstrate). They also reveal, in Haraway's words, that "we are in a [complex] knot of species coshaping one another" that makes possible "response and respect" as well as the search for more livable "other worlds." Under this new framework, and from the perspective of an oral historian, it can be argued that the young Rodríguez Pantoja was a reliable and insightful oral storyteller when he described his everyday embodied companionate relations with wild and domestic animals in Sierra Morena. Even if anthropomorphic metaphor, fantasy, and nostalgia might have played a role in the construction of his narrative, rigid anthropocentric sociocultural perspectives do not seem to explain effectively the complexity and indeterminacy that characterize historically situated human-nonhuman animal relations. Furthermore, I think that Marcos's remarkable narrative of his experience as an isolated goatherd very *politely* cohabiting with wild animals has a mundane, worldly, and everyday dimension, as well as a non-anthropocentric ethical significance, that most of the research, books and films on his life have obscured or deemphasized.[9]

To conclude this section, it is also important to note that what we know about Marcos Rodríguez Pantoja and his historical experience in Sierra Morena exceeds the written scientific archive about him. After many years suffering diverse degrees of social and economic marginalization, Rodríguez Pantoja now lives in Galicia with a well-off friend. However, I think he is still a marginalized person in many senses. When he was invited to appear in the Spanish public media, for example, his voice was always transcribed, interpreted or translated by "experts" (such as social scientists, journalists, or filmmakers) in order to have some value as symbolic capital in the public sphere. As the final sequence of *Entrelobos* shows (in which he marvelously howls and plays with tamed Iberian wolves), Rodríguez Pantoja seems to have conserved many of the skills and attitudes he developed in his years in Sierra Morena as a "wild child," and also his deep love and empathy to nonhuman animals. It is in this sense that Rodríguez Pantoja's own living and acting body, as well as his own registered performances, howls, bird-songs, narratives, and unorthodox ideas about animals, can be understood as what performance scholar Diana Taylor has called the "repertoire" of "embodied memory": a means to store and transmit subaltern knowledges through the spoken word, gesture, or performance which either can be an alternative to or work along with the archive.

Out of the Archive and onto the Screen

The film narrative of *Entrelobos* incorporates many of the episodes and anecdotes told by Rodríguez Pantoja and published as part of Janer's anthropological study on his case as a feral child. However, instead of adopting the scientific interpretation of these "real facts" and events, Olivares constructs his narrative mainly (but not exclusively) through the focalization of the fictional character of Marcos and, as we have seen, by adopting different film-genre narrative models. In this section I examine how the fictional character of Marcos and his non-anthropocentric worldviews are represented through film rhetoric; how fictional nonhuman-animal perspectives are integrated into the film narrative; and which are the intertextual connections between these cinematic representations and the worldviews of the historical Rodríguez Pantoja as a non-anthropocentric life storyteller in the 1970s. I argue that, at least in some moments of the film, Olivares makes an effort to translate by cinematic means Rodríguez Pantoja's oral narrative about his companionate relationships with nonhuman animals, as well as his non-anthropocentric views and ideas about animals. In this way, *Entrelobos* points out to a non-anthropocentric reinterpretation of the archive on "Marcos's case" as a "feral child." Nevertheless, the fact that this narrative translation of the archive is infused with popular-film animal imaginaries and contemporary urban public perceptions and sentiments about wild fauna mitigates the non-anthropocentric critical potential of the film. Rodríguez Pantoja's non-anthropocentric oral life history about his everyday companionate relations with different nonhuman animals translates, thus, into a sentimental man-and-canid relation of friendship and mutual aid developed amidst situations of danger.

According to Olivares, one of the aims of his film was to narrate Marcos' "story of love" for nature. As we have seen above, in the oral life history recorded by Janer, Rodríguez Pantoja describes affective, companionate relations with wild animals developed through everyday bodily interactions and material and communicative exchanges by means of vocal expression, gesture and touch. For Haraway, human-nonhuman animal encounters and relations of this kind constitute an open interspecific coshaping process—what she calls a "becoming with" in naturecultures—rather than the product of the scientifically predictable behaviors of pre-established biological entities.

In some of its scenes and sequences, and particularly in the central episodes of the film that depict Marcos's life as a child and teenager after his arrival to the valley, *Entrelobos* narrates the development of companionate relationships between Marcos and two animals: a ferret (whom he uses to hunt rabbits) and a young wolf. In Rodríguez Pantoja's oral life history there is no mention of a hunting ferret, so this relationship is the product of Olivares's imagination as scriptwriter. The relationship with the wolf, on the contrary, is freely based on experiences narrated by Rodríguez Pantoja in his memorialistic account. Nevertheless, Rodríguez Pantoja only talks about "wolves" or a "wolf pack" (who were his "neighbors" in the valley, who trusted him and respected his herd of goats, and who helped him when he got lost) without mentioning the existence of a particular wolf with whom he had a special relationship. In any case, the film represents the relationships of Marcos with the wolf and the ferret by means of the performance by human and nonhuman actors (trained animals, in the last case) of some of the actions that Rodríguez Pantoja narrates in his oral life history. These include petting and kissing wild animals, observing patiently and carefully their behavior, communicating with them through voice and gesture, feeding them, hunting with them, sleeping with them, helping and getting help from them, and playing with them. More particularly, some of the scenes that make up the wolf-child-relationship subplot of the film are based on episodes from Rodríguez Pantoja's oral life history: the first contact with the valley's wolves when Marcos starts to feed them in order to get their trust, an encounter with wolf cubs and his playing with them, and the sharing of a hunting prey with the wolves.

While Rodríguez Pantoja does not narrate his life experiences in chronological order (the structure of his oral life history might have depended on Janer's questions and Rodríguez Pantoja's spontaneous acts of remembering and verbalizing his memories), the film's discourse is ordered chronologically. In this manner, the film focuses on the evolution of the child character, on his gradual adaptation to the life in the valley, as well as on the development of Marcos's companionate relationships with nonhuman animals. *Entrelobos* shows how Marcos's initial feeling of fear toward the wolves and other nonhuman animals gradually develops into curiosity, mutual respect and trust, and companionate love. By selecting some of Rodríguez Pantoja's "real experiences" from the archive, and by inserting them into a fictional narrative of human coming-of-age among animals (or human *becoming with* nonhuman animals, if we use Haraway's terminology), *Entrelobos* suggests a non-anthropocentric reinterpretation of the anthropological archive. While Janer interprets with anthropocentric skepticism Rodríguez Pantoja's narrative of his relations with animals as the result of a psychological fantasy generated by his lack of socialization in total isolation, Olivares's film represents Marcos's experience in Sierra Morena as the result of social-like affective and embodied companionate relationships between human and nonhuman animals.

Furthermore, Olivares employs cinematic focalization both to represent Marcos's subjectivity and nonhuman animal cognition and perspectives. Marcos is the main focalizer in most of the scenes that show his observation of, and interaction with, wild animals. By means of this focalization (along with other rhetorical aspects of the film such as editing and sound) *Entrelobos* represents the child's lived experience of the natural world, as well as the evolution of his feelings and attitudes toward nonhuman animals: from fear and wonder at nonhumans to familiarity with them and companionate affection to some. We can see this type of focalization, for example, in the scene when Marcos wakes up drowsy by the river with a flock of vultures flying over him, when the child peeps at a wolf pack devouring prey, and in the sequence of the chase of the deer. In other scenes, the idea of companionate affection between Marcos and his animal partners is represented by means of sequences made up of close-up shots of the protagonist petting or kissing his nonhuman friends. These scenes are edited with a conventional musical soundtrack that puts the accent on emotion.

Olivares also employs focalization to generate the effect of a nonhuman animal cognition and to insert fictional nonhuman perspectives within the variable focalization of the cinematic narration. Within contemporary European fiction film, the technique of using animal actors as fictional nonhuman characters with their own perspectives on the world expressed through focalization was most famously used in Jean Jacques Annaud's 1988 box-office hit *The Bear*. Following this trend in nature fiction-film, in *Entrelobos* the wolves are focalizers recurrently throughout the narrative in their different roles as predators and companions. Occasionally, other nonhuman animals such as the ferret or different bird species are also focalizers in the film, which thus displays a variety of fictional cognitive perspectives from different species on the world of fiction.

By means of a selective use of Rodríguez Pantoja's oral life history as material for the script, and by representing human and nonhuman cognition and subjectivity through variable focalization, *Entrelobos* points out to a non-anthropocentric reinterpretation of the anthropological archive on "Marcos's case" as a "feral child." Notwithstanding, as we will see in the next section, this cinematic reading of Rodríguez Pantoja's story is also influenced by the film's genre and its intertextual relations as a popular film. In order to construct Rodríguez Pantoja's historical life experience as an adventure movie, Olivares resorts to narrative and iconographic traditions such as feral-heroes adventure narratives, wildlife documentary film, folk tales about animals, and the conventions of uncritical anthropomorphism. As we will see below, these aspects of the film make it problematic as a "translation" for popular audiences of Rodríguez Pantoja's oral life history and mutes its non-anthropocentric and nonspecieist critical potential.

The Framing of Rodríguez Pantoja's Life History by Popular Film Traditions

After an initial scene in which a pack of wolves chase a deer, the story of *Entrelobos* begins with a workday in the life of the children Marcos and Juanillo: two brothers who help their poor family as goatherds. Their herd suffers a wolf pack attack and, when they come back home, their stepmother violently scolds them for their neglect. When the father comes back home later we know that the family's goats and house are property of a powerful local landowner and that the family needs to pay for the killed animals to avoid being evicted. In order to pay the debt, the father sells Marcos as an unpaid laborer to the *señorito*, who sends the child to a remote valley to help herd goats. Marcos travels to his new workplace with the property owner's foreman, Ceferino. During the trip, they meet a patrol of the *Guardia Civil* who asks Ceferino for information about El Balilla, a local *maquis* who survives in the area since the end of the Civil War. Ceferino believes that El Balilla was responsible for his brother's death during the Civil War and obsessively seeks to avenge him by killing the *maquis*. When Marcos gets to his new workplace in the valley, he meets an old goatherd, Atanasio, with whom he will live in a cave. From this moment, an affective and pedagogical relation gradually develops between both characters and Marcos learns from Atanasio hunting techniques, medicinal uses of herbs, and how to get along with the wolves. Just before Atanasio dies from disease, he tells Marcos he is El Balilla's father. After Atanasio's death Marcos experiences the hardships of living alone in the valley but he gradually learns to take care of himself, to hunt, and to get the trust of wild animals. Particularly, a relationship between Marcos and a wolf cub unfolds as both the child and the animal grow. In successive moments of the narrative Marcos feeds and gets food from the animal, a pack helps him to escape from Ceferino, and he takes care of his wolf-friend after Ceferino's men shoot him. After this episode, the narrative takes a large chronological leap through ellipsis and Marcos appears as a young man perfectly adapted to his life in the valley. The adventure-film plot, however, resumes when Ceferino and the *Guardia Civil* chase after and shoot El Balilla. Even though Marcos manages to hide El Balilla, Marcos himself is captured. In the poor tenant's house where he spends the night, the tenant's daughter misinforms Ceferino of where El Balilla is hiding in hopes that, if the hated foreman goes out, the wolves might attack him. That is exactly what happens and the narrative closes when the Guardia Civil takes Marcos out of the valley while he dramatically howls to his wolf friends. As we mentioned, the film ends with a narrative coda that represents the contemporary Rodríguez Pantoja, acting as himself, visiting the valley where he used to live.

The purpose of this summary of the plot is to illustrate how the film's script recodifies some of the "real facts" and events of Rodríguez Pantoja's oral life history as an adventure film with all the "plot formulas, settings, character networks, icons, and conventions of action" that for Gehring "mark out the range of this category" (9). According to Gehring, adventure stories "are [ultimately] based on the structure of the romance of medieval literature" in which "a protagonist either has or develops . . . special skills and overcomes insurmountable obstacles in extraordinary situations" (10). For this author, "[i]n these films mankind is depicted as living in a world fraught with perils of all kinds, but through the exemplary effort, cooperation, and bravery of individuals and groups, the social order . . . is maintained" (10). This restoration of the social order, Gehring continues, contrasts with the fact that adventure films "give free play to activities which are outside the normal framework of the everyday world" (10). Gehring concludes that adventure films' characters and conflicts are located "in isolation from ordinary life" (in a romantic past or in exotic spaces, for example) "in order to maximize the dangers and thus to increase the magnitude of the efforts of the characters" (10). *Entrelobos* clearly operates as an adventure narrative, even though with some local and historical particularities with respect to the more archetypical examples of classic Hollywood cinema that Gehring had in mind. Because of his cruelty to humans and animals (manifested in his double persecution and shooting of wolves and the *maquis*) Ceferino represents a threat to a moral order that is both social and natural. Once Marcos saves both his wolf-friend and El Balilla and after the wolves kill Ceferino the moral order is partially restored through narrative closure.

This narrative of restoration of the moral order, as well as the overtly sentimental social drama of Marcos's abandonment by his family, operates also as a denunciation of the oppressive sociopolitical order of rural Andalucía during early Francoism. Anyone acquainted with the Spanish cinema of the last turn-of-the-century will find familiar elements in this treatment of Francoist past, such as the romantic *maquis* or melodramatic representations of class and political conflict. José Colmeiro and other critics have suggested that recent popular films dealing with the Spanish Civil War and Post-War period, "in spite of their noble goals of making the traumatic past of repression and resistance accessible and understandable to contemporary Spanish audiences, . . . have made it perhaps too palatable and too comfortable, thus neutralizing their potential as instruments for social intervention" (29). For Colmeiro, these films, "can appear to reinforce the idea of the pastness of the past, easily accessible but just as easily disconnected from present day historical realities and social and political concerns, [and] may suggest a certain aesthetic commoditization of the horrors of the past" (29). I think Colmeiro's argument applies to *Entrelobos* in the sense that Olivares transforms certain materials, drawn from Janer's work or the historiographical archive, into a fictional cinematic representation of early-Francoist rural Andalucía which is "accessible" and "commodified," at once politically progressive and politically unchallenging. In the film the quasi-feudal social stratification of post-Civil War rural Andalusia, the family abuse and forced labor to which many working-class children were subjected, and the violent persecution that the internal opposition to the fascist Regime suffered are largely represented by means of the cinematic conventions of Hollywood-style costume drama and adventure film without establishing apparent connections with the historical present of the audiences.

This construction of the film narrative as an adventure story and as a "commodified" picture of the rural working-class's hardships and struggles during early Francoism determines Olivares's portrayal of Rodríguez Pantoja's historical experience of cohabitation with wild animals. As we mentioned above, for Haraway, companionate relationships between historically situated human and nonhuman animals develop in ordinary, mundane, everyday material encounters through which both partners "become with" one another. Yet the protagonist of *Entrelobos* participates throughout the film in some "extraordinary situations" and "perils" (such as shootings, fights, and persecutions) that situate his experience not only "outside the normal framework of the everyday world" (Gehring 10) but also outside the "worldly" everyday life evoked by Rodríguez Pantoja in his oral life history. Furthermore, even ordinary situations in Rodríguez Pantoja's life in Sierra Morena such as hunting are constructed in the movie as a thrilling, exotic spectacle intertextually linked to popular cinematographic feral-child heroes such as Burroughs's Tarzan or Kipling's Mowgli. The clearest example of this Tarzan-like representation of the character is the scene of deer hunting which introduces Marcos as a young adult (interpreted by the popular Spanish actor Juan José Ballesta) to the audience after a big chronological leap in the narrative. This scene was shot with a wide variety of cameras, angles, and types of shots, edited with fast cutting technique and with a "primitive" percussive music score as soundtrack. Marcos chases down the deer through the forest along with his wolf friend, jumps from a ravine to the river, and swims with his knife among his teeth before slashing his prey. After watching this scene one understands the reasons that led some Spanish critics to head their press reviews of the film with titles such as "Tarzán visita Córdoba" (Sánchez) and "El libro de la sierra" (Arce).

In the spirit of global popular wildlife documentary film such as the wildlife series produced by National Geographic, but also drawing from local classics of the genre such as Rodríguez de la Fuente's immensely popular *El hombre y la Tierra*, many sequences of *Entrelobos* portray animal life as a visual spectacle for the spectatorial gaze. The connections here are more than merely textual since the director of the film's second unit for wildlife shooting, Joaquín Gutiérrez Acha, worked as a cinematographer for National Geographic wildlife documentaries before his collaboration with Olivares. In his study of North American nature documentaries, Alexander Wilson observed a shift from pastoralism to scientism in the representation of animals that culminated in the 1980s. For Wilson, contemporary environmentalist values and demands for more accurate scientific information came to challenge previous anthropomorphic representations of animals that constructed them as incarnating diverse human values. Nevertheless, for Wilson, "the demand for spectacle tended to lead to an obsessive interest in predation, usually enhanced by the exciting music and slow-motion, fast-edited sequences that audiences might expect of a thriller" (qtd. in Garrard). In some sequences of *Entrelobos* we find this type of dramatic, slow-motion representation of animal predation, which is the hallmark both of National Geographic aesthetics and Rodríguez de la Fuente's documentary style.[10]

Even though contemporary wildlife-documentary aesthetics play a role in the representation of nonhuman animals in *Entrelobos*, even more significant are the effects that diverse forms of anthropomorphism generate in the film. In the last decades, anthropomorphism has been the subject of an intense scrutiny and debate both among ethologists and among critics who study cultural representations of nonhuman life. In the zoological field of ethology, the "anthropodenial" of those behaviorist ethologists who consider anthropomorphism as the worst pitfall in the scientific study of animal behavior has been criticized from Darwinian and cognitivist positions. Frans de Waal, a leading primatologist, argues that "as soon as we admit that animals are far more like our relatives than like machines, then anthropodenial becomes impossible and anthropomorphism becomes inevitable and scientifically acceptable" (51). Still, according to de Waal, anthropomorphic interpretations of animal behavior are scientifically valid only when they are "put in the wider context of a species' habits and natural history" and "used to develop ideas we can test" empirically (52). Marc Bekoff, a prominent specialist in canid behavior and animal cognition, also defends the ethological validity of a critical anthropomorphism, as well as the need to extend the notions of justice, cooperation and moral behavior to nonhuman social species of the animal kingdom. From the field of cultural and literary criticism, several authors have drawn distinctions between different types of anthropomorphism. While Greg Garrard distinguishes between a "crude" and a "critical" one, literary critic John Simons proposed the term "strong" anthropomorphism for those anthropomorphic representations of nonhuman animals that either show "how the non-human experience differs from the human" or create "profound questions in the reader's mind as to the extent to which humans and non-humans are really different" (120).

Anthropomorphic understandings and representations of nonhuman animals play a variety of roles both in Rodríguez Pantoja's oral life history and in Olivares's film. Rodríguez Pantoja spontaneously used an anthropomorphic language to describe and explain his encounters, interactions and cohabitation with other embodied animals in a natural environment, as well as his extensive observations of animal behavior. Rodríguez Pantoja's narrative also suggests that some of these encounters with animals developed into interspecific relations of companionship, and that these relations depended on different forms of *naturecultural* acquired knowledges, material exchanges, and nonlinguistic communication.

As we already have seen in the previous section, some scenes of *Entrelobos* suggest a valuable effort to translate into the language of fiction cinema Rodríguez Pantoja's non-anthropocentric views and personal experiences of human-nonhuman animal companionship. Furthermore, in *Entrelobos* we find some examples of animal representation that may be considered "critical" or "strong" anthropomorphism. This is the case in two juxtaposed scenes. The first shows the abusive mistreatment to which Marcos is subjected in his own home. The second shows a female wolf in her den carefully feeding and looking after her cubs. The point made with this sequence—inspired by a personal reflection that Rodríguez Pantoja makes in his life narrative—might seem rather obvious. Yet, it is still consistent with ethological descriptions of canid behavior and rather effective in questioning specieist assumptions about the meaning of "human" and "inhuman" behavior in the animal kingdom.

Nonetheless, Olivares' film clearly moves away from Rodríguez Pantoja's non-anthropocentric oral life history in two aspects: the anthropomorphic design of individual animal characters with an active role in the adventure plot, and the uncritical reproduction of folkloric myths and cultural stereotypes about Iberian wild fauna. In the same vein of the stories of Tarzan and Mowgli, the wolf and the ferret are constructed as individual characters who help the protagonist in situations of danger. As such, they show an implausibly anthropomorphic comprehension of the human world that seems motivated by the structural needs of the adventure emplotment of the story and by the absorption of pop-culture imaginaries. We can observe the uncritical reproduction of folkloric myths and cultural stereotypes about wild animals in the sequence when the wolves kill Ceferino at the end of the story. In this sequence, set in a foggy full-moon night as if it were a horror folk tale, ages-old folkloric and literary myths and stereotypes about the wolf are activated with a "noble" purpose. As we already noted, the film recurrently represents human-wolf relations as a developing companionship. However, traditional myths and superstitions of European rural cultures that conceive the wolf as a mysterious, fearsome, and loathsome killer are triggered in order to give a narrative closure to the story.

By killing Ceferino, the Iberian wolf, mysterious once again, becomes the symbol of a moral order that is not only social and natural, but that also has mythical and otherworldly overtones. Like the *lobos* of Iberian folk tales that inspired the masterful short stories of Galician writer Ánxel Fole, Olivares's wolves kill humans in full-moon foggy nights, but in this case with the noble purpose of restoring the social-natural order threatened by Ceferino's violence.

Entrelobos is a hybrid popular film whose cinematic discourse operates dialogically by means of diverse generic and intertextual relations. Pop-culture imaginaries of feral-child heroes, adventure-film genre narrative structures, "commodified" representations of the oppressive sociopolitical realities of twentieth-century Spain, wild-life documentary aesthetics, and folkloric myths about local fauna, all interweave in the film along with facts, events and worldviews drawn from Rodríguez Pantoja's life history and Janer's study of his "case" as a feral child. In this manner, while *Entrelobos* seems to contest some of the anthropocentric Great Divides (such as human/animal, nature/culture or wild/domesticated) imposed by Western Modernity and still dominant in contemporary urban-centered Iberian cultures, it does so by resorting to familiar, unchallenging, or stereotyping cultural imaginaries that thwart the non-anthropocentrical critical potential of Rodríguez Pantoja's life history.

Conclusions

In his anthropological study on Rodríguez Pantoja as a "feral child case," Janer points out that it may be impossible to know "to what extent . . . Marcos establish[ed] real communications with the animals he lived with" (*Wild Child* 149). In spite of Janer's attempts to elucidate this point, it may be equally difficult to know to what extent nostalgia or imagination influenced Rodríguez Pantoja's oral account of his solitary life in Sierra Morena from 1943 to 1965. Still, and beyond these interpretative challenges, I believe Rodríguez Pantoja's narrative—made possible thanks to Dr. Janer's fieldwork in the mid-seventies—constitutes one of the most intellectually inspiring and anthropologically challenging working-class life histories ever documented in twentieth-century Spain. In order to understand the cultural value and significance of Rodríguez Pantoja's oral life history, though, we need to call into question anthropocentric, humanist, or urban-centered critical perspectives and languages that have dominated the study of the "cultures" and the "nature" of the Iberian Peninsula.

According to Janer, since Marcos lived in isolation during his childhood, he had to imagine meaningful relationships with animals in order to survive in solitude. By drawing from Haraway's materialist and non-anthropocentric critical theory, and by separating Rodríguez Pantoja's oral narrative from the anthropo-*logical* "archive" into which it was originally situated, I have proposed a non-anthropocentric critical reading of the text. This non-anthropocentric reading is not based on the ideological Great Divides (such as animal/human, nature/culture or wild/domestic) that have separated humans from the non-human since the inception of Western modernity, but on the idea that animal species (including humans) have been historically and materially entangled in a web of encounters and shape one another in multiple ways. From this perspective, we may interpret Rodríguez Pantoja's narrative as a *becoming with* animals in *naturecultures*: a material, embodied interspecific socio- and nature-cultural process through which humans and nonhumans affect and are affected by each other in different ways. We can observe this process in Rodríguez Pantoja's description of the everyday embodied companionate relationships (based on mutual recognition, behavioral adaptation, trust, and respect) he was able to develop with some of the wild animals he cohabited with in Sierra Morena.

Rodríguez Pantoja's oral life history was recorded as part of an anthropological archive on a "feral-child case." For Foucault, the "archive" regulates the formation and transformation of statements: what can and cannot be said, as well as how statements must be interpreted in order to be considered valid knowledge. The archive in this Foucauldian sense, however, cannot legislate statements and interpretations of archived materials if they are taken out from the archive to be used as intertexts in nonscientific spheres of the cultural field. It is true that the archive may relegate "illegitimate" uses of archival materials to the sphere of what does not count as true knowledge. Nonetheless, nonscientific intertextual uses of the archive, for example in popular fiction film, may contribute to the formation, or reflect the emergence, of new public sentiments and popular historical memories. Furthermore, they may put into question implicit assumptions of the dominant scientific discourse through the formation of what Raymond Williams has called "structures of feeling": "emergent" manifestations of a social "practical consciousness" (such as a more ethically sensitive or sympathetic understanding of animals) that can be traced in some of the literature and art of a particular historical period, and that "do not have to await definition, classification, or rationalization before they exert palpable pressures . . . on experience and on action" (Williams 132). This is, in my opinion, what *Entrelobos* does with respect to the anthropological archive on Marcos's case as a "feral child."

There is a subplot in *Entrelobos* that constitutes a worthy effort in representing (by means of film rhetoric and the use of human and nonhuman actors) Rodríguez Pantoja's account of embodied companionate relations across biological difference. Yet, the use of adventure-film narrative structures, conventional wildlife documentary aesthetics, and stereotyping pop-culture and folkloric imaginaries situate the story outside the everyday world Rodríguez Pantoja described, and covers up the non-anthropocentric critical potential of his life history. In this way, Olivares's translation of Rodríguez Pantoja's oral life history is infused with the anthropocentric visual languages and cultural codes that media and tourism industry commonly use to transmit images and ideas of "wild nature" to urban mass audiences. The nonhuman animals constructed through these languages and codes are exotic, other-worldly, extra-ordinary, a thrilling spectacle, or a "natural heritage." They are something other-than-human that only innocent children, primitive indigenous folks, too-radical greens, or too-scrupulous Darwinists can confuse with our own civilized human selves.

This understanding of "wild" nature and animals contrasts with Rodríguez Pantoja's non-anthropocentric oral life history. For the young Rodríguez Pantoja, who told his experience as an isolated child-goatherd to Janer, there was not a "nature" beyond "culture", nor a "wild" world beyond a "civilized" one. There was simply an everyday world with many different animal species, both "wild" and "domestic," human and nonhuman, entangled in a web of embodied encounters and relations. This world was neither a dreadful wilderness nor an Edenic, pastoral or Rousseauian Nature, but a cool valley with plenty of water, many plant and animal species, and a good number of food sources for a resourceful isolated goatherd with a deep knowledge of the environment where he lived. I think that Rodríguez Pantoja learned to inhabit this naturecultural Iberian world (is it not Iberian?) not only without specieist or anthropocentric prejudices, but also with a "worldly" politeness and a "mundane" considerateness to other animal species that might well inspire environmentally progressive ways of human–nonhuman cohabitation of the Iberian Peninsula.

Acknowledgments

I would like to express my gratitude to Prof. Kata Beilin, Prof. Juan Egea, and Belén Hernando of the University of Wisconsin-Madison, and to Prof. William Viestenz of the University of Minnesota for their critical and stylistic revisions of different versions of this article.

También me gustaría agradecer a Marcos Rodríguez Pantoja sus inspiradoras ideas e historias sobre su relación con los animales y al Doctor Janer Manila el haber hecho realidad el relato de Marcos a través de su labor investigadora.

Notes

1. Janer Manila, La Problemàtica Educativa Dels Infants Selvàtics, 260.
2. De Waal, "Are We in Anthopodenial?" 51.
3. Haraway, *When the Species Meet*, 19.
4. Janer Manila, La Problemàtica Educativa Dels Infants Selvàtics, 168.
5. Janer Manila, *The Wild Child of Sierra Morena*. Trans. Deborah Bonner. 30.
6. According to the web site Internet Movie Database the box office takings of *Entrelobos* was $4,224,250 in Spain by March 2010 ($922,522 the opening weekend) and $6,820,000 worldwide by August 2012.
7. Critical reviews in Spain and abroad were mixed in their assessment of the film: while some critics valued its technical and production quality, the relative novelty of its story, and its effectiveness as a genre movie (Batlle Caminal; Holland), others lamented the unsatisfactory narrative unfolding of the plot (Salvá), its moments of sentimental excess (Sánchez), and the fact that representations of "nature" and the "human" interfere with each other and make the film an unaccomplished hybrid product (Crespo).
8. Colin Allen and Marc Bekoff have traced a historical account of the zoological discipline of ethology from the "classic" studies of Konrad Lorenz and Nikolaas Tinbergen in the first half of the twentieth century, to the dominance of a behaviorist scientific paradigm, and the contemporary emergence of a cognitivist and evolutionary model defended by the authors. The work of Bekoff and the research of the primatologist Frans de Waal on animal mind and behavior are paradigmatic of the new cognitivist and evolutionary approaches and methodologies in ethology (as well as of the ethically-committed stance adopted by many ethologists in relation to animal wellbeing and human-animal relations). Bekoff's and de Waal's academic and popular-science works deal with innovative research topics such as animal's play, emotions, morality, and conflict resolution that have contributed to expand our understanding of nonhuman animal life beyond the methodological restrictions and biases of behaviorism.
9. Apart from his anthropological study on "Marcos's case," published in Catalan in the late 1970s and reprinted in a short version in 2010, Janer Manila (who is also a recognized and prolific fiction writer in Catalan) published in 2010 a young-adult novel based on Marcos's life, *He llugat amb ells llops*. This work was also published in Spanish translation that same year.
10. For some of the reviewers of *Entrelobos*, these wildlife-documentary style sequences do not fit seamlessly into the film's main narrative (Crespo; Sánchez).

Works Cited

Allen, Colin, and Marc Bekoff. *Species of Mind: The Philosophy and Biology of Cognitive Ethology*. Cambridge, MA: MIT Press, 1997.

Alonso, Agustín. "En el rodaje de 'Entrelobos', una película basada en hechos reales." *RTVE.es*. 18 Apr. 2010. Web. 19 May 2012.

"Among Wolves." *IMDb*. 26 Nov. 2010. Web. 2 June 2013.

Arce, José. "'Entrelobos': El Libro de La Sierra." *LaButaca.net*. 24 Nov. 2010. Web. 20 May 2012.

Batlle Caminal, Jordi. "Entrelobos: Buen Cine de Aventuras." *La Vanguardia*. N. p., 26 Nov. 2010. Web. 19 May 2012.

Bekoff, Marc. *Wild Justice: The Moral Lives of Animals*. Chicago: University of Chicago Press, 2009.

Belinchón, Gregorio. "El buen salvaje entre lobos." *EL PAÍS*. N. p., 26 Nov. 2010. Web. 19 May 2012.

Benzaquén, Adriana S. *Encounters with Wild Children: Temptation and Disappointment in the Study of Human Nature*. Montreal: McGill-Queen's University Press, 2006.

Crespo, Irene. "Críticas: Entrelobos." *Cinemanía*. 26 Nov. 2010. Web. 19 May 2012.

De Waal, F. B. M. "Are We in Anthopodenial?" *Discover* 18.7 (1997): 50–53.

"Días de cine: 'Entre lobos.'" *RTVE.es*. 26 Nov. 2010. Web. 19 May 2012.

"Especiales: Entre Lobos." *El Mundo*. 2010. Web. 19 May 2012.

Fole, Anxel. *Contos de Lobos*. 2a. ed. Vigo: Edicións Xerais de Galicia, 1988.

Foucault, Michel. *Archaeology of Knowledge*. New York: Routledge, 2002.

Garrard, Greg. *Ecocriticism*. 2nd ed. New York: Routledge, 2012.

Gehring, Wes D. *Handbook of American Film Genres*. Westport, CT: Greenwood Press, 1988.

Haraway, Donna Jeanne. *When Species Meet*. St. Paul: University of Minnesota Press, 2008.

Holland, Jonathan. "Film Reviews: Among Wolves." *Variety* 7 Dec. 2010. Web. 19 May 2012.

Janer Manila, Gabriel. *He llugat amb els llops*. Barcelona: La Galera, 2010.

______. *La Problemàtica Educativa Dels Infants Selvàtics: El Cas de "Marcos."* 1. ed. Barcelona: Laia, 1979.

______. *Marcos, The Wild Child of the Sierra Morena*. Trans. Deborah Bonner. London: Souvenir, 1982.

Janer Manila, Gabriel, and Maria Aurèlia Capmany. *Entre llops: l'infant salvatge de Sierra Morena*. Barcelona: Celesa, 2010.

Latour, Bruno. *We Have Never Been Modern*. Cambridge, MA: Harvard University Press, 1993.

National Geographic: The Wildlife Collection. National Geographic Video, 2006. Film.

Ocaña, Javier. "Crítica: La lucha por la vida." *El País* 26 Nov. 2010. Web. 19 May 2012.

Olivares, Gerardo. *Entrelobos*. Cameo, 2010. Film.

_____. "Como Conocí a Marcos." Especiales: Entre Lobos. *El Mundo*. 2010. Web. 19 May 2012.

Rodríguez de la Fuente, Félix. *El Hombre y La Tierra: Serie Ibérica*. RTVE, 1975–1980. Film.

Salvá, Nando. "'Entre Lobos', Una Odisea Desperdiciada." *El Periódico* 26 Nov. 2010. Web. 19 May 2012.

Sánchez, S. "'Entre Lobos': Tarzán Visita Córdoba." *La Razón* 3 Dec. 2010. Web. 19 May 2012.

Simons, John. *Animal Rights and the Politics of Literary Representation*. New York: Palgrave, 2002.

Taylor, Diana. *The Archive and the Repertoire: Performing Cultural Memory in the Americas*. Durham, NC: Duke University Press, 2003.

Williams, Raymond. *Marxism and Literature*. Oxford: Oxford University Press, 1977.

Wilson, Alexander. *The Culture of Nature: North American Landscape from Disney to the Exxon* Valdez. Toronto: Between the Lines, 1991.

11

The Bull Also Rises: The Political Redemption of the Beast in *La pell de brau* by Salvador Espriu

William Viestenz

This essay approaches the Catalan poet Salvador Espriu's *La pell de brau* (The Bull's Skin) (1960) as a type of bestiary that poeticizes an inventory of animals, notably the bull, as figurations of the political. As a response to both the ideological tenets of Francoism and the federalist critique of the unitary development of the Spanish State in the nineteenth century, *La pell de brau* proposes a model of divided sovereignty that redeems Iberia's beasts—*las bestias negras* (the black beasts) denounced by Francoism—through a relational rapprochement that rethinks the taxonomic strictures of what constitutes a political animal (*zoon politikon*) on the Peninsula.[1] In Espriu's figuration of the eponymous bull, the politicity of the redeemed beast springs forth from a covenental pact amongst Iberia's *diversos homes i diverses parles* (diverse peoples and diverse speeches), an agreement consecrated by the grace conferred by the methodical dedication to work and the ability of formerly incomprehensible languages to respond to a political vocation. With the assistance of the late thought of Jacques Derrida, I will postulate that the capacity to respond and enter into pacts elevates the beast to the category of political animal by opening up the possibility of self-thesis.

Espriu was born in 1913 in Santa Coloma de Farners, near Girona, but spent summers at his family's vacation home in Arenys de Mar, a locale that gave birth to the anagram Sinera, which he featured in many of his writings. In addition to penning prose novels and short stories, he spent his late teens and early twenties studying law at the Universitat de Barcelona and was pursuing studies in Egyptology when the Spanish Civil War broke out in 1936. After the war, Espriu focused almost exclusively on poetry and, in defiance of the Francoist regime, his writings were composed entirely in Catalan. The arc of Espriu's postwar literature begins with a cosmologically abandoned time of destitution, the Spanish Civil War, and therefore works such as *Cementiri de Sinera* (1946) and *El caminant i el mur* (1954) dwell on the loss of language's plasticity, literature's connection to a dynamic reading public, and the type of poetic subjectivity required to exit such a moribund state. This poetic arc arrives at an articulation of Iberianism in *La pell de brau*, published in 1960, as Espriu steps outside of the hermetic enclosure of his previous verses and details how the sacred communion of each nation on the peninsula connects with a universal project of coexistence.

La pell de brau, more precisely, details the painstaking work that goes into substituting the proper names "Spain" and "Portugal" with a new, and yet very old, common signifier; namely, throughout the collection Espriu refers to the peninsula as Sepharad, the ancient Hebrew term for Hispania. As regards semiology, I argue that Espriu's concept of "Sepharad" functions according to the philosopher and linguist Wilhelm von Humboldt's understanding of language. In basic terms, Humboldt theorizes language as a dynamic process (*energeia*) and not a calcefied product (*ergon*). Sepharad, corresponding to *energeia* under Humboldt's rubric, would be a sign-in-becoming defined in the act of its production; i.e., by the activities and principles of action of the collective that comprises the volume's poetic voice (nearly all of the verses use the first person plural verbal conjugation).[2] Against the *energeia* of language, Humboldt proposes the term *ergon*: discourse as a fixed and finished product or a *Werk*.[3] Language "in itself is no product (*Ergon*), but an activity (*Energeia*). Its true definition can therefore only be a genetic one" (*On Language* 49). Essentially, I will assert that for Espriu the terms "Spain" and "Portugal" have ceased to be acts of dynamic production and are instead the ergonic products of a past littered with violence and antagonism.

Sepharad, on the other hand, replaces the etiolated proper names formerly attached to the Iberian Peninsula and genetically produces a new structure of identification through the mode of poetic production. Through his own symbolic substitution, Espriu enunciates the principles underlying the vital *energeia* of the term Sepharad, which is a political construct only viable as a processional state of activity opposed to the concluded history inscribed within Iberia's previous identitary signs. Humboldt postulates that the very "bringing-forth of language" makes possible the development of "mental powers and the attainment of a world-view, to which man can attain only by bringing his thinking to clarity and precision through communal thinking with others" (27). In *La pell de brau*, the constancy of a communal dialogue amongst the diverse speeches of Sepharad vitalizes a political pact that attains a pan-Iberian worldview (*Weltanschauung*) that reapportions the centralization of sovereign power, a model that is eminently transportable to other fields of power—a maneuver I will initiate at the end of this essay with respect to academic fields of study.

Already in *La pell de brau*'s opening verses does the Humboldtian theory of language foreground the relational subversion of Iberia's political antinomies. The collection begins with a description of a bull charging a stretched out hide and raising it up as a flag: "El brau, en l'arena de Sepharad / envestia l'estesa pell / i en fa, enlairant-la, bandera" (The bull, in the sands of Sepharad / charged the stretched out skin / and rising it up, transforms it into a flag).[4] In this poeticization of the Ancient Roman geographer Strabo's comparison of Hispania to a bull's hide, Espriu establishes the geopolitical delimitation of Sepharad, an arena of historical thrownness implicating any of the peoples whose contingent histories led to and from the Peninsula and occasioned a *trepitjar* (stepping onto) of the stretched out hide. The conclusion of the first poem reaffirms this geo-political enclosure, which places the setting of the verses "sota la closa eternitat del cel" (under the closed eternity of the sky). Poem II solidifies that the hide raised in the beginning verses bears a syntagmatic relation to the political body of Sepharad: "Ets estesa pell de brau, / vella Sepharad" (You are a stretched out bull's hide, / old Sepharad). Thinking of Sepharad as a "political body" designates the people in biological terms, which Poem I stresses through its focus on the corporeality of the bull, from whose horns the flag hangs.

The stretched-out skin encompasses the entirety of the Peninsula, Portugal included, and the ersatz flag bearing the mark of past violence and subjugation lifted by the bull in Poem I symbolically references multiple centers of national identification within Iberia. The poem continues: "Contra el vent, aquesta pell / de toro, del brau cobert de sang, / és ja parrac espesseït per l'or / del sol" (Against the wind, that skin / of a bull covered in blood, / is now a tattered rag / thickened by the gold / of the sun). The bloodied bull's hide, as a finished symbolic product ([*ergon*]) communicates the consequences of the Francoist Regime having placed its *cara al sol con la camisa nueva, que tú bordaste en rojo ayer* (face to the sun wearing the new shirt, that you embroidered in red yesterday), as the infamous hymn goes. Moreover, the confluence of the scarlet blood of the bull and the gold of the sun on the hide gestures toward the *Rojigualda*, the distinguishing emblem of one of the signs that Sepharad comes to displace, "Spain." The accumulated product of Franco's centralization of sovereignty, with all of its attendant exclusions and expropriations, is a ragged piece of clothing caked with blood.

At the same time, the bull—the beast from whose body the skin was violently shorn—betrays an autonomous agency that appropriates with a defiant mien the monopoly of violence that scarred the hide-cum-flag. The *energeia* of the bull's movement thus contrasts with the enervated, sclerotic *Werk* that Espriu attaches to "el domini del flac amblador" (the dominion of the feeble ambler) in poem LIII. Effectively, the collection's opening sets the stage for the poeticization of the principles underlying the potentiality of Sepharad's body politic and declares finished the calcified product—the coagulated, caked blood—of a retrograde administration whose course is run. In order for Sepharad to succeed as a political project, all inscriptions of previous models of sovereign mastery must be lifted up, like the hide, and discarded for the purposes of discovering those parts of the peninsula previously veiled in silence. Sepharad is thus no mere appendage to an existing political structure, nor is it an awakening of the subaltern within the habitus of an extant nation-state's social field.

The active *energeia* of articulating the breadth and logic of Sepharad, as a sign, produces new sounds that express a more appropriate correspondence to the objects "out there"—for Espriu, this is the enclosed geo-political arena of the Iberian peninsula. A consequence of the shift to a collective language-in-becoming—from Spain and Portugal to Sepharad—as the poetic voice notes in poem LIII, is that "el gran nombre / après i ordenat de les paraules / es perd lentament en el silenci" (the great learned and ordered / number of words / is lost slowly in silence); while the collective poetic voice's superior cognizance of "els camps" (the fields), "els boscos" (the forests), "la remor de les fulles i les fonts" (the rustle of the leaves and fountains) of Sepharad in the final poem of the work permits a new form of writing: "anem escrivint / en aquesta pell estesa, [. . .] a poc a poc el nom / de Sepharad" (we go about writing / in that stretched out hide, [. . .] little by little the name / of Sepharad). The use of the gerund is worth noting, and the active "writing" of the name Sepharad parallels the bull's gerundial *enlairant* of the ersatz flag in the first poem of the collection. The writing into existence of Sepharad functions as a processional act that nevertheless occurs within a standalone geopolitical milieu under the closed eternity of the sky. This concept mirrors the work of the gerund: an active verb functioning as a standalone noun; a set-apart term that is simultaneously relational. The work of bringing Sepharad into being is thus energeiac and possesses the processional contingency that Humboldt perceives in the development of language.

The final inscription of "Sepharad" onto the bull's hide thus comes to replace the testimony of the animal's subjugated history represented by coagulated blood. This self-naming is a prepotent declaration of sovereignty. As Agamben trenchantly stresses, "an act is sovereign when it realizes itself be simply taking away its own potentiality not to be, letting itself be, giving itself to itself" (*HS* 46). Jacques Derrida echoes Agamben's position, arguing that "the concept of sovereignty will always imply the possibility of this personality, this thesis, this self-thesis, this autoposition of him who posits or posits himself as *ipse, the (self-)same, oneself*" (67). Inscribed within the notion of sovereignty is the naming of oneself into existence and the possibility of self-thesis. Indeed, the willing into existence of the name "Sepharad" in the final poem of the collection returns one to the bull's opening obstreperous act of *embestir* Iberia's stretched out skin. This poignant image of a bull rising up from a state of inexistence—a life rendered bare by the loss of its outer hide—asserts an ontological potentiality; a letting a bull be as a bull ought to be. Similarly, the inhabitant of Sepharad, with the freedom to join a covenental pact amongst *diversos homes*, freely gives itself to itself by writing the peninsula's new name onto the bull's skin at the end of the collection. In both cases, the two forms of animal morph into beings-in-themselves and cease being instruments through which the Sovereign derives mastery and power.

Espriu thus directly links the poeticization of Sepharad to a reconceptualization of sovereignty and a conversion of the bull, as a political figuration of all of Franco's *bestias negras*, into a political animal. Unlike previous iterations of mastery, however, the collective poetic voice of Sepharad refuses to define itself through a symmetrical relationship with subjugated bare life. In Agamben's thought, such a symmetrical relationship defines the link between the Sovereign and the *homo sacer*, the criminal bare life that inhabits a state of exception embedded within the confines of the nomos. "What unites the surviving devotee, *homo sacer*, and the sovereign in one single paradigm is that in each case we find ourselves confronted with a bare life that has been separated from its context and [. . .] is for this very reason incompatible with the human world" (100). In *La pell de brau*, animal life, such as the bull, and the *bestia negra*, a human being stripped of its compatibility with the polis, are no longer the negative electrodes that give sovereignty a positive charge. Espriu's interdiction of the *domini del flac amblador*, of course, refers to the Franco Regime, but critiques any mode of sovereignty that maintains zones of ontological indistinction within which any form of life—human or animal—can be reduced to a *bête noire*.[5] In the prologue to the *La pell de brau*'s first edition, Espriu precisely notes that his conception of the peninsula and his ideal social and political organization of the State "contempla una Espanya federal composta per quatre nacions: la nació castellana, la portuguesa . . . la basca i la catalana" (contemplates a federalist Spain comprised of four nations: the Castilian, Portuguese . . . Basque and Catalan nations) (cited in *OC XII* 243.5). In 1976, Espriu further explains that *La pell de brau* was originally a contestation to "unes paraules d'Ortega, que també els homes de la perifèria peninsular érem capaços d'entendre el complexíssim conjunt dels essencials problemes ibèrics" (243) (some words by Ortega [y Gasset], that the peoples of the peninsular periphery were also capable of understanding the complex array of essential Iberian problems).[6] Espriu directs his critique at the historical project of Ortega, but with a cognizance of its consistency with the reactionary ideology of Francoism.

The counterproposal of Statist organization to which Espriu signals, tied to a Federalist relationality, directly cites the work of a cohort of Catalan intellectuals in the nineteenth century. One obvious name, especially in reference to the periphery's ability to contribute to overarching Iberian problematics, is Valentí Almirall and his philosophy of *particularisme*.[7] Almirall, in addition to penning, amongst others works, *Lo Catalanisme* (*Catalanism*) in 1886, also founded *El diari català* (1879–1881), a landmark newspaper in the early history of Catalan journalism. In *Lo Catalanisme*, Almirall argues that a particularist State, what he calls an "Estat compost" (compound State), contests the centralization of sovereignty typical of unitary political configurations: "la base del sistema d'associació d'Estats és la divisió de la soberania" (176) (the base of the system of association of States is the division of sovereignty). Not swayed by the ineffective and fraudulent Alfonsan *turno pacífico*, Almirall puts forth that the devolution of sovereignty to Iberia's myriad nations is ultimately "un remei" (23) (a remedy) to the poverty of the nineteenth-century Spanish nation-state. For both Almirall and another crucial figure of the period, Francesc Pi i Maragall, this division is regulated under the auspices of collectively agreed upon principles negotiated through political pacts.

In *La pell de brau*, the poetic voice conceives the union of Sepharad's diverse peoples as a pact underpinned by adherence to structurally fixed principles of action.[8] The two most frequently cited principles underlying the Iberian social pact are love and freedom. In Sepharad, "diversos són els homes i diverses les parles, / i han convingut molts noms a un sol amor" (poem XXX) (diverse are men and diverse are their speeches, / and many names are capable of expressing a single love). Moreover, the eternal hope of the collection's poetic voice is "que Sepharad visqui eternament / en l'ordre i en la pau, en el treball, / en la difícil i merescuda llibertat" (poem XLV) (Sepharad live eternally / in order and peace, in work / in difficult and deserved liberty). Freedom is *prima facie* deserved, not at all a quality that a subject progressively gains. As a right inherent to the constitution of Sepharad *qua* Sepharad, autonomy cannot be sacrificed in order to enter into a social contract; therefore, a State cannot deprive freedom and retain its politicity. Any contractual pact possesses an important precondition: a respect for the right of each adherent to the pact to maintains its legal charters, political institutions, shared histories, modes of communication, and so on.

The notion of work in poem XLVI also emphasizes the processional nature of the term Sepharad and sets the parameters for inclusion within the refashioned body politic. Gesturing toward Sepharad's eternal life confers a metaphysical dimension to the polity's existence, which elevates the stretched out hide from a merely physical marker to the status of symbolic shell or placeholder. Against the contingency of the passage of history, idealizing particular principles holds Sepharad together on a metaphysical register. These principles include political love, liberty, promoting dialogue over violence, and attempting to comprehend and esteem the variegated cultures linked to the history of the Iberian peninsula. The poetic voice ties the notion of work to the persistent dedication to translating those ideal principles to political existence on the ground; in a sense, work is the conduit through which one receives grace and the blessing needed to continue *trepitjant* (stepping) over the geopolitical space demarcated by the stretched out skin of the bull.[9] Resurrecting Sepharad from its period of decadence precisely involves constructing "el lent temple / del teu treball" (poem XLVII) (the slow temple / of your work), which explicitly ties methodical work to the conferral of sacred grace.

Thinking of the hide-cum-flag in poem I in metaphysical terms gestures toward a splitting of the Sepharadian political body into a noetic and physical dimension. Ernst Kantorowicz's landmark study of the King's two bodies begins with an analysis of the use of the term "body politic" in Elizabethan England. There, the King was believed to embody—literally—two capacities: a body natural and a body politic. The body natural corresponded to the King's finite mortal existence whereas the body politic was the eternal, sacred container tied to the institution of sovereignty. Kantorowicz notes that the body politic remained static while the finite natural bodies of various kings succumbed to mortality. This "continuity of repetitive incarnation of the body politic in exchangeable bodies natural suggests a 'noetic' interpretation of kingship" (18). A noetic take on kingship posits a metaphysical facet to sovereignty and allows for geo-political spaces of power to exist beyond the lifespans of its individual members. Essentially, sovereign power is a sum worth more than its finite parts. For Espriu, Sepharad can only live eternally if the newfangled flag raised in the first poem ties the polity together around universal principles that the "bodies naturals" on the ground enunciate through their collective acts. The raised up hide, in other words, represents the noetic half of Sepharad's political body, through which the polity "lives eternally," whereas the charging bull, with its display of fleshy corporeality, gestures toward the *corpus naturale* of the peninsula's finite bodies that will come and go.

The division of sovereign power amongst various national centers highlighted by Almirall and present in Espriu's ideal conception of the State reappears in Francesc Pi i Maragall's concept of revolution. The second President of the First Spanish Republic argues that revolution maintains as an end goal the representation of power "pero tiende a dividirlo; no mata aún la fuerza, pero le clava el puñal hasta donde sabe y puede" (*La reacción* 244) (but it tends to divide it; it does not even dispel force, but drives the dagger as far as it knows how and is able). Compare this sentiment to the definition of the nature of *treball* (work) that goes into the active *energeia* of articulating the symbolic reference of the sign "Sepharad" in poem XLII of *La pell de brau*: "clavem el punyal / del crit al duríssim / cor de Sepharad" (we drive the dagger / of the cry into the hardened / heart of Sepharad). The echo of Pi i Maragall's discourse on revolution in Espriu's poeticization of Sepharad emphasizes further that the metamorphosis of sovereignty in the work is a radical alteration that wipes away, in one fell stroke, all previous iterations of the political.[10] As poem XLVII stresses,

> lluny del port de refugi
> rentaràsen
> aigües d'esperança
> tota la sang
> d'aquesta trepitjada
> pell de brau
>
> (far from the port of refuge
> you will cleanse
> in waters of hope
> all of the blood
> on that treaded on
> bull's skin).

Such a cleansing of the carmine-tinged content of the flag, far from the golden rays of the sun, represents an inimical disavowal of the emblematic symbol formerly attached to the Spanish State. In other words, the advent of Sepharad, in terms of Niklas Luhmann's thought, is not the mere "realization of higher-level relations among relations" within a closed social system (23). Espriu's poetics, against this notion, erects an entirely new system with a distinct logic of relationality amongst its networked nodes—the *diverses parles* of Sepharad.

For Pi i Maragall, as in a host of other writers of federalism such as Proudhon, Montesquieu, and Alexander Hamilton, the principles of free election and of *agapé* for *diversos homes i parles* are a precondition for any political pact that divvies up sovereign power. As Pi i Maragall notes in an appendix to the third edition of *Las nacionalidades* (The Nationalities), "no es autónomo el mancebo a quien se impone una esposa. No lo es el hijo mientras vive bajo la potestad del padre [. . .] No lo son los pueblos que se agreguen por la violencia o la autoridad de los príncipes" (493) (the bachelor onto whom a wife is imposed is not autonomous. Neither is the son while living under the authority of the father [. . .] Neither are those nations that come together through violence or the authority of princes). Along nearly identical lines, in *La pell de brau* the political is formed "en la llei i en el pacte / que sempre guardaràs, / en la duresa del diàleg / amb els qui et són iguals" (poem XLVII) (in the law and in the pact / that you will always keep / through the constancy of dialogue / with those who are your equals). For Espriu, the political is only legible under these conditions; therefore, "no pot escollir príncep / qui vessa sang" (poem V) (he who spills blood / cannot choose a prince) and the "dies d'odi" (days of hatred), inscribed within the old, ergonic symbols such as the blood-stained bull's hide, are eras of "desgovern" (poem XLII) (disgovernment). A government ceases to exist as such whenever it fails to define itself through its acts of production, instead falling back on antiquated and outmoded sources of identity. By the same token, a pact is only truly political when formed by the constancy of a mutual dialogue in which all parties freely assent to participation. Thus, whoever resorts to violence exhibits apolitical behavior and forfeits the right to choose a prince.

However, in order to create a political space of pacific dialogue amongst diverse peoples, the poetic voice of *La pell de brau* first must appropriate the *flac amblador*'s unilateral hold on the exercise of violence. As poem XLII instructs, "si vols el poder, / fes que sigui teva" (if you want power / make it yours). Displacing power from the *flac amblador* to the Sepharadian collective secures the freedom required in order to wash away the bull hide's scarred past and inscribe onto its materiality the name "Sepharad," imputing the sign with the principles pursuant to a reordered State. Already in the first poem the bull's charging and impaling of the stretched out hide symbolizing Hispania denotes the force necessitated by any attempted division of sovereignty. The *solar de la llibertat* (plot of liberty) on top of which the inhabitants of Sepharad construct a temple late in the collection is only cleared by a sacrifice; that of *l'ocell solar* (the sky bird) in poem XVI.[11] In that poem, the collective poetic voice representing the *zoon politikons* of Sepharad string up and hang the bird: "I esdevindrà / l'esglaiós sacrifici d'imperial / captiu que va per segles senyorejar / el cims, el cel, els somnis de Sepharad" (The frightful sacrifice of the imperial captive / will come to pass / Captive that for centuries ruled / the peaks, sky, the dreams of Sepharad). Elsewhere, Espriu identifies the *l'ocell solar* as "L'Àguila de Sant Joan, heràldica a l'escut d'Espanya" (qtd. in *OC XII* 303) (the Eagle of St. John, heralded on the Spanish Coat of Arms). Based on these remarks, Delor i Muns has argued that the *ocell* prevails as the political figuration of both Monarchical Spain and the Castile-centric vision of the nation that Espriu references in the 1976 prologue of *La pell de brau* apropos Ortega y Gasset (cited in *OC XII* 303 n.10). The *ocell*, it might be said, is the sovereign counterpart to the bull reduced to a state of bare life. The bull rises up from a beastly state whereas the poetic voice construes the sky bird as a type of sovereign.

In the final set of seminars that Jacques Derrida presented to the EHESS in Paris shortly before his death, the French philosopher delves into the common occurrence of filtering a discourse on the exercise and division of sovereignty through the figure of the animal. Derrida distinguishes the beast from the animal, which he views as already political in an Aristotelian sense as a *zoon politikon*.[12] The beast exists outside of the political, stripped of *bios*, that part of life dependent on inclusion within a group, and possesses a symmetrical relation to the Sovereign all the while paradoxically being his antimony: "Sharing this common being-outside-the-law, beast, criminal, and sovereign have a troubling resemblance [. . .] [they] strangely resemble each other while seeming to be situated at the antipodes, at each other's antipodes" (17). The symmetricality between beast and Sovereign is seen in Espriu's poetry in the correspondence between the skinned bull and the *ocell solar*. The Sovereign ability to will oneself into existence and declare law legitimate by maintaining a zone of indistinction beyond the political is only possible by simultaneously regulating, beyond the *nomos*, a bestial excess that is denied speech, representation, and responsibility to others.

That Espriu would percolate a particular pattern of Spanish sovereignty through still another figuration of the animal, the sky bird, bears logical consistency with Derrida's thought. The Sovereign and the beast, both beyond the law, bear an identificatory resemblance related to the animal. Political figurations of the animal, however, tend to designate the beast as something less-than-human and the Sovereign as superhuman or monstrous. On the latter point, Derrida points to Hobbes's imagining of the artificial nature of sovereignty through the monstrous dragon referred to as Leviathan in the Book of Job. Because sovereignty is created artificially by humankind, and not God, so must be the animal that represents the institution of power. In Espriu's poetry, *l'ocell solar* may not betray the monstrous, fantastical qualities of a dragon, but certainly does possess the inhuman capacity for flight, which links the firmament and the ground. It is for this reason that the *ocell* for centuries ruled the peaks, skies, and dreams of Sepharad.

The bull, on the other side of the equation, fulfills the role of terrestrial creature, a quality emphasized in the opening lines of the first poem designating its limited space of movement as *l'arena de Sepharad*. Despite their similarities, the beast, as a base creature, is "ignorant of right" while the sovereign *ocell* possesses "the right to suspend right, to place himself above the law that he is, that he makes, that he institutes, as to which he decides, sovereignly" (Derrida 32). The ornithological ability to rise above the terrestrial makes the *ocell* a fitting symbol for the sovereign right to suspend and rise above the rules that keep other animal forms tied to the Earth. Espriu, through poetic symbolism, attaches two figurations of the animal to two distinct classes of human beings. On one hand, the outlawed humans that have suffered the brunt of the tyrant's sovereign force equate with the beastly bull. On the other, the Sovereign assumes the mantle of *ocell solar*. The *plaza de toros* thus becomes a perfect metaphor for the state of exception that Agamben posits at the center of the *nomos*, a liminal space in which the outlawed human—the *homo sacer*—is like the wolf-man, a monstrous hybrid of human and animal. The *barrera* surrounding the bullring is that "threshhold of indistinction and of passage between animal and man . . . the life of the bandit is the life of the *loup garou*, the werewolf, who is precisely *neither man nor beast*, and who dwells paradoxically within both while belonging to neither" (*HS* 105). On a plane that resembles Bataille, Agamben posits that the *homo sacer* and the wolf-man cannot be sacrificed because they constitute, along with the Sovereign, the originary political relation. For Espriu, the *ocell solar*'s death is a sacrifice, banishing the concept of *homo sacer* and rupturing the immanence of the two poles of the sovereign exception. *La pell de brau* thus proposes a form of power without exclusion by paradoxically banning all relationality with the previous Sovereign.

In 1946, 1966, and 1968, the Catalan painter and sculptor Joan Miró produced a total of four sculptures cast from the same mold entitled "Ocell solar."

The nearly identical aesthetic objects—the first two were made of bronze, the latter two of white marble—possess a heavy materiality that emphasizes the bird's conversion from an energeiac, vital status to the polished completeness of a finished sculpted work. In Espriu's poetics, the *ocell solar* suffers a martyrdom that also leaves it immobile: "Estiraven les ales de l'ocell solar, / per la façana el pugen cap a dalt / El claven prou enlaire . . . el deixen ja / fix en el mur, immòbil" (They spread the sky bird's wings, / and push it up the facade and nail it high in the air . . . they leave it fixed to the wall, immobile). The immobility of the bird contrasts with the free range of movement of the bull in the first poem of the work. Going a step or two beyond Miró's aesthetic interpretation of the *ocell*, Espriu, through the motif of sacrifice, denies the *ocell* participation in the world of relational objects. Georges Bataille argues in particular that the end goal of sacrifice is to abrogate the "thingness" of the victim: "Sacrifice essentially turns its back on real relations. If it took them into account, it would go against its own nature, which is precisely the opposite of the world of things on which distinct *reality* is founded. It could not destroy the animal as a thing without denying the animal's objective reality" (44–45). In order to clear the *solar de llibertat* of past vestiges of the sky bird's mastery over Sepharad, the collective poetic voice of *La pell de brau* must obliterate the objective reality of the animal and preclude further relations with it. In the wake of the sacrifice, then, the inhabitants of Sepharad "oblidem a poc a poc / i acabàven / ells balls i les cançons / del vell cel / de Saron" (forget little by little / and the dances and songs / of the old sky / of Sharon / are forgotten). The leitmotif of forgetting results from the *ocell*'s sacrificial turning away from the world of objects, a shift that the poetic voice marks through the conclusion of the rituals linked to the "old sky" of Sharon. This is the sky, of course, over which the *ocell* formerly possessed mastery.

The offing of the *ocell* is quite literally the removal of a false idol—the "ídol que vares alçar, imatge del teu mal" (poem VI) (the idol that you raised, image of your wickedness)—and the disavowal of an epoch of *desgovern*. Moreover, substituting the *domini del flac amblador* with a political pact formed through law and the constancy of dialogue legitimates the rule of Sepharad through what Max Weber calls a "belief in the validity of legal *statute* and the appropriate (*sachlich*) juridical 'competence' founded on rationally devised rules" (*Political Writings* 312). This particular mode of legitimating rule is one of three possibilities that Weber outlines, with the other two being: "the authority of 'the eternal past,' of *custom*" and "the authority of the exceptional, personal '*gift of grace*', or charisma, the entirely personal devotion to, and personal trust in, revelations, heroism, or other qualities of leadership in an individual" (312).

Franco, of course, attempted to adopt both of these latter two variations, being at once *caudillo por la gracia de Dios*—the charismatic leader chosen by a call to service issued by God—and the inheritor of a glorious, eternal tradition, epitomized by the Catholic Kings, of purifying Spain of its non-Catholic elements.

Dividing political power through a pact that normalizes relations amongst the different cultural centers of Iberia through adherence to particular principles of action deliberately shuns the individualistic sovereignty and historical exceptionalism inherent to Francoist ideology. In poem XLII, the great "esforç" (effort) of the progressive inscribing of the name "Sepharad" is to produce a land "sense ramat / ni pastor" (without flock / nor shepherd). That is to say, Sepharad's division of sovereignty leaves in its wake a political organization not directed by a single, charismatic leader in charge of a homogenous herd. *La pell de brau*'s concept of "deserved freedom" amongst diverse peoples declares null and void the brand of decisionistic sovereignty that permits any single individual to decide, in Schmittian fashion, on the exception.[13] Legitimating rule through normative law and rationally devised principles also sets in motion a rising up of the *bête* to an intermediary position above an ontology of bare life outside the nomos but below the onmipotent, individualistic stature of the Sovereign as a stand-in for God on Earth. Derrida anoints the *zoon politikon* (the Aristotelian political animal) at this interstitial nexus: "politicity, the being-political of the living being called man, is an intermediate between those two other living beings that are beast and god, which, each in its own way, would be 'apolitical'" (25–26). Espriu, too, would posit that a manifestation of sovereignty buttressed by a synchronous relationship between the *domini del flac amblador* (dominion of the feeble wanderer) and the voiceless, stripped-of-bios bull in the sands of Sepharad inaugurates a period of apolitical "disgovernment."

Important for the concerns of this essay, Derrida argues that the Sovereign *and* the beast are beyond conventions—the former because it is impossible to negotiate with God and the latter due to a lack of language. "If one cannot make a convention with the beast, any more than with God, it is for a reason of language. The beast does not understand our language, and God cannot respond to us [. . .] in both cases, there could not be an exchange, shared speech, question and response, proposition and response, as any contract, convention, or covenant seems to demand" (55). The beast, in other words, is radically passive, which is a counterweight to the Sovereign's monopoly on force and access to "worlding" language. Already in Sophocles's *Antigone*, a play that Espriu himself adapted in 1939, does this asymmetry arise in a comment directed at Kreon: "But this is one of kingship's many blessings—that it can both act and speak just as it wishes to" (53). For Derrida, *la bête*, like the subjugated bull in

Espriu's work, is a being-in-itself with a materiality already wrought and closed to other kinds of relation or determination. The political animal, the *zoon politikon*, is at once a form of animality but an elevation above the beast due to the ability to respond, propose, and enter into covenental pacts.

For this reason, the political animalization of *las bestias negras* of Sepharad in Espriu's poetry hinges on the ability of the *diversos homes* to congregate according to a called response: the "diverses homes" with "diverses parles" participate, in their own unique ways, in the universality of love ("un sol amor"). Mutual intelligibility also becomes a factor; the diverse peoples are guided "cap a la unitat" (poem XLII) (toward unity); but it is commanded that "siguin segurs els ponts del diàleg / i mira de comprendre i estimar / les raons i les parles diverses dels teus fills" (poem XLVI) (the points of dialogue be secure / try to understand and respect / the reasoning and diverse speeches of your sons). This, of course, is not the first "Adéu, Espanya" that arises due to a lack of understanding the speech of one of the State's *fills*.[14] Here, the poetic voice bids Spain adieu in favor of Sepharad, and this dissolution and repartition of sovereignty precisely germinates at the point that those who worship at the altar of *l'ocell solar* finally hear "la veu atronadora" (the booming voice), to borrow from Joan Maragall, of the State's excluded beasts.

The covenental nature of the pact formed by the diverse peoples of Sepharad deepens the meaning of the final inscription of the name "Sepharad" onto the bull's skin. The inscription, more than a mere substitution of one proper name in favor of another, carries the solemnity and responsibility of a political oath. Agamben notes that the oath is "situated at the intersection of religion and politics" and gains prominence whenever a political pact "enters into crisis or turns to renew itself in diverse forms" (*Sacrament* 1). The sacred and the secular indeed intersect in the very outset of *La pell de brau*: the poetic conceptualization of Sepharad is referred to as "oració nostra / i blasfèmia nostra" (our prayer / and our blasphemy) and the *poble*'s recovery of its voice culminates in a resounding "Amén" at the conclusion of poem XXXVIII.

The question then becomes how to theorize any kind of utterance that takes the form of a political oath or testament. For Agamben, the matter revolves not around the affirmation of factual phenomena—in this case the clearing away of space in order to erect a new palace of congregation on the sands of Sepharad—but with the semiotic power of language to compel its speakers to adopt certain poses toward the social and accept responsibility for words uttered. "The testimony that is in question in the oath must therefore be understood in a sense that has little to do with much of what we normally understand by this term. It concerns not the verification of a fact or an event but the very signifying power of language" (*Sacrament* 33). The evolution of *La pell de brau*, from an accounting of the cruelty embodied within Iberia's enervated identitary signs, to the clearing away of the bull's hide through sacrifice and the formation of a political pact attached to the metaphysical ideals of plurilingualism, freedom, and political love, all leads up to the creation of an alternative sociopolitical organization of the peninsula. Channeling Humboldt, the term "Sepharad," over the course of the book, is a sign-in-becoming gradually taking shape; on the other side of the semiotic equation, the actions of the poetic voice enacts the methodical evolution of a referent. At the work's conclusion, all that remains is the need to ratify, through the sacrament of language, the validity of the semiotic term through an oath. Every oath, of course, entails accepting a grade of responsibility according to the terms of the agreed-upon political pact. The notion of a sacred response via the oath is embedded within a subject's capacity for *responsibility*. The signature of each participant in the construction of Sepharad—an oath taken through the deployment of "molts noms" for "un sol amor" -thus completes the evolution of the black beasts of Francoism to the status of political animal.

The collections of poetry that Espriu published prior to *La pell de brau*, such as *Les hores* (The Hours), *El caminant i el mur* (The Walker and the Wall), and *Final del laberint* (End of the Labyrinth), meditate at length on the disappearance of language's elasticity and musicality and deplore the impotence of the words of poetic creation. The poetic voce of "Ofrenat a Cerber" (Offered to Cerberus) in *Les hores*, for example, sinks his hands into "l'or misteriós / del meu vell català" (the mysterious gold / of my old Catalan) but raises them "sense cap guany, blanques de cendra" (without any profit, whitened by ashes). In the concluding lines of the final poem of *El caminant i el mur* titled "Sentit a la manera de Salvador Espriu" (Felt/Heard in the Manner of Salvador Espriu), the poetic voice declares himself to be "sense missatge, sol, / enllà del cant / enmig dels oblidats / caiguts amb por, només un somni fosc" (without message, alone, / beyond song / amongst the forgotten / fallen in fear, nothing more than a black shadow). This lyric, undoubtedly an allusion to Dante's walking shades, also establishes a reciprocity amongst a forgotten, beastly caste consigned to an oblivion made possible, first, by the lack of a message, and secondly through spatial placement within a zone "enllà del cant." The elevation of the beast to political animal in *La pell de brau* thus proceeds from Espriu's prior poeticization of a mute state of Being beyond the possibility of a covenental pact; a description of an existence that could never congregate in a common *solar de llibertat* and respond to a vocation to participate in an Iberian political project predicated on universal principles.

La pell de brau thus issues a call to us, as readers, to denounce the subjective gradation of life that allows the neighbor or the enemy more generally to become the other than human; the beast. It also instructs the reader to be wary of the anthropocentric limiting of the definition of response to *logos* alone; the opening image of the bull regaining its range of motion and reasserting its right to be is the recuperation of an innate freedom that the taurine arts undeservedly abrogate. The bull, as a form of matter, regains its vitality and vibrancy not within the enclosed *Umwelt* of *la plaza de toros* but within the open space, *l'arena*, of Sepharad. The *plaza*, in this sense corresponds to the closed-off state of exception, beyond the nomos, in which the beast and the Sovereign exist at each other's antipodes. In the sands of Sepharad, no such sinister exception exists just beyond the boundaries of the polis. In terms of human-animal life, this signifies that Espriu promotes an ontology of human and animal matter that mirrors the horizontal equality of the diverse peoples of Sepharad. The elevation of the *bestias negras* of Iberia to the status of political animal concomitantly obliterates the hierarchical ordering of political status that even permits for a categorical distinction between beast, human political agent, and God. This reverberates within the ontological realm—with the elimination of the denigrating category of beast, the Sepharadian animal, in this case the bull, discloses its unique Being autonomously without symbolic appropriation as an unethical stand-in for a class of human beings denied political participation. In accordance with the thought of John Frow, the animal, by existing on a horizontal plane of movement whose space is shared by human agents, makes it possible for, "these two sets of kinds to exchange properties" (283). Under such an arrangement, the human thing (the inhabitant of Sepharad) and the nonhuman (bull) can be "read horizontally as a juxtaposition rather than vertically as a hierarchy of being" (283). In *La pell de brau*, *energeia* is the property that yokes these two classes together. *Energeia* both pervades the active writing of the term "Sepharad" and also seeps into the bull's recuperation of its shorn hide in the act of charging the stretched out hide in the first poem. Alongside the elevation of the Sepharadian human to political animal, the bull, ironically, is elevated from the status of beast to a register of ontological authenticity: the beast becomes a bull.

Jordi Pujol, the former leader of the center-right *Convergència i Unió* party and president of the Catalan Generalitat from 1980 to 2003, has remarked that "Espriu era el poeta que calia a la generació que havíem nascut sota el signe de la derrota" (qtd. in Walters 127) (Espriu was the poet that the generation born under the sign of defeat needed). Armed with the uncritical security of being outside the political limelight, a retired Pujol has recently declared the mission of translating Sepharad into a political project a failure, despite his own best efforts. Pujol's tone has sharpened in the wake of the Tribunal Constitucional's abrogation of the most recent Catalan Statute of Autonomy in 2010, but even prior to this event Pujol noted in 2009 that "hi ha hagut gent—aquí, entre nosaltres, i sobretot fora d'aquí, en altres indrets de Sepharad—que no ha cregut mai en tot això" ("El fracàs") (there have been people—here, amongst us, and above all on the outside, in other locales of Sepharad—that have never believed in any of that). With the arrival of the fiscal crisis, the relations between the center and the "diverse peoples" of the North have grown more tenebrous and made the notion of Espriu's Sepharad seem even more out of step with the zeitgeist of the twenty-first century. On the occassion of Espriu's centenary in 2013, Edgar Illas cogently put forth that Espriu's concept of dialogue amongst Iberia's diverse cultures "ja no ens serveix" (no longer is useful to us) amidst "una guerra financera i legal amb l'Estat" ("Vigència") (a legal and financial war with the State).

In light of my analysis, the apparrent implausibility of Sepharad as a model of sovereignty in present-day Spain speaks less to the conceptual validity of the political theory worked out in *La pell de brau* than to a failure to understand the revolutionary scope and nonnegotiable terms of the pact that brings Espriu's ideal vision of the State into being. Though this is fodder for another essay entirely, one could question the extent to which Pujol truly managed to wipe away the inherited framework of the Francoist State and push the vestigial *ocells solars*—those not eligible to *escollir príncip*—outside of all political relationality. As I have repeated throughout this essay, Espriu's Sepharad, unlike other forms of Catalan pactism, is not merely a higher form of relations within an already established social field; i.e., it is not a model for the gradual increase of peripheral autonomy within the strictures of the Spanish State. Furthermore, Espriu's political pact demands that each individual peninsular party submit to mutually agreed upon terms, a precondition that has certainly not always been present.

Of course, the failure of Sepharad's *implementation* in the post-Francoist State does not signify that the concept itself, in abstract terms, is irremediably flawed. On that level, one could push Illas's concluding assertion that "si cal llegir Espriu . . . no és per gaudir del valor present o universal de la seva obra" ("Vigència") (if one must read Espriu . . . it is not to enjoy the present or universal value of his work). In conclusion, I submit that Espriu's theorization of the term Sepharad provides a model for harnessing the *energeia* of any word-in-becoming that emphasizes a revolution of relationality. As an example, I will discuss briefly the recent rise of Iberian Studies. Iberian Studies is a burgeoning field of research representing a methodological shift to the traditional approaches of Hispanism. Concretely, Joan Ramon Resina has asserted that Iberian studies is "una propuesta federativa para el ámbito cultural" (a federalist proposal for the cultural field) that challenges the monolingualism and canonical exclusions of traditional Hispanist discourse.

The question of divided sovereignty remains patent within this pedagogical shift. As a discipline, Iberian Studies is "un espacio academico todavía virtual donde se afirma la memoria de la humanidad por respeto a todas las lenguas. Un espacio en que a ninguna se le disputa la condición de 'dominante' dentro de su propio ámbito social" (206) (a still-virtual academic space in which the memory of humanity is affirmed through respect for all languages. A space in which no one's 'predominance' within their own social ambit is disputed). Like the divided sovereignty of Sepharad, an Iberian Studies paradigm precludes that a single language or cultural tradition possess a unilateral hold on the symbolic capital that determines the visibility or silence of other social ambits without a voice.

Espriu's thinking through Sepharad as *energeia* denotes a state of motion that is at once an actuality and a concept laden with potentiality for change—recall that Sepharad's eternal future is partly assured by a dedicated to work, to active renovation, and *Bildung*. I filtered this concept through the use of the gerund in the writing of "Sepharad" onto the bull's hide. Similarly, Resina places an interdisciplinarity approach at the root of the ethical parameters for Iberian Studies (159–60). Interdisciplinarity, however is combined with a cultural specification determined by the contingent development of the Iberian Peninsula, giving the discipline an adroit permutability while maintaining specialization in myriad cultural histories. In other words, Iberian Studies thinks, in disciplinary terms, of the field of Iberian and Latin American cultural discourse as a confluence of diverse peoples and speeches that each reserve a modicum of freedom. Indeed, the political failure of Espriu's concept certainly provides a clear set of warning signs that could derail the implementation of the new Iberian Studies paradigm. In the end, thinking through this disciplinary evolution by way of Espriu's Sepharad begs the following question: Could a political failure become a pedagogical success?

Notes

1. These *bestias* often included Freemasons, Basque and Catalan separatists, homosexuals, Communists, Jews, Protestants, and Republicans.
2. The concept of *energeia* has Aristotelian roots. In *The Metaphysics*, Aristotle associates energy (*energeia*) with an object's actuality, which is determined by its first causes. He compares the term to the growth of a child into a man: "But energy is an end, and on account of this is potentiality assumed; for not in order that they may have the power of vision do animals see: but they have the power of vision that they may see" (191). Aristotle does not oppose *energeia* to *ergon*—work—whereas von Humboldt would view the term as a state of existence with a processionality pregnant with potential for change.
3. Many writers have compared this division to Ferdinand de Saussure's notion of *langue* and *parole*. *Langue* refers to the socially determined rules of a language; it is "both a social product of the faculty of speech and a collection of necessary conventions that have been adopted by a social body to permit individuals to exercise that faculty" (*Course* 9). *Parole* represents the concrete execution, on an individual level, of the conventions of the *langue*: "Execution is always individual, and the individual is always its master: I shall call the executive side *speaking* [*parole*]" (13). Von Humboldt's concept of language as *energeia* encompasses both *langue* and *parole* and presupposes that linguistic conventions are contingent and only valid in their acts of execution.
4. All of the poems from *La pell de brau* are taken from Salvador Espriu, *Obres completes II* (Barcelona: Edicions 62, 1963). Translations are my own.
5. This notion that the Sovereign is he who decides on the exception to the law, existing both in a state in which the *nomos* is suspended and within the normative strictures of the political, is central to Agamben's theory of the state of exception (*HS* 30–38).
6. In *España invertebrada*, Ortega surmises that if the Catalans and Basques had been given the responsibility of developing the Iberian Peninsula instead of the Castilians, "lejos de arribar a la España una, habrían dejado la Península convertida en una pululación de mil cantones" (41) (far from arriving at a unified Spain, they would have left the Peninsula multiplied into a thousand cantons).
7. Almirall calls his ideology *particularisme* in lieu of federalism due to the latter term's connotations of anarchism and other radical interpretations of Pierre Joseph Proudhon principles of federation.
8. Edgar Illas convincingly locates pactism at the core of several important ideological premises and theories of Catalonia as a nation, including Jaume Vicens i Vives's *Notícia de Catalunya* and Josep Ferrater Mora's *Les formes de la vida catalana*. Illas asserts that pacticsm traditionally argues for a redefinition of Catalonia's relationship with the State that "supersedes the current subordination or colonial bond" (94, 102) and is not related to separatist discourse. I argue that Espriu, though taking his cue

from these ideological undercurrents, puts forth a new type of relationality only in the wake of an abrogration of Spain-qua-signifier.

9. Max Weber, in his explication of the Protestant ethic, has argued that in the most ascetic forms of the religion, such as Calvinism and Puritanism, work is "the *end* and the purpose of life commanded by God [. . .] Unwillingness to work is a symptom of the absence of the state of grace" (107). In previous collections of poetry, Espriu has expressed a fondness for the Northern European ethic, certainly in reference to Protestant worldviews. Take, for example, the "Assaig de càntic en el temple" (Rehearsal of a chant in the temple): "Oh! què cansat estic de la meva covarda / vella, tan salvatge terra, / i com m'agradaria allunyar-me'n, / nord enllà" (*El caminant i el mur* 51) (Oh! How weary I am of my cowardly / old, so savage land, / and how I'd like to distance myself, / toward the north).
10. The metaphor of the dagger, especially in reference to revolution, harkens back to Donoso Cortés's discourse on dictatorship. Declaimed in January of 1849, a year after a wave of revolts gripped the European continent, Cortés compares a sovereign state of exception to a popular revolt, calling the former a "dictadura del sable" (dictatorship of the sabre) and the latter "la dictadura del puñal" (dictatorship of the dagger). "Yo escojo la dictadura del sable, porque es más noble" (261) (I choose dictatorship of the sabre, because it is more noble).
11. Note the homophonic connection between the *solar*, as a plot of land, and the *ocell solar*, whose solar nature links the animal to the golden rays that have coagulated the blood on the bull's hide. Poem LI details that the collective dream of the new Sepharad includes a "nova casa / alçada en el solar de la llibertat" (new house / erected atop the plot of liberty).
12. Aristotle posits that what defines man as a political animal [*zoon politikon*] is speech, chiefly because mastery of the polis is over "free and equal persons": "speech serves to reveal the advantageous and the harmful, and hence also the just and the unjust. For it is peculiar to man as compared to other animals that he alone has a perception of good and bad and just and unjust" (*Politics* 37). I will argue at the end of the essay that Espriu, in contrast with Aristotle, argues that the defining political quality of the human being is in fact dialogue but other animals, such as the bull, have their own means of political disclosure not reliant on logos.
13. Carl Schmitt famously asserted that "sovereign is he who decides on the exception" (5) and "it is precisely the exception that makes relevant the subject of sovereignty, that is, the whole question of sovereignty" (6). Espriu's federalist proposal still allows for decisions on the exception to normal legal order and the validity of the political pact; however, such decisionistic sovereignty can never be the sole business of a "he."
14. In Joan Maragall's "Oda a Espanya" (1898), the poetic voice—"la veu d'un fill que et parla en llengua – no castellana" (the voice of a son that speaks to you in a non-castilian language) – questions at the end of the piece, "No sents la meva veu atronadora? / No entens aquesta llengua – que et parla entre perills? / Has desaprès d'entendre an els teus fills? Adéu, Espanya!" (Do you not hear my booming voice? /

Do you not understand this language, that speaks to you amidst perils? / Have you forgotten how to understand one of your sons? Goodbye, Spain!)

Works Cited

Agamben, Giorgio. *Homo Sacer: Sovereign Power and Bare Life*. Trans. Daniel Heller-Roazen. Stanford: Stanford University Press, 1998.

_____. *The Sacrament of Language: An Archaeology of the Oath*. Trans. Adam Kotsko. Stanford: Stanford University Press, 2011.

Almirall, Valentí. *Lo Catalanisme*. Barcelona: Edicions 62, 1979.

Aristotle. *The Metaphysics*. Trans. John H. McMahon. Amherst, NY: Prometheus, 1991.

_____. *The Politics*. Trans. Carnes Lord. Chicago: University of Chicago Press, 1985.

Bataille, Georges. *Theory of Religion*. Trans. Robert Hurley. New York: Zone, 1992.

Cortés, Donoso. "Discurso sobre la dictadura." *Ensayo sobre el catolicismo, el liberalismo, y el socialismo. Otros escritos*. Barcelona: Planeta, 1985. 241–61.

Derrida, Jacques. *The Beast and the Sovereign I*. Trans. Geoffrey Bennington. Chicago: University of Chicago Press, 2009.

Espriu, Salvador. *El caminant i el mur*. Barcelona: Edicions 62, 1983.

_____. *Obres Completes II. Poesia*. Barcelona: Edicions 62, 1963.

_____. *Obres completes XII. Edició crítica*. Ed. Olívia Gassol i Bellet. Barcelona: Edicions 62, 2008.

Frow, John. "A Pebble, a Camera, a Man." *Critical Inquiry* 28.1 (2001): 270–85.

Illas, Edgar. *Thinking Barcelona: Ideologies of a Global City*. Liverpool: Liverpool University Press, 2012.

_____. "Vigència d'Espriu." *El Punt Avui*. El Punt Avui, 18 March 2013. Web. 29 April 2013.

Kantorowicz, Ernst. *The King's Two Bodies*. Princeton: Princeton University Press, 1997.

Luhmann, Niklas. *Social Systems*. Trans. John Bednarz, Jr. and Dirk Baecker. Stanford: Stanford University Press, 1995.

Ortega y Gasset, José. *España invertebrada*. Madrid: Revista de Occidente, 1962.

Pi i Maragall, Francesc. *Las nacionalidades*. Madrid: Akal, 2009.

_____. *La reacción y la revolución*. Barcelona: Anthropos, 1982.

Pujol, Jordi. "El fracàs de l'Espriu." *Centre d'Estudis Jordi Pujol*. Centre d'Estudis Jordi Pujol, 24 November 2009. Web. 6 May 2011.

Resina, Joan Ramon. *Del Hispanismo a los Estudios Ibéricos*. Madrid: Biblioteca Nueva, 2009.

Saussure, Ferdinand de. *Course in General Linguistics*. Trans. Wade Baskin. Collins: New York, 1981.

Schmitt, Carl. *Political Theology*. Trans. George Schwab. Chicago: University of Chicago Press, 2005.

Sophocles. *The Theban Plays*. Trans. Ruby Blondell. Newburyport, MA: Focus, 2002.

Von Humboldt, Wilhelm. *On Language*. Trans. Peter Heath. Cambridge: Cambridge University Press, 1999.

Walters, D. Gareth. *The Poetry of Salvador Espriu: To Save the Words*. Woodbridge: Tamesis, 2006.

Weber, Max. *Political Writings*. Cambridge: Cambridge University Press, 1994.

_____. *The Protestant Ethic and the 'Spirit' of Capitalism*. Trans. Peter Baehr and Gordon C. Wells, 2002.

12

Animals in Contemporary Spanish Newspapers

John Beusterien

From 2012–2013, the Spanish press contributed to the invention of wild spaces on the Iberian Peninsula through the publication of articles on the resurgence of nearly extinct indigenous Spanish animals, including many types of birds, the brown bear in Asturias, and the Iberian wolf. In addition, Spanish print media reported that foreign species, such as the raccoon, had invaded those wild spaces. Media not only fabricated a sense of the wild Spanish space through articles that depicted invasive species and the return of nearly extinct animal populations, but also released an article on the project Paleolítico Vivo (Living Paleolithic). This project explicitly established a protected zone for wild fauna in the north of Spain.

Reporting on this unique reserve for protected species, the Spanish newspaper *El Mundo* features a full display picture of two enormous European bison scheduled to arrive to Atapuerca from Poland in May of 2013. The article on the Living Paleolithic Project discusses how it seeks to reinvent a 30,000-year-old ecosystem by letting bison and other large animals roam a region in northern Spain (Corral, "El Paleolítico"). Editors include an image of the two bison with another picture of a herd of a different large bovine animal. It resembles the bull specially bred for the Spanish bullfight, but it is actually a newly created version of the extinct aurochs, ancestor of modern cattle, specifically the taurine variety of cattle. The aurochs in the image, butting heads, *look* strikingly similar to the modern variety of the bull bred for Spanish bullfights. Is the creation of aurochs in this animal park in northern Spain some kind of national repentance for its bull slaughter spectacle or is it, in turn, a celebration of the animal that gave rise to its celebrated national pastime?[1]

But the creation of the animal park in the north of Spain stimulates even broader questions: What is the nature of contemporary attitudes toward animals that informed the creation of an environmental preserve that includes bison, aurochs, and other animals from Spain's deep history? With respect to my own profession, the teaching of Spanish-speaking cultures, what impact does a better knowledge of animals from contemporary Spain have on how critics conceive their academic field? Although some newspapers in Spain are written in Basque, Catalan, and Galician, this essay focuses on contemporary print media written in Spanish, a language that in the context of the region of West Texas where I teach and research is a minority language. Moreover, I live in an area of Texas that did not harbor permanent human settlement until around 1900. Prior to the twentieth century, millions of bison for tens of thousands of years had lived on the southern high plains, also known as the Llano Estacado, where the university where I teach is located. One only has to imagine life a little over a hundred and fifty years ago—not 30,000 years ago as in the case of Spain—to visualize a radically different landscape. The imagination of new landscapes is a particularly necessary undertaking given that the current agricultural economy of the Llano Estacado, a land with hundreds of miles of cotton farms, no longer will be viable in the near future.

I carry out the following description of a heightened contemporary Spanish consciousness toward animals by signaling these two important contexts—the teaching of the Spanish language at Texas Tech, a university where English-speaking is dominant, and where an urgent need to reexamine the environment exists, especially in light of diminishing water sources and the relatively recent disappearance of the bison and many other animals. With the intent of offering a possible future scenario of reshaped attitudes for my region's ecosystem, the essay focuses on print in Spanish since the history, the language, and culture of Spanish speakers are underrepresented at my university and through a Spanish-speaking lens I can inspire my students to act.

Spanish attitudes toward animal life have received attention outside of Spain and, despite traditions despised by animal rights groups such as the bullfight, many scholars of Spanish culture and Animal Studies scholars cite the case of the Spain as a leading model in its conception of the animal, especially based on its proposal to give human rights to apes.[2] The following examines a series of articles from Spanish newspapers published in 2012 and 2013, such as the one with the picture of the bison and the aurochs from *El Mundo,* in order to more fully assess Spain's contemporary relationship with the animal. I signal a few characteristics of Spanish animal consciousness from the articles: they emphasize animals as entrapped by using images of cages and the representation of the body of the animal as described in the articles is also accompanied by a description or image of advanced technology. Moreover, the articles indicate that increased attention to animals often depends on the grafting of global animal awareness models onto a local or regional site of urgency in Spain.

Symptomatic of a desire to see the animal free to roam, many articles about animals highlight an image of it trapped in a cage. In "Piensos con carne de perro y oveja para alimentar animals de consumo humano" (*El Mundo*, February 1, 2013) (Food made with dog and sheep meat for animals destined for human consumption), the accompanying image to the article shows a group of dogs of different breeds looking at the lens of the camera from behind a cage made of iron wire. With the boundaries of the cage out of focus in the forefront of the image, the picture focuses on the background image of the dogs with the caption, "Perros encerrados en una de las instalaciones, situada en Vincios (Pontevedra), implicadas en la trama" (49) (Dogs enclosed in one of the installations involved in the case located in Vincios [Pontevedra]). The article cites the regional newspaper *La voz de Galicia* and situates the problem within the context of European dietary norms (the meat was sold not only in Spain but also in Portugal and Holland) and within the international context of fears of prionic illness such as bovine spongiform encephalopathy (mad cow disease) because of the threat of sheep eating sheep meat.

The article appeals to the emotions primarily through the image of a group of dogs destined for human consumption trapped within what appears to be a salvage yard. The dogs all look out from behind their enclosure to the newspaper reader, who occupies the free space on the other side of the cage. In contrast, another article published in *El Mundo*, "Así se 'reciclan' los animals en la perrera" (November 3, 2012) (This is the way animals are "recycled" in the pound), uses the image of the cage as a way to communicate an emotionally upbeat message. The image shows a long series of cages that appear to be empty and in the forefront of the picture a young man squats down holding a medium-sized dog's paws in his hands. The door of the cage is open and the image suggests that all the dogs were liberated from their cages except for this one. Indeed, the article indicates that the majority of animals in this pound in Madrid (Centro de Protección Animal del Ayuntamiento de Madrid) have not been killed, but that 80 percent are reported as having found a family.

Every year in the summer, before many Spaniards begin a July or August vacation, newspapers often report a rise in the abandonment of domestic animals. Such articles have continued in recent years and, particularly salient, they have included reports about the abandoning and sacrificing of horses due to the economic crisis.[3] Nonetheless, the portrayal of the dog liberated from its cage indicates a growing Spanish embracing of the dog as domestic pet. "El auge de los galgos" (October 21, 2012) (The rise of greyhounds), an article in the Cataluña-based newspaper *La Vanguardia*, indeed, shows images of happy owners with their greyhounds. The article argues that, despite economic hardship and the downturn in the economy, the caring for pets, in this case the greyhound, is on the rise. Articles in a similar vein occupy the pages of other newspapers. *El Mundo* reports the following headline: "La primera pastelería para perros" (The first pastry shop for dogs) on November 3, 2012. *El País* in "La industria de los mascotas pasa el corte" (March 6, 2013) (The pet industry makes the cut March 6, 2013) discusses how the 20 million domestic animals in Spain afford some 13,300 jobs to a veterinary industry that is growing.[4]

The image of the dog freed from its cage reflects a broader Spanish interest in the animal as a sentient being that deserves certain rights. In this respect, *El Mundo* published images of great apes behind the bars of a cage. In "El simio ya no es cobaya" (February 2, 2013) (The ape is no longer a guinea pig), *El Mundo* reports that the Spanish government prohibited experimentation with great apes (chimpanzees, bonobos, gorillas, and orangutans). It also established extremely strict norms not just for primates, but also for dogs and cats. To illustrate the new Spanish codes, the editors chose a close-up image of the face of a chimpanzee behind bars. Its shiny eyes, fingers clutching the bars, and thumb in mouth with a piece of fabric, emphasize both its entrapment and the pedomorphosizing of the chimp. The sucking of thumb suggests the action of a human infant and is intended to stimulate a response that triggers what many scientists consider to be an inherent nurturing need in humans. This "cute response" is a common goal in which breeders have exaggerated physical traits in animals so that their appearance suggests those of children and infants.[5] The title of the article also implicitly suggests that the ape is not an animal upon which experimentation is justified, unlike the guinea pig.

Studying the images of the animal and its treatment in the Spanish press fills a gap in animal studies scholarship. While many cite the Spanish one as an especially unique case, few scholars have examined the complexity of the treatment of the animal in Spain, particularly through the lens of contemporary print media. World renowned for the bullfight that meets the chagrin of animal rights groups, Spain is also, as the editors of this special issue on "Debates on Life in Contemporary Spain" suggest, at the forefront of ideas behind progressive legislation on behalf of the animal. *El Mundo* directly engages the question of human rights when it opts for the metaphor of the place known for its blatant lack of basic human rights, Guantánamo. The author of the article, Martín Mucha, conflates the detention facility in Cuba with that of an experimentation animal facility in Spain. The article, "El Guantánamo español de los simios" (February 10, 2013) (The Spanish Guantánamo for apes) devotes an entire page to two images of cages. The first shows an image of multiple cages. The dark image of about fifteen macaques is seen behind the cages. The caption reads: "Su cárcel. Ubicada en Camarles (Tarragona), puede albergarse más de 3,500 primates" (11) (Their jail. Located in Camarles [Tarragona] where more than 3,500 primates can be lodged). The second image shows a close-up image of one of the macaques, whose head is angled in the same position as the chimp's in the previous image of the caged primate. The caption reads: "Primer plano" (A close-up). The caption also claims that this article was an exclusive: the first time that any Spanish news organization shows images of the facility. It directs its reader to its Web page, on which one will find more than a hundred additional images of this Spanish Guantánamo.

The cage is a photographic motif used in these images of dogs and primates in order to create the sense that the viewer looks into a restricted space, thereby suggesting that the animal belongs elsewhere, either within a human home or in an abstract wild. In the case of primates in Spain that are not in experimental facilities, others arrive to domiciles through illegal trade, circuses, and zoos. Many of these animals are abandoned, and one article from the conservative-leaning newspaper *La Razón* published an article entitled "Los rescatadores de primates" (March 16, 2013) (Primate rescuers) that highlights a facility in Madrid dedicated to the rehabilitation of primates providing "un hogar tranquilo y seguro para el resto de sus vidas" (44) (a peaceful and secure home for the rest of their lives).

The article "Los rescatadores de primates" is found in a section of the paper called *La Razón Solidaria* (*La Razón* in solidarity). Companion animals are the principal focus of this special section of *La Razón* dedicated to reader participation and debate. The context of the article on a page devoted to domestic animals indirectly suggests that apes are another humanized companion animal like the dog. On the same page as "Primate Rescuers" the newspaper features a winning photo of one of its readers' dogs. The page also includes an advertisement for Pedigree dog food and a dog toothpaste with the caption "Los perros no quieren dentaduras postizas" (Dogs don't want false teeth).

The image that the editors chose to accompany the article "Primate Rescuers" takes up the center of the page and shows what appears to be a large chimp lying face up on a stretcher. It is covered by blankets and has various tubes connecting to it. Two appear to be intravenous and another large tube delivers it oxygen. The animal's mouth is open and its eyes half open in a coma-like state. The image from the primate rescue facility evokes an image of a hospitalized character from the one of the *Planet of the Apes* series, but it also, given its context on a page devoted to companion animals, suggests that apes not only find a home, but also medical care in Spain. No pretense is made that the abandoned animal will end up in some idyllic Tarzan-type African wilderness. The technology of animal life support in the photo thereby evokes another significant theme found in a survey of contemporary Spanish newspapers. It underscores that animals are both visualized and defined through the lens of human technology.

An increasing number of articles depict the resurgence of certain species, such as the sparrow and partridge ("El gorrión y la perdiz, entre las aves en peligro"; *El Mundo*, April 24, 2013) or the brown bear ("Osos despiertos pese al invierno"; *La Vanguardia*, March 7, 2013, and "Los osos pardos causaron daño por 150,000 Euros en 2012: La Junta de Castilla y León edita un manual para fomentar su conservación"; *El Mundo*, February 2, 2013). Often authors of these articles describe the human technology that enables both the viewing of the animal and its return. In "El lobo vuelve a criar en Madrid" (February 26, 2013) (The wolf returns to breed in Madrid) the newspaper *El País* displays an Iberian wolf cub (*canis lupus signatus*) looking into the lens of the camera and out at the newspaper reader. The article describes how researchers put out an installed camera near the River Lozoya in August of 2011 and within two weeks the wolf cub appeared. A page of *El Mundo* that contains an article on the return of the Imperial Iberian eagle entitled "El águila ibérica más amenazada remonta al vuelo" (December 13, 2012) (The extremely threatened Iberian eagle returns to flight) also contains a companion article on the same page dedicated to how researchers were able to observe the eagle. This article "Siete días en un 'hide'" (Seven days in a "hide") describes how the photographer Andoni Canela was able to achieve unprecedented views of the eagle's behavior and track its recovery from near extinction. "Siete días en un 'hide'" shows a photo of one of the viewing sites that Canela used and describes how he took refuge in the tent-like "hide," a space of one square meter in which he stayed for seven days taking photographs.

In contrast to the animals that the press celebrates as making a comeback, other articles signal the entrance of invasive species. *El País* treated its readers to an informative overview on the raccoon in 2012. Many Spaniards at the beginning of the twenty-first century adopted this animal indigenous to the Americas as a household pet. People abandoned the raccoon as a fashion pet a few years later and the article entitled "El mapache no es bienvenido en Doaña" (November 5, 2012) (The raccoon is not welcome in Doaña) describes how it turned into an invading varmint, reproducing out of control, especially in the areas around Guadalajara and Madrid. Employing medical rhetoric and metaphor of the epidemic, the article fits into a journalistic generic mold that describes invasive species, be they animals, bacteria, or viruses. Aside from the news story about the increasing invasiveness of the autochthonous American species on Spanish lands, it provides the following zoological primer: the raccoon becomes aggressive while in heat and smells bad; it is a voracious, indiscriminate eater; and it transmits diseases to other animals and humans. This introduction to the raccoon was the background information for the report of the plague of raccoons on the Iberian Peninsula, most notably in Doaña, the protected national park in Andalucía. One veterinarian explained the animals were detected through photo-traps. The camera located a family of raccoons and set off an alarm, enabling the capture of the raccoons from Doaña.

Manuel Planelles, the author of the article, not only signals this technology and the regional areas in Spain where the raccoons increasingly appear, but also connects the plague of the raccoon with a similar situation in Germany and points to a recent article from the German newspaper *Der Spiegel* that reported an invasion of some 400,000 specimens and the capture and sacrifice of some 67,700. (As opposed to Spain, where they only recently arrived as exotic pets, in Germany they arrived in the 1930s for the fur industry and to enrich the variety of fauna in the country.) Beyond citing the German example, Planelles also places the problem of the raccoon within an international context when he refers to ideas of autonomous species conservationists and when he cites the English academic journal *Biological Invasions*, devoted to the control of exotic species invasion.

Much of the move toward animal consciousness in the Spanish press is guided indirectly and directly by international attitudes toward animals, generating from Anglo-American ecological and zoological influences. In 2012, for instance, California's ban on foie gras received heavy media attention. For instance, the *New York Times* published "Some in California Skirt Ban on Foie Gras" (August 12, 2012). In the Spanish press the case of foie gras in California is generally not mentioned, but the United States ban finds its own incarnations in local debates published in the Spanish press. In one striking photo, the Basque owner of a farm force-feeds one of her geese, the center image for the article "Adiós al 'foie gras' de Mugaritz" (*El Mundo*, December 13, 2012) (Goodbye to Mugaritz's foie gras). Basque health officials found the company did not comply with hygiene standards. But given the image and the article's tone—it mentions the complaints of an animal rights group, *Igualdad Animal*—the article's main objective goes beyond the story of sanitation at a local foie gras provider and touches on the animal rights ideology of international groups such as those behind the California case published a few months earlier in the *New York Times*. *El Mundo* recognized that the article fit the mold of groups that want to see the prohibition of foie gras and published another article expressing the contrary point of view alongside "Adiós al 'foie gras' de Mugaritz," entitled "Pecado venial" (A venial sin). In this opinion piece, Juan M. Bellver, one of Spain's most renowned food critics, argues that the use of foie gras was an inconsequential sin, a *peccadillo*. The famed Basque cook, Andoni Luis Adoriz, who was using it at his restaurant, should not be blamed in any way. In his comments on the case, Bellver ignores the issue of animal abuse so vividly pronounced in the photo of the woman force-feeding the goose and approaches the issue from the point of view of Andoni's craftsmanship as chef. Bellver explains:

> Del rigor de cirujano de Andoni en la cocina puedo dar fe porque conozco a pocos cocineros más maniáticos que él respecto al higiene alimentario y el cuidado del producto. Si su proveedora de ánades incumple alguna normativa, que le hagan la correspondiente inspección técnica. Pero denunciar al cliente como si fomentara las malas prácticas alimenticias suena oportunista y zafio. (47)
>
> (I can guarantee Andoni's surgeon-like attention to detail in the kitchen because I know few cooks more fastidious with respect to food hygiene and care. If the goose provider is not following code then inspectors should punish the provider. But punishing its client by suggesting that he encouraged wrongful food practice is nothing but opportunistic and crude.)

The juxtaposition of the article suggesting animal sympathy with that of the opinion of a Spanish chef makes manifest contradictory attitudes toward the treatment of animals: both sides of the debate are presented, showing Spanish recognition of the California animal rights model as well as the French culinary one.

Other articles reveal relatively new animal rights realities, suggesting the presence of an animal consciousness that flatly contradicts stereotypical and traditional assumptions about Spain and animals. "El santuario para animales de granja" (The sanctuary for farm animals) in the science pages of *El Mundo* (May 6, 2013), under the heading "Protection of Species," describe a group of vegans who care for abandoned and injured farm animals and who argue that no living being should be exploited or consumed. One member stated that if we cannot free the 70 billion animals reared for food each year, then at least we can raise consciousness (Guerrero). The vegan sanctuary takes in all kinds of farm animals, including geese. Across Spain, restaurants and bars hang legs of cured ham with pride, and one of the most prized *serrano* hams is that from the type known as *cerdo ibérico*. But this article reports how the caretakers of the sanctuary rescued one of these pigs when it was a suckling *cerdo ibérico* and raised it. A picture of this pig, Rapunzel, is placed along a picture of Gary the ram, Félix the injured goat, and Giorgio the rooster. No doubt many Spaniards find the content of this article about vegans and their pig Rapunzel extremely comical, given the popular attitude that ham should be savored, especially from the highly esteemed *cerdo ibérico*.

Sometimes articles with an international focus do not mention their direct connection to animal rights issues within the Spanish context. For instance, the question of the legitimacy of the Spanish monarchy linked to the question of animal rights when, in the spring of 2012, the Spanish King Juan Carlos came under scathing criticism for hunting Botswana elephants. In an unprecedented act of royal contrition, Juan Carlos apologized to the country over the incident. Two articles, "Cuernos con glamur" (Fashionable horns) and "El lujo asiático expolia África" (Asian luxury markets pillage Africa), suggest the king's action to the Spanish reader even though they do not mention his act. Published side by side in the Catalan-based *La Vanguardia* on March 4, 2013, the articles indirectly allude to the royal hunting incident, particularly through their references to the World Wildlife Fund (WWF) and elephant hunting. Many readers of the article will have known that the Spanish branch of the World Wildlife Fund dismissed the king as honorary head shortly after the hunting incident. The article, however, never mentions the king's hunting mishap and, in contrast to the economic emphasis of the original story (the king wasted state funds on a hunting trip while Spaniards suffered in an acute economic crisis), treats the questions of animal rights and illegal commerce on an international level. In the article, then, the Catalan newspaper directly dismisses the king by not mentioning him at all. Instead, it appeals to the Catalan reading public's sense of belonging to an international reading community interested in the illegal trade of elephant tusks and rhinoceros horns from Botswana and other African countries to China and Thailand. Beyond the theme of the article, its sources are international—a report on the United Nations Conference on the international commerce of endangered animals and the incorporation of quotes of authorities such as Julia Marton-Lebèvre and Mike Knight from the International Union for the Conservation of Nature. Affirming an international bent, the two articles on the illegal commerce of endangered animals also quote the journal *Science* on the growing insatiable demand for the tusks and horns. On one hand, news about the animal in contemporary Spanish newspapers demonstrates an increased communication and transparency between international news organizations in which news on animals of an international scope is swiftly translated to the Iberian context. Moreover, the use of sources such as *Science* highlights the degree of the international dissemination of information since Spanish newspapers increasingly take advantage not just of animal rights causes elsewhere in the world, but also scholarly international studies on the animal that support those causes.

Even the relatively low-circulating *Cinco Días* referred to studies published in *Science* (and also *Nature Communications*) when it reported on the European Union's vote to prohibit the use of three pesticides produced by Bayer CropScience and Syngenta (April 30, 2013) because of the damage they caused to bee populations. (The prohibition was tabled in part due to Bayer CropScience and Syngenta's argument that studies on the effect on bees of their pesticides lack scientific rigor.) Opposed to *La Vanguardia*'s treatment of the illegal animal trade, the article in *Cinco Días* firmly positions the bee problem in a national context. The bee article not only cites the deleterious global catastrophe of a diminished bee population due to crop loss from the lack of pollination, but also spotlights a collective national Spanish identity by using a first person plural: "nuestro país es el primer productor de miel de la Unión Europea" (11) (our country is the principal producer of honey in the European Union).

In an act that speaks directly to the creation of a collective national space, Spain, as mentioned, brought bison, aurochs, and wild horses to its lands. The proposal seeks to repopulate parts of the north with animals from the period when the earliest humans arrived to Europe. One might argue that the decision in Spain to restore an ancient ecosystem is fundamentally a project that concerns itself with national cultural identity. Spain's oldest and most famous texts are cave paintings in the Cantabria region in Altamira in northern Spain that include the oldest images of bison (known as steppe bison or *Bison priscus*).

Indeed, studies have concluded that images found in the Altamira caves are Europe's oldest dated art by at least 4000 years and most likely the oldest reliably dated cave art anywhere in the world. The image of the depiction of extinct bison in different poses from the amazing polychrome ceiling of the cave is one of its most remarkable features and the autonomous government of Cantabria already had created a logo in which it used a version of one of the images of bison to promote tourism to the region. Now, another region, Atapuerca (also a UNESCO World Heritage Site because it is considered to contain the earliest human remains in Europe) not far from Altamira in the north, has designated a landscape for bison.

The enlivening of an iconic Spanish image exemplifies how the animal connects to the technological and artificial. Something artificial, the image or text of the Altamira bison, reconstitutes itself in roaming, flesh and blood animals on some 3000 hectares in Atapuerca in a reconstituted European landscape from the Quaternary time period, about 12,000 years ago. In this Spanish version of a Jurassic Park, people see animals like they would on an African safari. The hope is that tourism will revitalize an economically depressed and abandoned rural part of the country. The fabrication of a wild ecosystem also produces economic profits due to increased meat production. One of the architects of the plan compares the arrival of bison with Ted Turner's success in the United States, arguing that Turner earned more money with the sale of bison than from his profits as owner of the news channel CNN and that Spain will take advantage in the same way of its abandoned lands. The *El Mundo* article that features the image of the bison discusses the project as part of a project that sought to make Spain wild, even though the project depends on a concerted employment of human technology in creating wilderness. The bison will be transported from a region in Poland. The aurochs was only rescued from extinction due to genetic advances. But advanced human technology does not only create fauna on the landscape. The wild space also provides a place for the advancement of the technology that fuels the global economy. The article concludes,

> Se trata de situar el medio rural en el epicentro de la economía. Que aquellos de la *crème de la crème* de la Sociedad de la Informacón que sean emprendedores con vocación por vivir in el campo, puedan traslocar su puesto de trabajo vía internet a aldeas semiabandonandas en espacios naturales y allí iniciar y desarrollar una nueva economía rural, mixta y multifuncional entre el sector terciario y el primario. (36)

> (The project proposes to situate the rural landscape as epicenter of the economy. Those of the *crème de la crème* of the Information Society who are entrepreneurial with a calling to live in the country, can have their office space in a semi-abandoned country town in a natural landscape with internet access. There they will develop a new rural economy, a hybrid, multifunctional between the tertiary and primary economic sectors.)

The present essay has reviewed the treatment of animals in Spanish newspapers. It shows the complex, sometimes contradictory, nature of Spanish social consciousness toward the animal in a variety of print media sources. Often, this consciousness depends on a perspective about the animal taken from international animal rights groups or conclusions from scientific studies published in the United States or England and then applied to the situation of a local area in Spain. Nonetheless, the present study taken as a whole demonstrates a uniquely Spanish sensibility toward the animal with respect to local concerns in which the press fostered a growing notion of ecological awareness through perpetuating a paradoxically artificial, but also wild, Spanish ecological space.

Beyond offering a richer perspective of contemporary Spanish attitudes toward the animal, this essay provides material for a class that I teach in Spanish that examines the history and future of animals on the Llano Estacado. Teaching the class in Spanish using Spanish texts helps provide academic legitimacy to a minority language. It gives students pride in speaking Spanish since in West Texas, until quite recently, elementary and secondary students were punished for speaking it in school. The class describes the history of bison in the area by studying how they formed an important part of Native American Comanche and Spanish-speaking New Mexican economy before they were slaughtered by white settlers. After examining the past, the class inspires students to imagine new futures for the region. The class walks and tours the world's only playa-lake classroom to discuss the native grasses in the region and the way in which playas, ephemeral ponds that hold water after rain events, help to replenish the Ogallala Aquifer and water in the region.[5] At the playa-lake classroom, we discuss the variety of animals that lived on the grasses, most especially the bison. The class discusses the possibility of the realization of the idea of the Buffalo Commons, an idea that seeks to bring bison back to the region, and it also visits the park at Caprock Canyon to see a restored herd.

As part of the discussion on the Buffalo Commons the class examines animals in contemporary Spanish newspapers to help better understand another culture's attitudes toward animals and how it looks to restore bison on the landscape. The use of contemporary Spanish print media is a pedagogical exercise in showing students something different. The material is written in a different dialect of Spanish from that spoken in West Texas and the landscape and animals are different from North America (indeed, American species like the raccoon are invasive in the Spanish context). Even the European bison is different in many ways from the American Plains bison. Nonetheless, after processing these differences, the class can arrive to a new vision of the present. They leave the class with an enriched understanding of a minority language through which they discover a new sense of place. The inclusion in this course of Spain's relationship to animals in print media will thereby have aided students articulate their own ecological alternatives for the West Texas landscape of the future.

Notes

1. Katarzyna Beilin's argument that strong criticism existed regarding the legitimacy of the bullfight ever since its inception paradoxically means that "yes" could plausibly answer both questions.
2. For instance, see Wolfe.
3. See, for instance: "Aumenta el abandono y el sacrificio de caballos por la crisis económica" (*El Mundo*, September 18, 2012) (The abandonment and sacrifice of horses rises due to the economic crisis) and "Lista de espera en el matadero: Los sacrificios de caballos" (*El País*, January 27, 2013) (The waitlist for the butcher: sacrificing horses). These articles about horse killing, in contrast to those of many French regions that eat horse meat as part of their diet, look with condemnation on the butchering of horses. *El País*, in fact, in a panegyric to horse, devotes a feature length article to the importance of a special type of chariot-type horseracing in the Balearic Islands and focuses on the brutal beating to death of one these horses by its irate owner ("Cuando el galope es mortal," January 13, 2013).
4. The press is feeding a growing consciousness of the dog as domestic companion in Spain (this article suggested that the Spanish economy related to the domestic pet has plenty of room to grow since it still has not reached the level of France, Italy the United Kingdom or Germany). In contrast to articles showing happy Spaniards with their dogs, one article attributes illegal dog fighting to gypsy gangs ("Jóvenes y clanes gitanos organizan peleas de perros clandestinas en el sur de Madrid," *ABC*, February 25, 2013). The article suggests a racial overtone in the question of dog ownership because, in contrast to articles that portray Spaniards as caring for dogs, gypsies raise them for brutal, bloody, and illegal fights.
5. In her study on animals in the Middle Ages, Joyce Salisbury explains the terms *pedomorphosis* and *cute response*: "the main characteristic of all domestic animals is

pedomorphosis, that is, the retention of juvenile characteristics, both in body shape and in personality characteristics, such as whining and submissiveness. The most extreme example of the retention of juvenile traits comes in toy dogs: in addition to their small size, they have disproportionally broad heads, small limbs, large eyes, and smaller noses and mouths. All these are characteristics of human infants and thus evoke what Nobel Prize-winning ethologist Konrad Lorez defines as the "'cute response'" (116).

6. Darryl Birkenfeld might be considered a West Texas Don Quijote since he is not looking for immediate gains, but heads up Ogallala Commons, a nonprofit community development network, nestled among the Llano Estacado's acres of cotton crop. Near his home in Nazareth, a town in Castro County in the Panhandle Plains, he has an outdoor classroom at the site of a playa lake.

Works Cited

Beilin, Katarzyna. Unpublished manuscript. 2014.

Bellver, Juan M. "Pecado venial." *El Mundo* 13 Dec. 2012: 47. Print.

Benvenuty, Luis. "El auge de los galgos." *La Vanguardia* 21 Oct. 2012: 4–5. Print.

Cerrillo, Antonio. "El lujo asiático expolia África" and "Cuernos con glamur." *La Vanguardia* 4 Mar. 2013: 26–27.

Colomba, Marta G. "Adiós al 'foie gras' de Mugaritiz." *El Mundo* 13 Dec. 2012: 47. Print.

Corral, Miguel G. "El águila ibérica más amenazada remonta el vuelo" and "Siente días en un 'hide.'" *El Mundo* 13 Dec. 2012: 49.

_____. "El gorrión y la perdiz, entre la saves en peligro." *El Mundo* 24 April 2013: 47.

_____. "El Paleolítico revive en Atapuerca." *El Mundo* 22 April 2013: 35–6.

_____. "Piensos con carne de perro y oveja para alimentar animales de consume humano." *El Mundo* 1 Feb. 2013: 49.

_____. "El simio ya no es cobaya." *El Mundo* 2 Feb. 2013: 48.

Espinosa, Pedro. "Lista de espera en el matadero: Los sacrificios de caballos." *El País* 27 Jan. 2013: 81.

Guerrero, Teresa. "El santuario para animales de granja." *El Mundo* 6 May 2013: 42.

Herraiz, Pablo. "La primera pastelería para perros." *El Mundo* 3 Nov. 2012: 59.

Hoyo, Chema M. del. "Los rescatadores de primates." *La Razón* 16 Mar. 2013: 44.

"Los osos pardos causaron daño por 150,000 Euros en 2012: La Junta de Castilla y León edita un manual para fomentar su conservación," *El Mundo* 2 Feb. 2013: 7.

Manresa, Andreu. "Cuando el galope es mortal." *El País* 13 Jan. 2013: 9.

Matos, María. "La Unión Europea, al rescate de las abejas." *Cinco Días* 30 Apr. 2013: 11.

Mucha, Martín. "El Guantánamo español de los simios." *El Mundo* 10 Feb. 2013: 10–11.

Planelles, Manuel. "El mapache no es bienvenido en Doñana." *El País* 5 Nov. 2012: 32–33.

Ricou, Javier. "Osos despiertos pese al invierno." *La Vanguardia* 7 Mar. 2013: 34.

Rivas, Tatiana G., and Víctor Lerena. "Jóvenes y clanes gitanos organizan peleas de perros clandestinos en el sur de Madrid." *ABC* 25 Feb. 2013: 58–59.

Romero, Ana. "La industria de las mascotas pasa el corte." *El Mundo* 6 March 2013: 31.

Salisbury, Joyce E. *The Beast Within: Animals in the Middle Ages*. Second edition. London: Routledge, 2011.

Sánchez, Ésther. "El lobo vuelve a criar en Madrid. *El Páis* 26 Feb. 2013: 57.

Treceño, Jaime G. "Así se 'reciclan' los animales en la perrera." *El Mundo* 3 Nov. 2013: 58.

Wolfe, Cary. "Human, All Too Human: 'Animal Studies' and the Humanities." *PMLA: Publications of the Modern Language Association of America* 124 (2009): 564–75.

◆ 13

Accounting for Violence, Counting the Dead: The Civil War and Spain's Political Present

Sebastiaan Faber

Introduction

What is the weight of a human life? Its value is often construed as infinitely large and therefore fundamentally incomparable: as the truism has it, every unnatural or premature death is one too many. Yet in practice human deaths are regularly quantified in coldly straightforward terms in order to assess their relative importance. The trope of the scale—the balance—is a popular one in this context. Speaking on the United States drone program—in which unmanned, remote-controlled aircraft roam the globe in search of suspects to kill, regardless of their citizenship and without any semblance of due process—U.S. President Barack Obama acknowledged that drone strikes were to blame for the loss of innocent life, although he immediately went on to invoke mitigating factors. "[I]t is a hard fact that U.S. strikes have resulted in civilian casualties, a risk that exists in all wars," Obama said on May 23, 2013;

> For the families of those civilians, no words or legal construct can justify their loss. For me, and those in my chain of command, these deaths will haunt us as long as we live, just as we are haunted by the civilian casualties that have occurred

> through conventional fighting in Afghanistan and Iraq. / But as Commander-in-Chief, I must weigh these heartbreaking tragedies against the alternatives. To do nothing in the face of terrorist networks would invite far more civilian casualties—not just in our cities at home and facilities abroad, but also in the very places—like Sana'a and Kabul and Mogadishu—where terrorists seek a foothold. Let us remember that the terrorists we are after target civilians, and the death toll from their acts of terrorism against Muslims dwarfs any estimate of civilian casualties from drone strikes. (Obama 2013)

Obama invokes the notion of weight twice here, though only once explicitly. The President believes it is his duty to "weigh these heartbreaking tragedies against the alternatives." At the same time, he states that the innocent deaths his orders have caused will never cease to haunt him; they will never stop weighing on his conscience. The value of a human life, then, appears in Obama's speech as both infinite *and* quantifiable: we are asked to accept the paradoxical notion that the President's decisions are based on cold, realistic calculation but also, at the same time, on empathy and the unflinching acceptance of personal, ethical responsibility. The loss of innocent life cannot be justified, Obama says; and yet he goes on to justify it, in the very next paragraph, as the tragic, unintended but unavoidable consequence of a decision meant to prevent a greater loss of innocent life in the future.

What is remarkable about Obama's speech is not just the contradictory nature of his statements, but also their level of sophistication. Unlike most of his predecessors, Obama resists dehumanizing the victims of the violence he has set in motion, and he explicitly justifies his decision by arguing that it will help save *non-American* lives. Judith Butler has called attention, by contrast, to the manner in which the Israeli government has consistently explained away its targeting of civilians by rhetorically turning every Palestinian—including women and children—into a potential weapon (for attacks on Israel) or a potential shield (to defend terrorists), marking civilians as a legitimate object of Israeli military violence wielded in self-defense.[1] In *Precarious Life* (2004) and *Frames of War* (2009), Butler argues that subjecting a life to violence, as for example in war, is more difficult once that life is apprehended *as life,* which for Butler means that it is apprehended—or, more actively, recognized—as precarious and grievable: "Without grievability, there is no life, or, rather, there is something living that is other than life" (Butler, *Frames* 15). This apprehension and recognition are in turn mediated through processes of representation such as photography. Thus, during the war in Vietnam, she writes in *Precarious Life,*

> it was the pictures of the children burning and dying from napalm that brought the U.S. public to a sense of shock, outrage, remorse, and grief. . . . The images furnished a reality, but they also showed a reality that disrupted the hegemonic field of representation itself. Despite their graphic effectivity, the images pointed to somewhere else, beyond themselves, to a life and to a precariousness that they could not show. It was from that apprehension of the precariousness of those lives we destroyed that many U.S. citizens came to develop an important and vital consensus against the war. (Butler, *Precarious* 150)

"We read about lives lost and are often given the numbers," Butler notes in *Frames of War*,

> but these stories are repeated every day, and the repetition appears endless, irremediable. And so, we have to ask, what would it take not only to apprehend the precarious character of lives lost in war, but to have that apprehension coincide with an ethical and political opposition to the losses war entails? Among the questions that follow from this situation are: How is affect produced by this structure of the frame? And what is the relation of affect to ethical and political judgment and practice? / To say that a life is precarious requires not only that a life be apprehended as a life, but also that precariousness be an aspect of what is apprehended in what is living. Normatively construed, I am arguing that there ought to be a more inclusive and egalitarian way of recognizing precariousness, and that this should take form as concrete social policy regarding such issues as shelter, work, food, medical care, and legal status. (Butler, *Frames* 13)

Obama gestures toward the ethical position that Butler advocates inasmuch as he explicitly, and personally, grieves for the non-American, civilian victims of his actions, emphasizing the humanity and innocence of the victims, and the profound tragedy of their deaths. This change in political discourse arguably aligns with a more general cultural shift in attention—in historiography, legal studies, and ethics as much as in popular culture and the media—from the figure of the hero or perpetrator to that of the victim (Judt 828; Sarlo; Bilbija & Payne 18–19; Naqvi 1, qtd. in Gómez 94). If Obama consciously sets himself apart from other world leaders in his acknowledgment of non-American victims *as victims*, moreover, he implicitly draws on a version of U.S. exceptionalism that, in the recent past, has helped justify military and other interventions around the globe—one that posits the United States as exceptional because it takes on the moral burden of protecting the lives of all citizens of the world, not just its own (Levine; Neier 94–95).

Another remarkable feature of Obama's speech is his assumption of accountability, even if it is on a moral and not a judicial level. To be sure, Obama and his attorney general have been notoriously unwilling to hold the Bush administration accountable for war crimes and torture committed as part of the so-called "war against terrorism." Still, the President's acceptance of *personal* accountability for his own administration's actions can be argued to reflect the effects of what Kathryn Sikkink has called the "justice cascade": the rapid rise of the idea that political and military leaders can and should be held individually and criminally accountable for violating human rights. Since the mid-1970s, Sikkink writes, "there has been a shift in the *legitimacy of the norm* of individual criminal accountability for human rights violations and an increase in criminal persecutions on behalf of that norm" (5). (Ironically, in this particular speech the assumption of personal responsibility serves Obama to justify actions that, many argue, actually violate principles both of the U.S. Constitution and international law. The privatization of political accountability, in other words, may have a downside.)

How do we move from questions of representation—questions about the public narration and depiction of violence—into a politics and ethics of life? It is one of the questions that propels Butler's project, and one that has a particular urgency for societies dealing with the memory of internal violence. Let's move some 75 years back in time, to the Spanish Civil War, when international public indignation over the deaths of innocent civilians as a result of military violence, particularly city bombardments, reached unprecedented heights—in large part thanks to the work of war photographers like Robert Capa and David Seymour (Capa et al.)—and when, in spite of this indignation, the majority of democratic powers, including Roosevelt's United States, decided against intervention. The Spanish Civil War is interesting in historical terms (how did public opinion and authorities frame the loss of innocent life at the time?), but also in diachronic terms (how has this framing of the events in Spain changed over time?).

Any discussion about the ethics of life in the Spanish context is obliged to take into account the tremendous loss and mutilation of life in the violent five decades between 1931 and 1981, as well as the ways in which democratic Spain has positioned itself in relation to that violence. In recent years, the precise numbers of dead caused by the Spanish war and its aftermath have returned to occupy a central place in the continuing debates about the conflict. And if counting the dead has been one problem for historians of the Civil War and Francoism, another, more complicated one has been to account more generally for the unspeakable violence that Spaniards inflicted on each other between 1936 and 1975. In what follows I want to consider this double problematic—determining the number of victims and understanding the nature of the violence—in relation to an ethical and historiographical question: What is the proper attitude to adopt, from the present, vis-à-vis past acts of violence and their victims? I am particularly interested in positions that frame these four questions through notions of national or cultural identity. In reality, though, other more fundamental questions precede these four; fundamental questions that, in the Spanish context at least, are anything but resolved: Should there be a public conversation (discourse, dialogue, debate) about past violence in the first place? And if so, who should have it and what should its purpose be? Should the main goal of this conversation be to understand violence, clinically isolated, in its proper historical context? Or should it serve to inform a debate about forms of violence that continue to be operative in the present? Should the conversation about violence result in an assignation of guilt or responsibility, and a concomitant condemnation of the perpetrators and a recognition or reparation of the victims? Or should it resist the temptation of judicialization? I'll return to these larger questions at the end of this essay.

The connection between violence, cultural identity, and the Spanish Civil War is less far-fetched than it seems. A couple of instances immediately come to mind. The violence inflicted by Spaniards on each other between 1936 and 1939 has long been framed through culture: Spaniards, the argument went, are simply more prone to extreme violence than other peoples. Versions of this narrative, rooted in the so-called Black Legend, were prominent in the dictatorship's justification for the 1936 coup and the regime's subsequent harsh rule (Aguilar 35), but also made their appearance in anti-Francoist accounts such as Gerald Brenan's *The Spanish Labyrinth*. ("The troubles of Spain," Brenan wrote, "come from the belief, shared by almost every element in the country, in violent remedies"; 148). Further, as Preston has most recently shown, the ruthlessness of Nationalist violence toward particular sections of the Spanish population was partly rooted in perceptions among the leaders of the military revolt, who had enjoyed long careers in Spain's colonial army, that Spanish peasants did not share these leaders' identity: they were seen as racially more mixed and therefore inferior.[2]

It can be also argued that, at the level of international politics, the connection between cultural identity and collective positions on violence have become increasingly explicit. Tony Judt has shown, for example, that the collective identity of Europe today is closely linked to the idea that a nation or a people cannot be fully European if they are not able or willing to come to terms with shamefully violent episodes in their own history—primarily, but not exclusively, the Holocaust (803–804). "Coming to terms" here stands as a shorthand phrase for a series of actions and attitudes that range from the purely discursive and performative to the more practically political and judicial: they might include the creation of monuments to victims, the issuing of public apologies for crimes committed in the past, the persecution of war criminals, or the granting of reparations to victims or their descendants (Judt 826–29). Within the Spanish context, the enthusiastic embrace by the Generalitat Catalana of the recovery of historical memory movement (Fernández) and anti-bullfight legislation (see also other contributions in this volume), in explicit contrast with the much more reluctant positions adopted by the Spanish national government in Madrid, provides another example of the link between cultural identity and positions on violence. In this context we should also remember that, for post-Franco Spain, "Europe" and "European" have long functioned as synonyms for a much-desired modernity and normality in political, economic, and even literary terms.[3] As I have argued elsewhere, the same generation of intellectuals and politicians who orchestrated the transition to democracy, which failed in important ways to "come to terms" with the country's violent past (at least as that process is being interpreted internationally now), and which celebrated the transition as Spain's entry into European modernity, is ironically forced to adopt something akin to an exceptionalist argument to resist calls for a judicial accounting for that violence, for example by means of a truth commission (Faber, "Raising"). Meanwhile, developments in Spain over the past fifteen years, particularly in the wake of numerous exhumations of previously untouched mass graves from 1936–1939, have fundamentally transformed the ways in which the violence of the years 1931–1978 is being discussed, represented, and processed in the Spanish public and private spheres (Ferrándiz). And while Antonio Gómez is right to state that "for at least some demographic groups, the debate about the Spanish Civil War is also and primordially the debate about (some aspect of) twenty-first century Spain," efforts to translate the renewed interest for a painful past into a viable ethics and politics for the present have encountered significant roadblocks (Gómez 89). In this regard it is telling that one of the most visible recent calls for a moral reveille, by the prominent novelist Antonio Muñoz Molina, ends up dismissing Spain's widespread interest in the country's violent twentieth-century past in the years 2000–2008 as a *distraction* from the profound political and ethical problems facing Spain in the present, which were brought to light by the 2008 economic crisis (Muñoz Molina 12, 202).

Counting the Victims

Statistics of casualties have been a source of controversy since the very outbreak of the fighting in Spain. As early as 1961, Hugh Thomas adjusted the total number of mortal casualties from the war down from the proverbial *millón de muertos* to less than half that number (Thomas, *Spanish Civil War* 632). Yet for a surprisingly long time, the numbers of victims employed by historians have been strangely fluid. In 1996, for example, Michael Richards wrote that "It has been estimated that as many as 200,000 men and women were executed in the post-war period alone" (179)—a number that Preston, in 2012, set at 20,000: a full order of magnitude less (*Holocaust* xi). This confusion about precise numbers of victims has practical explanations, to be sure. These include the distorted nature of official Francoist historiography, the systematic destruction of documents, and the fact that key archives were opened very late or remain inaccessible (Preston xvi). At the same time, the fluidity of the numbers indicates something about their rhetorical value a well.

Starting even before the war, during the mounting violence and political polarization in the years of the Second Republic (1931–1936), victim statistics had become a political tool. Once the war broke out, both sides felt the urgent need to discredit their opponents and garner international support. They were thus naturally tempted to inflate the numbers of victims caused by their enemies and downplay those resulting from their own actions. The Nationalists were particularly brazen in this regard. As Paul Preston writes, "[o]n 18 July 1938 in Burgos, Franco . . . claimed that 54,000 people had been killed in Catalonia. In the same speech, he alleged that 70,000 had been murdered in Madrid and 20,000 in Valencia. On the same day, he told a reporter there had already been a total of 470,000 murders in the Republican zone" (*Holocaust* xix). Conversely, the Nationalists systematically minimized or flatly denied their own responsibility for civilian deaths. Even today, for example, pro-Franco historians continue to claim that the death toll of the bombing of Gernika, in the spring of 1937, was minimal (Irujo 239–80; Viñas xxi; Southworth 507–31).

Another factor that has granted new relevance to the victims of violence is what has been termed the ethical turn in historiography. In the past couple of decades, historians have moved from a descriptive and explanatory mode, in factual and causal terms, to a mode of writing that, in addition to establishing facts and causes, seeks to pass judgment and apportion moral responsibilities.[4] This trend has also affected professional historians' accounts of the Spanish Civil War, and has coincided with the rise of the grassroots movement calling for a recovery of historical memory (Ferrándiz). As Santos Juliá explained some years ago, the generation of Spanish historians that grew up under Francoism and lived the transition to democracy as adults was convinced that the recent past should be approached as aseptically as possible. He and his contemporaries were "sobre todo interesad[os] no en recordar, sino en conocer el pasado, en investigar, en saber qué pasó" (above all, interested not in remembering the past but knowing it, investigating it, finding out what happened);

> Aquellos jóvenes prefirieron, pues, no fiarse de la memoria; más aún: optaron por echar la guerra al olvido en un sentido muy preciso: la consideraron como historia, como un pasado clausurado, algo que había afectado a sus padres, pero de lo que era preciso librarse si se quería desbrozar el único camino que podía reconducir a la democracia, a la libertad. . . . La guerra era sencillamente historia, objeto de conocimiento, no de memoria; su herencia no era bien venida. . . . De modo que nos pusimos a investigar recusando simultáneamente la presunta memoria recibida y las narrativas que les habían servido de vehículo, para volver la vista hacia la política, el sistema de partidos, la acción de los sindicatos, el movimiento obrero, la crisis económica, la lucha de clases, el papel de la Iglesia. De modo que, mientras España mudaba de instituciones, un montón de jóvenes investigadores se asomaba a ese pasado con el propósito de conocer, de identificar lo que había ocurrido. . . . aquella gente, por sus biografías, estaba más interesada en saber del pasado que en saldar alguna cuenta pendiente con ese pasado: no hicieron historia con el propósito de acumular pruebas para llevar ante un juez que dictaminara sobre culpas. (Juliá, "Bajo el imperio" 12, 14–16)
>
> (Those young [scholars] preferred, then, to not trust memory; what is more, they opted to confine the war to oblivion in a very precise sense: they considered it as history, as a closed-off past, something that had affected their parents but from which they needed to liberate themselves if they wanted to clear the only road that could lead back to democracy, to freedom. . . . The war was simply history, an object of knowledge but not of memory; its legacy was not welcome. . . . This is why we began to do research rejecting, at the same time, the supposed memory that had been handed down to us and the narratives that had been its vehicle, in order to turn our gaze to politics, the party system, the actions undertaken by the labor unions, the labor movement, the economic crisis, the class struggle, the role of the Church. This is why, precisely when Spain itself was changing its institutions, a lot of young researchers turned to that past with the goal of knowing it, of identifying what had happened . . . given their own life trajectories, those people were more interested in knowing about the past than in settling some account or other with

that past: they did not practice history with the goal of accumulating evidence to bring before a judge in order for him to pass a guilty verdict.)

What Juliá expresses here, I would argue, is his generation's desire for a genealogical break. By treating the violent Spain that was lived and shaped by the generation of his parents as a past that is *not* welcome at the table of the present, his generation asserted its right to *disavow* any claim from the past on Spain's future: it assumed no duty, no debt, and no obligation (to continue the struggle, for instance, or to settle accounts).[5] Nor did it accept the notion of (political) violence as the Spanish curse. Instead, it expressed its commitment to building a Spain *not* defined by its own past. This process had an affective dimension as well: If the desire to be part of a modern, European Spain was driven by a measure of embarrassment, the transition to democracy that Juliá and his generation saw as their greatest accomplishment was a source of considerable pride.

What to Juliá and his contemporaries felt as a liberation, however, younger generations experienced as a void, unfinished business, and—ironically—a source of embarrassment. In contrast with their teachers, historians born in the 1960s have preferred to frame their work in ethical and political terms. In this trend they have no doubt been influenced by the historiography of the Holocaust, which has been framing the historian's task in part as a duty to remember and do justice to the dead, or, in Dominick LaCapra's words, "not only remembering what happened in the past but actively recognizing the fundamental injustice done to the victims as a premise of legitimate action in the present and future" (LaCapra 21, qtd. in Dintenfass 11). Juliá, who himself disagrees with this position, summarizes his younger colleagues' views, not inaccurately, as follows:

> [D]ijeron: aquí no se ha hecho justicia porque se ha olvidado a los vencidos, a los que combatieron por la causa de la República y sufrieron luego depuraciones, cárceles, fusilamientos. Nadie se ha acordado de ellos, nadie ha reivindicado su memoria, la legitimidad de su causa, nadie les ha organizado un homenaje. La indagación en el pasado recuperó así su carga moralista inspirada en lo que Ginzburg ha denominado modelo judicial. Los crímenes fueron tan monstruosos que pretender una neutralidad valorativa, quedarse sólo en la comprensión y en la explicación, mantener lo que Bloch consideraba como máxima perversión del oficio de historiador, esto es, convertirse en juez, se criticó como una abdicación del oficio. Tarea del historiador tendría que ser, por tanto, recuperar la memoria, o sea, no exactamente conocer esos pasados, sino exigir justicia. ("Bajo el imperio" 17–18)

> (They said: justice has not been done here because the defeated have been forgotten, those who fought for the Republican cause and later suffered purges, prison, executions. No one remembers them, no one has defended their memory, the legitimacy of their cause, no one has organized any tribute to them. In this way, the investigation of the past recovered its moralistic charge, inspired by what Ginzburg has called the judicial model. The crimes were so monstrous that to strive for a neutrality of judgment, to stick to understanding and explanation, to maintain what Bloch considered as the greatest perversion of the historian's craft, that is, to become a judge, was criticized as an abdication of that craft. Hence it became the historian's task to recover memory, that is, not exactly to know those pasts, but to demand justice.)

For Juliá, the work of the academic historian should be confined to the disinterested search for the truth, narrowly defined as a *factual* truth (Mate 21). Representations of the past that are informed by political, emotional, affective, or ethical concerns, by contrast, fall under the notion of "memory" ("Presentación" 17; see also Faber, "Debate" 177–80). In response to Spain's adoption of the so-called Law of Historical Memory, moreover, Juliá has argued that the State has no business legislating on representations of the past, which in his view is tantamount to the imposition, from above, of one uniform version that, moreover, will tend to be self-serving. A number of his colleagues in the profession, however, have emphatically disputed both positions. Pedro Ruiz Torres and Pablo Sánchez León have argued against a rigid distinction between history and memory (Ruiz Torres; Sánchez León "Objetividad"). Ricard Vinyes, for his part, suggests that a democratic state does indeed have the obligation to help shape the narrative of a nation's past through a *política pública de la memoria*, precisely in order to "reconocer en qué se basa el sedimento ético de las instituciones que tenemos, del sistema de convivencia que nos hemos dado" ("Memoria" 57) (recognize the ethical foundation of our institutions, of the system of coexistence that we have adopted):

> Es el reconocimiento de este patrimonio y la demanda de trasmisión de este patrimonio, su valoración positiva, lo que *constituye* la memoria democrática con toda su pluralidad, y la instituye en un derecho civil —no en un deber moral— que funda y basa un ámbito de responsabilidad en el gobierno. Y el deber político generado en la administración es garantizar a los ciudadanos el ejercicio de este derecho con una política pública de la memoria, no instaurando una memoria pública. ("Memoria" 58; qtd. in Faber, "Usted" 20)

(It is the recognition of this legacy and the demand that it be transmitted, that it be given a positive value, which *constitutes* democratic memory in all its plurality, and institutes it as a civil right—not a moral duty—that founds, and lays the basis for, a climate of responsibility in government. And the political duty that the practice of governance generates is to guarantee to all citizens the exercise of that right, and to do so with a public memory policy, not by implementing a particular public memory.)

Vinyes's position brings us back to the link between cultural or national identity, on the one hand, and a collective position vis-à-vis a violent past, on the other. For Vinyes, a state-sponsored *política pública de la memoria* can serve to cement a collective identity in particular modes of public representation, ranging from school curricula to museums and monuments (Vinyes, *Asalto* 11–24, 65–72).

Accounting for Violence

This dispute among professional historians is a part of a wider series of recent debates waged in the public and academic spheres of Spain, in which historians, intellectuals, and citizens have disagreed sharply over a set of basic questions related to ethics and judgment vis-à-vis the Spanish Civil War, the Franco dictatorship, and the transition to democracy. At stake in these debates have been the duty or debt of the present generations toward those of the past—and whether it is proper to speak of such a duty or debt at all; the differences between the violence perpetrated on the Republican side versus that on the Nationalist side; and the distinctions between history and memory. Statistics have figured prominently in some of these debates. In addition to the numbers, however, three other components have played a key role in the ethical judgment of the past: the characterization of the victim, the imputed motivations of the perpetrators, and the level of knowledge and complicity of those in command on either side.

Among professional and amateur historians, several conflicting arguments have emerged that are worth a tentative attempt at categorization. Predominant among historians like Paul Preston, Ángel Viñas, and Helen Graham—scholars often identified as "pro-Republican"[6]—has been the argument that the differences between the two groups of victims—those who died in Republican territory and those who died in Nationalist territory—are of a qualitative and quantitative kind. Quantitatively, many more people died on the Nationalist side than on the Republican side. (The numbers according to Preston stand roughly in a proportion from three to one [xvii–xviii].) Qualitatively, most of the victims on the Republican side died as a result of uncontrolled outbursts of violence that the Republican authorities tried to quell and eventually succeeded in quelling. The victims on the Nationalist side, by contrast, were killed as a result of an explicit strategy consciously designed by the rebel leadership to sow terror among their political enemies.

> The repression carried out by the military rebels was a carefully planned operation to eliminate, in the words of the director of the coup, Emilio Mola, "without scruple or hesitation those who do not think as we do." In contrast, the repression in the Republican zone was hot-blooded and reactive. . . . Unlike the systematic repression unleashed by the rebels as an instrument of policy, this random violence took place despite, not because of, the Republican authorities. Indeed, as a result of the efforts of successive Republican governments to re-establish public order, the left-wing repression was restrained and was largely at an end by December 1936. (Preston xiii–xiv)

Historians in this camp also tend to adopt a position of solidarity and identification with the victims of Nationalist violence as representatives of the modern ethical and political values that we still recognize as our own: separation between church and state, women's emancipation, cultural diversity, or workers' rights (Graham 22). At the same time—much like Obama—they explicitly disassociate themselves from all forms of violence, clarifying that they do not aim to justify the violence committed on the Republican side. Others have rejected the argument that the violence committed by Republican sympathizers was largely uncontrolled and not part of an explicit tactic. The filmmaker and amateur historian Carlos García-Alix, for example, has argued that a conscious strategic use of terror was also part and parcel of factions aligned with the Republic: "Habría que plantearse si . . . existió una política de usar el terror como arma de disuasión frente al enemigo en el bando republicano. Y la respuesta es que sí la hubo." (Contratiempo 11:09) (The question that should be asked is whether . . . the Republican side, as a matter of policy, used terror as a weapon to deter the enemy. And the answer is that it did).[7]

A third group of historians believe that it is useless to think in terms of relative blame. Santos Juliá, for example, has argued that when it came to the *motivations* for violence, it is important to recognize that both sides were driven by ideologies that twenty-first century Spaniards can no longer recognize as their own. Those who murdered in the name of the Republic, Juliá writes, "consideraban la muerte de los representantes del viejo orden social como una exigencia de la revolución" (considered that the revolution demanded the death of the representatives of the old social order) and "creían que una verdadera revolución —que es una conquista violenta de poder político y social— solo puede avanzar amontonando cadáveres y cenizas en su camino" (believed that a true revolution—which is a violent conquest of political and social power—can only advance if it leaves heaps of dead bodies and ashes in its wake) ("Duelo"). Spaniards today, by contrast, should think of "los crímenes entonces cometidos en zona republicana como una tragedia por la que todos tendríamos que hacer duelo" (the crimes that were then committed in the Republican zone as a tragedy for which we should all mourn). For Juliá, the killings "nada tenían que ver con su defensa ni con los valores por ella representados, sino con el comienzo de una revolución social que, entre otras catástrofes como acelerar la derrota, significaría, de triunfar, el fin de la misma República" (had nothing to do with the defense of the Republic or a defense of the values that it represented, but rather with the beginning of a social revolution that, in addition to the fact that it accelerated the Republic's defeat, among other catastrophes, would, should it have succeeded, meant the very end of the Republic):

> Cuando se comparan los crímenes de los rebeldes con los de los leales, al modo en que Ossorio se lo decía a Azaña: ellos comenzaron; o se insiste en que fueron menos: ellos matan más; o se reducen a desmanes de incontrolados: ellos planifican; lo que se olvida es que esos crímenes obedecieron a una lógica propia, reiteradamente publicitada desde discursos de líderes anarquistas, comunistas y socialistas, repetidos cada vez que se cometía un crimen masivo: que era preciso destruir desde la raíz el viejo mundo, prender fuego a sus símbolos y proceder a la limpieza de sus representantes. ("Duelo")
>
> (When comparing the crimes of the rebels with those of the loyalists, it is like Ossorio told Azaña: they started it; or people underscore that [the loyalists'] crimes were fewer: "they kill more"; or [the loyalists' crimes] are reduced to uncontrolled excesses: they plan [their crimes]; what people forget is that those crimes obeyed their own logic, which was publicized repeatedly in speeches by anarchist, communist, and socialist leaders and reiterated every time a massive crime was committed: that it was necessary to destroy the old world from its roots, to set fire to its symbols and proceed to purge its representatives.)

Along similar lines, the novelist and literary critic Andrés Trapiello contends that the sheer intensity of the violence on both sides—the *qualitative* similarities, in other words—trump any considerations of the *quantitative* dimension. "Los crímenes, de una y otra zona," he writes, "fueron, ciertamente, equiparables" (*Las armas* 14) (the crimes committed in both zones were certainly comparable); or, as one of his alter egos in the novel *Ayer no más* states: "Hubo cierta equivalencia en el horror de los crímenes cometidos en la Guerra. Y seguiría habiéndola igualmente si en una balanza hubiese doscientos mil y en la otra diez mil, si la naturaleza del crimen fuese la misma cuando hablamos de tal monstruosidad. La cifra exacta puede ser lo de menos" (140) (There was a certain equivalency in terms of the horror of the crimes committed during the war. And that equivalency would still exist even if one side of the scale had 200,000 crimes and the other 10,000, because when we talk about that level of monstrosity, the nature of the crime is the same. The exact numbers might well be less relevant). Trapiello, who has long been interested in the cultural production of intellectuals associated with the Nationalists, also attempts to complicate the binaries that are usually applied to the history of the Spanish Civil War. While it is true that it was the Republic that, by and large, represented the values of the Enlightenment, he writes, in practice there were liberals and anti-liberals, democrats and anti-democrats, on both sides (*Las armas* 14). For Trapiello, as for Julia, if there is anything worth rescuing from the war it is the position of those few, on either side, whose views and attitudes most closely match today's liberal-democratic consensus.

In Butler's terms, Juliá explicitly invokes the notion of grievability ("hacer duelo"), and seems to call for an end to *differential* grievability: the notion that *some* Spaniards should grieve for *some* victims ("their" victims), and other Spaniards for others. For Juliá, this need for a collective grieving beyond political divisions is closely linked with the idea that the violent political struggles from the past should never be translated into present politics. As he puts it, the legacy of those struggles is "not welcome" ("*su herencia no [es] bien venida*"). The aseptic shield that holds the legacy back is, for Juliá, the professional discipline of the historian—a discipline that those calling for a recovery of historical memory lack. I'll return to the role of professional competency in a moment.

A fourth, more unusual and solitary position is that of the American historian Michael Seidman. In two books on the Spanish Civil War, *Republic of Egos* and *The Victorious Counterrevolution,* Seidman has argued that historians have systematically overrated the ideological motivations of the individual actors involved. Most people just wanted to save their skin and protect their loved ones. Moreover, the eventual victory of the Nationalists over the Republicans had as much to do with material factors—each side's ability to provide food, comfort, and protection to those under their control or serving in their ranks—

as with political or ideological factors. Seidman thus ends up imputing a new kind of rationality to those involved in the war, including those responsible for acts of violence. The repression and extortion in the Nationalist rearguard, for example, becomes merely an effective way to maximize social order and pro-Nationalist fund raising (Seidman 117–27, Faber, "Seidman" 89–92).

What makes the explanation and understanding of Spanish Civil War violence particularly complicated in Spain today is the intersection of historiographical, ethical and political factors. At stake is not only what exactly happened, and why, more than seventy-five years ago. All of the academic authorities who adopt positions vis-à-vis the violent events of the past are also, at the same time, positioning themselves ethically and politically in the present. This is why, as I have argued elsewhere, Juliá's rejection of "memory" as inherently more politicized than "history" (as practiced by professional academic historians) does not hold water (Faber, "Raising" 120). (In that sense, Ricard Vinyes's position, outlined above, is more logically and politically consistent.) Pablo Sánchez León has contended, from an entirely different position than Juliá's, that this complication has not been beneficial to the rigor and quality of historical scholarship on the Spanish Civil War. In fact, he suspects there had been nothing less than a "collapse of the scholarly edifice built by the efforts of at least two generations of specialists" and a regression to attitudes that were prevalent in the 1950s that have led to "a collective dead end" ("Violence"). His argument adopts a wide and original angle, and is worth presenting in some detail.

Sánchez León describes an increasingly polarized scholarly landscape that "has begun to resemble a Hobbesian state of nature where different groups of scholars and amateurs fight for their own living space by aggressively attacking the rest" ("Violence"). Sánchez León identifies two main camps. On the hand, scholars like Preston and Viñas defend a reading of the Civil War that blames the revolting military for the outbreak of the war; emphasizes the planned, deliberate, and massive nature of Nationalist violence; ascribes the violence committed in Republican territory to uncontrollable and criminal elements associated with revolutionary groups, particularly the Anarchists; and defend the repression of those groups by the Republican government as the only option available given the need to galvanize the Republican war effort. On the other hand, there are less clearly pro-Republican historians like Manuel Álvarez-Tardío and Fernando del Rey Reguillo, who express doubts about the commitment to democracy among the leaders of the Republic.

Beyond their many disagreements, however, Sánchez León argues that both groups share a number of basic features. Both present themselves as examples of scholarly rigor, which in both cases is equated with a dispassionate and exhaustive reliance on documentary evidence, while both discredit each other as blinded by political prejudice. In the end, both also revert back to the "discourse of shared responsibility" and a blanket rejection of violence on either side of the war, while "excesses and misdeeds on the Republican side are exclusively attributed to autonomous groups and networks of more or less formally Anarchist or revolutionary Communist allegiance." In addition to their shared "reluctance to theorize," Sánchez León writes, "both these pro- and anti-Republican lines lack a hermeneutical foundation. Far too often, Preston, Rey and Viñas simply equate documentation with information, and information with truth" ("Violence"). In the end, Sánchez León argues, both groups of historians can be considered conservative.

The kind of gridlock and regression that Sánchez León describes, as well as the tendency to sharply mark authorities and competencies with the purpose of excluding or delegitimizing competing narratives, has been typical of the public exchanges around historical memory. Elsewhere I have characterized these exchanges as a *diálogo de sordos,* and speculated what should happen for the public dialogue about the past to be more productive (Faber, "Debate"). It is worth exploring, however, what *productive* might mean in the Spanish case. Many critics and commentators, myself included, have almost automatically assumed that a productive dialogue would lead to a form of *resolution* of the gridlock: the identification of a shared set of values—associated with notions of tolerance, respect, and democracy—that would allow for a proper recognition of all victims' suffering, of all forms responsibility, and the possibility of moving on. In short: a form of reconciliation, understood as an overcoming of conflict.

Antonio Gómez López-Quiñones has convincingly pointed to the fallacies in this set of expectations, however. In a dense but lucid essay he calls attention to the many ways in which the lack of productivity of the historical memory debate in Spain can in fact be ascribed to a *depoliticization* of the issues at stake—a depoliticization that has severely limited the political potential, in Spain's present, of narratives and other representations of the past. The processes that have contributed to this depoliticization are several. They include the attempts, mentioned already, to limit the discourse about the past to particular professional guilds (historians, judges), or to shield Spanish history from other histories (the invocation of exceptionalism). But they also include, Gómez argues, an uncritical assumption of liberal-democratic commonplaces; the idea, for example, that the Spanish public sphere simply should be able to accommodate many different memories of the past in a peaceful juxtaposition. Gómez thinks this is a fallacy: "we should not conceive the Spanish historical memory as a pluralistic market of recollections where there are potential consumers (so to speak) for each one" (91).

The invocation of human rights and international law, along with the judicialization of the accounting for the violent past, have been effective in some respects—for example, in countering exceptionalist positions, forcing Spanish government action, and garnering support from international organizations. But in other respects, these strategies have contributed to a sapping of political potential. By invoking supposed pre- or extra-political or notions of humanitarianism and common decency (values assumed to transcend any political difference), they have also encouraged a privatization and individualization of the suffering of the victims and their families (Gómez 93). The victims' associations themselves have been a part of this process inasmuch as they

> treat the crimes of Francoism as tragic events that fundamentally concern the affected families and their private suffering. The public and political facets of these offenses, their trans-individual and non-family dimension, their inevitable "correlation with" and "implication in" the trajectory of some political parties, and their ideological connection with some radical movements are usually dissolved or lost. (93)

Similarly, Gómez argues that the centrality of victimhood has become the basis for a new kind of identity politics delinked from the (revolutionary, emancipatory, modernizing) politics that motivated the victims to be targeted by right-wing violence in the first place. The groups that have emerged in this process seek social recognition, to be sure; but they are united *not* by "pre-existing political programs or ideological attachments." What unites them instead is "a common identity that a past experience of human rights violations confers unto them. Their politics are a causal extension of this identity and this identity is the underlying base upon which their petitions are molded and predicated" (Gómez 96). Finally, and most importantly, the hegemony of liberalism in the dialogue about the violent Spanish past has served to incorporate "non-liberal political aspirations (that is, political goals that are inimical to the liberal project)" but at the price of "deactivat(ing) their inassimilable and most unsettling content" (88).

So while political landscape appears to have changed since the years of the Transition—critics and activists routinely refer to the "breaking" of the "pact of silence" around the year 2000—it could be argued that the aseptic barrier between past and present actually remains largely in place, functioning as a filter that removes the sting from any political projects from the past that might unsettle the legitimacy of the status quo. Going back to Santos Juliá's account of the attitude that his generation of historians adopted toward the previous generation's involvement and engagement in violence, it seems clear that the persistence of this filter is closely related to the blanket repudiation of political violence that the country's elites, bound and determined as they were to avoid a second civil war, adopted as an imperative for post-Franco Spain (Aguilar). Interestingly, even the more radical Left—the Left most critical of the Transition—has trouble accounting for the violence that its forebears embraced in the 1930s. (Explaining this properly would take more space than I have available here, but the persistence of terrorism and state-sponsored counter-terrorism in post-Franco Spain is doubtless a central factor; it was in that context that the notion of the *victim* as a central category first gained its prominence in the country's public sphere and political discourse.)

It is true, of course, that the widespread disavowal of violence as a legitimate political tactic—even, most recently, on the part of ETA—can be seen as a genuine accomplishment in Spain today. What Gómez and Sánchez León show, however, is that the blanket, almost knee-jerk, repudiation of violence has led to a series of distortions—both in the historiography of the Civil War, where historians have been too quick to attribute political violence to "rogue," non-State actors, and in political life, where alternative political projects are disqualified out of hand because of their past association with violent strategies. This critical reading of the way in which historical violence has been processed in post-Franco Spain aligns to some extent with the equally critical reading of Transition Culture (*Cultura de la Transición* or *CT)* developed by Guillem Martínez, Amador Fernández-Savater and others (Martínez). These authors argue that the post-Francoist status quo has been marked by a fear of popular political mobilization beyond the party system and the ballot box; a steady hold on power by political and economic elites, with significant support from a largely co-opted intelligentsia; and a deep-rooted reluctance to considering other forms of political and social organization than those enshrined in the 1978 Constitution (Martínez).

The twentieth-century history of violence in Spain is painful and complicated. Yet it could serve as an important point of departure for an intellectual or scholarly discussion about the precise uses and semantic changes in the *concept* of violence as it appears in public discourse between, say, 1931 and 2014; and for a broader, *political* discussion about major questions facing the country today. What does it mean, for example, to be Spanish in a Europe that prides itself on an open attitude toward difficult or violent pasts and a willingness to be held accountable for them? What forms of violence continue to operate in the country—vis-à-vis the unemployed, for instance, in the foreclosure crisis, or with respect to illegal immigrants, animals, or the environment? (The fact that these forms of violence are less immediately visible or driven by law, policy, or regulation; or the fact that they are directed at nonhuman life, does not necessarily make them any less objectionable or identifiable as violence.) Does the violence with which, decades ago, certain groups and individuals defended particular values and projects—social justice, emancipation, workers' rights, cultural autonomy, an alternative economic system—invalidate those values and projects altogether in the present?

That this discussion is not or only very marginally held is, I would argue, due in part to the unwillingness among those in positions of authority—academic, political, or cultural—to allow the past to inform the present in politically meaningful ways. This unwillingness is not just manifested in self-propagating attitudes and behaviors, moreover; it is actually embodied in the *institutional structures* that shape public discourse and civic life in Spain. These include the media, the universities, the judiciary, the electoral system, and the internal organization of political parties. In recent years only the first and last of these categories (the media and party organization) have seen the timid emergence of some serious attempts at lasting reform. The immediate, refreshing results of these courageous but marginal attempts at institutional change—media outlets like *eldiario.es, La Marea,* Contratiempo (contratiempohistoria.org), or *Mongolia;* new coalitions within the smaller parties on the Left—seem to warrant a modest amount of optimism.

Returning to the larger questions posed at the beginning of this essay, it is clear that the answer to the first—should there be a public conversation about past violence in the first place?—is, in my view, affirmative. Participation, moreover, should not be limited to experts or authorities but instead be broadly civic. Its purpose should, indeed, be to understand violence in its proper historical context *but also* to include forms of violence operative in the present. And while a judicial or quasi-judicial framework—such as a Truth Commission—may be productive, it will likely not be sufficient.

Notes

1. "The Israeli government and the media started to say that everyone who was killed or injured in Gaza was a member of Hamas; or that they were all being used as part of the war effort; that even the children were instruments of the war effort; that the Palestinians put them out there, in the targets, to show that Israelis would kill

children, and this was actually part of a war effort. At this point, every single living being who is Palestinian becomes a war instrument. They are all, in their being, or by virtue of being Palestinian, declaring war on Israel or seeking the destruction of Israel. So any and all Palestinian lives that are killed or injured are understood no longer to be lives, no longer understood to be living, no longer understood even to be human in a recognizable sense, but they are artillery. And of course, the extreme instance of that is the suicide bomber, who has become unpopular in recent years. That is the instance in which a body becomes artillery, or becomes part of a violent act. If that figure gets extended to the entire Palestinian population, then there is no living human population anymore, and no one who is killed there can be grieved" (Aloni; qtd. in Kellogg 79–80).

2. "The leaders of the rebellion," Preston writes, ". . . regarded the Spanish proletariat in the same way as they did the Moroccan, as an inferior race that had to be subjugated by sudden, uncompromising violence" (xii).
3. Thus, Jordi Gracia and Domingo Ródenas celebrate Spain's attainment of literary and intellectual modernity as "el despliegue diverso y desatado de una europeidad moderna, al fin normalizada" (5) (the diverse and unbridled unfolding of a modern Europeanness, finally normalized).
4. "[I]f history is unavoidably an ethical activity," Alun Munslow writes in his description of the ethical turn in history, "historians must recognize the moral preferences they have. The ethical turn suggests moral lessons do not materialize *from* the past; they are embedded in history as we construct it" (197). See also Dintenfass.
5. As Žižek reminds us, there is such a thing as excessive historicization (670).
6. The label is problematic; for example, historians of World War II are not usually labeled as "pro-Allied" (Faber, "Reframing").
7. Recent investigations are producing a much more nuanced picture; see Maria Thomas and Ledesma.

Works Cited

Aguilar, Fernández P. *Memory and Amnesia: The Role of the Spanish Civil War in the Transition to Democracy*. New York: Berghahn Books, 2002.

Aloni, Udi. "Judith Butler: As a Jew, I was taught it was ethically imperative to speak up." *Ha'aretz* 24 Feb. 2010. Web. 20 Jan. 2014.

Bilbija, Ksenija, and Leigh A. Payne. "Introduction: Time is Money: The Memory Market in Latin America." *Accounting for Violence: Marketing Memory in Latin America*. Durham: Duke University Press, 2011. 1–40.

Brenan, Gerald. *The Spanish Labyrinth: An Account of the Social and Political Background of the Civil War*. Cambridge: Cambridge University Press, 1990.

Butler, Judith. *Precarious Life: The Powers of Mourning and Violence*. London: Verso, 2003.

_____. *Frames of War: When Is Life Grievable?* London: Verso, 2009.

Capa, Robert, Gerta Taro, David Seymour, and Cynthia Young. *The Mexican Suitcase: The Rediscovered Spanish Civil War Negatives of Capa, Chim and Taro.* Göttingen: Steidl, 2010.

Contratiempo: Historia y Memoria. "Muertos civiles e interpretación histórica." Broadcast 7 May 2013. Web. 20 January 2014.

Dintenfass, Michael. "Truth's Other: Ethics, the History of the Holocaust, and Historiographical Theory after the Linguistic Turn." *History and Theory* 39.1 (2000): 1–20.

Faber, Sebastiaan. "'¿Usted, qué sabe?': History, Memory, and the Voice of the Witness." *Revista Canadiense de Estudios Hispánicos* 36.1 (2012): 9–27.

_____. "Actos afiliativos y postmemoria: Asuntos pendientes." *Pasavento. Revista de Estudios Hispánicos* 2.1 (2014): 137–55. Web. 20 January 2014.

_____. "Raising the Specter of 'Argentinization': The Temptation of Spanish Exceptionalism." *Memory and Its Discontents: Spanish Culture in the Early Twenty-First Century.* Ed. Luis Martín-Estudillo and Nicholas Spadaccini. *Hispanic Issues On Line* 11 (Fall 2012): 117–36.

_____. "Reframing Our View of the War in Spain." Rev. of *The Spanish Holocaust,* by Paul Preston. New York: Horton, 2012. *The Volunteer* 19.2 (2012): 13–15.

_____. "The Debate about Spain's Past and the Crisis of Academic Legitimacy: The Case of Santos Juliá." *The Colorado Review of Hispanic Studies* 5 (2007): 165–90.

_____. Rev. of *The Victorious Counterrevolution: The Nationalist Effort in the Spanish Civil War,* by Michael Seidman. Madison: University of Wisconsin Press, 2011. *Journal of Spanish Cultural Studies* 14.1 (2013): 89–92.

Fernández, Antonio. "Cataluña dedica a recuperar la memoria histórica tanto como todo el Estado español," *El Confidencial,* 8 Jan. 2010. Web. 18 Jan. 2014.

Ferrándiz, Francisco. "Exhuming the Defeated: Civil War Mass Graves in 21st-Century Spain." *American Ethnologist* 40.1 (2013): 38–54.

Gómez López-Quiñones, Antonio. "A Secret Agreement: The Historical Memory Debate and the Limits of Recognition." *Memory and Its Discontents: Spanish Culture in the Early Twenty-First Century.* Ed. Luis Martín-Estudillo and Nicholas Spadaccini. *Hispanic Issues On-Line* 11 (2012). 87–116. Web. 20 January 2014.

Gracia, García J., D. Rodenas de Moya, and José-Carlos Mainer. Historia De La Literatura Española: 7. Barcelona: Crítica, 2011.

Graham, Helen. *The War and Its Shadow: Spain's Civil War in Europe's Long Twentieth Century.* Brighton: Sussex Academic Press, 2012.

Irujo, Ametzaga X. *El Gernika De Richtofen: Un Ensayo De Bombardeo De Terror.* Gernika: Gernikako Bakearen Museoa Fundazioa, 2012.

Judt, Tony. *Postwar: A History of Europe Since 1945.* New York: Penguin Press, 2005.

Juliá, Santos. "Duelo por la República española." *El País* 25 June 2010: 31.

_____. "Bajo el imperio de la memoria." *Revista de Occidente* 302–303 (2006): 7–19.

_____. "Duelo por la República española." *El País* 25 June 2010: 31. Web. 18 Jan. 2014.

_____. "Presentación." *Memoria de la guerra y del franquismo*. Ed. Juliá. Madrid: Taurus, 2006. 15–26.

Kellogg, Catherine. "Walter Benjamin and the Ethics of Violence." *Law, Culture and the Humanities* 9.1 (2011): 71–90.

LaCapra, Dominick. "Lanzmann's *Shoah*: 'Here There Is No Why.'" *Critical Inquiry* 23. 2 (1997): 231–69.

Ledesma, José Luis. "Total War behind the Frontlines? An Inquiry into the Violence on the Republican Side in the Spanish Civil War." *"If You Tolerate This..." The Spanish Civil War in the Age of Total Wars*. Ed. M. Baumeister and S. Schüler-Springorum. New York: Campus, 2008. 154–68.

Levine, Andrew. "The Moralization of American Exceptionalism." *Counterpunch* 4–6 Oct. 2013. Web. 20 January 2014.

Martínez, Guillem, coord. *CT o la cultura de la transición: Crítica a 35 años de cultura española*. Madrid: Delbolsillo, 2012.

Mate, Reyes. "From History and Memory—and Back: Factuality, Knowledge, and Morality." *The Holocaust in Spanish Memory: Historical Perceptions and Cultural Discourse*. Ed. Antonio López Gómez-Quiñones and Susanne Zepp. Leipzig: Leipziger Universitätsverlag, 2010. 15–30.

Muñoz, Molina Antonio. Todo lo que era sólido. Barcelona: Seix Barral, 2013.

Munslow, Alun. *Deconstructing History*. 2nd ed. London: Routledge, 2006.

Naqvi, Fatima. *The Literary and Cultural Rhetoric of Victimhood: Western Europe, 1970–2005*. New York: Palgrave Macmillan, 2007.

Neier, Aryeh. *The International Human Rights Movement: A History*. Princeton, NJ: Princeton University Press, 2012.

Obama, Barack. "Remarks by the President at the National Defense University." Washington, DC: White House Office of the Press Secretary, May 23, 2013. Web. 20 Dec. 2014.

Preston, Paul. *The Spanish Holocaust: Inquisition and Extermination in Twentieth-Century Spain*. New York: W.W. Norton & Co., 2012.

Richards, Michael. "'Terror and Progress': Industrialization, Modernity, and the Making of Francoism." *Spanish Cultural Studies: An Introduction: The Struggle for Modernity*. Eds. Helen Graham and Jo Labanyi. Oxford: Oxford University Press, 1995. 173–82.

Ruiz Torres, Pedro. "De perplejidades y confusiones a propósito de nuestra memoria histórica." *Hispania Nova. Revista de Historia Contemporánea* 7 (2007). Web. 1 Dec. 2010.

Sánchez León, Pablo. "La objetividad como ortodoxia: Los historiadores y el conocimiento de la guerra civil española." *Guerra Civil: Mito y memoria*. Ed. François Godicheau and Julio Aróstegui Sánchez. Madrid: Marcial Pons, 2006. 95–136.

_____. "Violence and the History and Memory of the Spanish Civil War: Beyond the Crisis of Inherited Narrative Frameworks." *Memories of the Second World War in Europe: Redefining the Self and the Nation*. Ed. Manu Bragança et al. London: Berghahn Books, 2015. Forthcoming.

Sarlo, Beatriz. *Tiempo pasado. Cultura de la memoria y giro subjetivo. Una discusión.* Buenos Aires: Siglo XXI, 2005.

Savater, Fernando. "Recuerdos envenenados." *El País* 22 June 2010. Web. 20 January 2014.

Seidman, Michael. *The Victorious Counterrevolution: The Nationalist Effort in the Spanish Civil War.* Madison: University of Wisconsin Press, 2011.

Sikkink, Kathryn. *The Justice Cascade: How Human Rights Prosecutions Are Changing World Politics.* New York: W. W. Norton & Co, 2011.

Southworth, Herbert R. La Destrucción De Guernica: Periodismo, Diplomacia, Propaganda E Historia. Granada: Comares, 2013.

Thomas, Hugh. *The Spanish Civil War.* New York: Harper Collins, 1963.

Thomas, Maria. *The Faith and the Fury: Popular Anticlerical Violence and Iconoclasm in Spain, 1931–1936.* Eastbourne, UK: Sussex Academic Press, 2013.

Trapiello, Andrés. Ayer No Más. Barcelona: Ediciones Destino, 2012.

_____. *Las Armas Y Las Letras: Literatura Y Guerra Civil (1936–1939).* 3rd ed. Barcelona: Destino, 2010.

Viñas, Ángel. Prefacio a la presente edición. La Destrucción *De Guernica: Periodismo, Diplomacia, Propaganda E Historia.* By Herbert R. Southworth. Granada: Comares, 2013. xi–xxiii.

Vinyes, Ricard. "La memoria del Estado." Trans. Juan Carlos Gentile Vitale. *El Estado y la memoria.* Ed. Ricard Vinyes. Madrid: RBA, 2009. 23–66.

_____. "Presentación." *El Estado y la memoria.* Ed. Ricard Vinyes. Madrid: RBA, 2009. 15–19.

_____. *Asalto a la memoria. Impunidades y reconciliaciones, símbolos y éticas.* Barcelona: Libros del Lince, 2011.

Žižek, Slavoj. "Tolerance as Ideological Category." *Critical Inquiry* 34 (2008): 660–82.

Afterword

Spain: Taking the Alternative?

Martín López-Vega and Luis Martín-Estudillo

This volume offers a multidisciplinary look at the politics and ethics of life in contemporary Spain, highlighting the role of literary and visual culture in the mediation and configuration of a number of key debates. The contributions form a coherent and engaging unity even as they come from different fields, including history, literary criticism, media studies, economics, law, and film studies, converging in a composite analysis of some of the most pressing collective concerns of the last decades. The essays relate to a global dimension the peculiarities of the Spanish manifestations of several issues, including the ecological crisis, abortion, genetic engineering, animal rights, and some of the effects of the so-called "Great Recession" which started around 2008 and still lingers for many. Benefitting from the pioneering work of scholars as varied as Donna Haraway, Bruno Latour, Timothy Morton, and Jorge Riechmann, the volume effectively exemplifies the scholarship done mainly from the humanities—albeit some may be in fact from a posthumanist perspective—that is emerging as an essential counterpart to the voices that present an unproblematic and often enthusiastic view of certain scientific applications and political trends. Many of these have been shaping life on our planet for some time and are certainly going to have a growing importance in the near future. The timeliness of the volume at hand is unquestionable.

The tension between the optimism of those who entrust our destiny to the success of technological initiatives and those who caution against an excessive faith in what is usually labeled as "reason" is of course not a new one. The ideology of progress that has been pervasive since the nineteenth century is not as discredited as some may think. In fact, scientific advancement has become a new taboo, in the sense that critics of technical innovations are often dismissed as retrograde, fundamentalist, obscurantist, superstitious, irrational, etc. But now more than ever, as developments in disciplines such as biotechnology make possible interventions that may radically alter the sphere of "life" as we have traditionally known it, those critical positions are increasingly necessary. The same can be said of policies oriented to the regulation (or de-regulation) of economic and social relations that are presented as a natural, unavoidable path. Examining their implications is part of the task of a revitalized humanism, of which ecocriticism and other related perspectives present in this volume are fundamental components.

The beginning of modern Spanish ecocriticism may be traced back to 1975, when Miguel Delibes delivered his acceptance speech on the occasion of his induction into the Real Academia de la Lengua. In his lecture, entitled "Mi Credo" (My Creed), Delibes discards modern ideas of progress and explains how in his novels nature revolts against humankind as a rejection of growing mechanization. He proposes the need to slow down development as the only way to avoid collective suicide: "ahora que hemos visto suficientemente claro que nuestro barco se hunde . . . ¿no sería progresar el admitirlo y aprontar los oportunos remedios para evitarlo? (now that we have seen clearly enough that our ship is sinking . . . would it not be actual progress to admit that it is happening and to prepare the remedies that would prevent it?). Delibes provides specific examples directly related to the series of environmental crises that were already visible to anyone who had the courage to see them, including the poisoning of natural resources, overpopulation, and the extinction of species. His views were characteristic of a liberal Catholic who enjoyed the outdoors as a hunter and a sportsman. Although from today's perspective some of them may sound naïve, they found resonance in a younger generation of intellectuals and activists who had a fundamental role in the establishment of environmental currents in post-Franco Spain.

Delibes's position as a reflective, bourgeois leisure hunter can be connected to Ramachandra Guha's idea that "the enjoyment of nature is an integral part of the consumer society" (571). As several essays in this volume make clear, most debates about wilderness (and nature in general) cannot be isolated from social and economical issues—actually, they are inextricably associated. One could reflect on the role of Spain's national parks (and more recently the so-called "biosphere reserves") as a commodification of nature by means of transforming it into a recreational space, and also on how this governmental intervention has had an impact on the life of shepherds and peasants. The opening of new business related to the exploitation of nature as a recreational commodity, for instance, has forced them to abandon their traditional livelihood and transformed not only their occupations but also the landscape they inhabit, including crops and animals.

The indissoluble link between ecological and economic issues is addressed by a number of essays in this volume. One of the volume's main concerns is that of revealing the semantic traps set by an emerging industry of marketing and public relations experts apt at washing their clients' image and presenting it as one that is environmentally conscious. Luis I. Prádanos in this volume appropriately mentions Joan Martínez-Alier's denunciation of a "gospel of eco-efficiency"; other authors, such as Lynn Keller, underline the problems around similar terms like "sustainability." Keller alerts us to the dangers of this notion, noting that "the readiness which businesses throughout the United States and Europe are embracing sustainability makes clear that in popular discourse sustainability is understood as fitting within the established patterns of capitalist consumption" (580). These terms—eco-efficiency, sustainability—are often merely carpets under which capitalism hides its inconvenient truth: that the overpopulation of our planet is becoming increasingly unsustainable, and so is the dramatic reduction of key resources. Debates about official regulations regarding birth control and population resonate in Pablo de Lora's essay in this volume. De Lora deals with the issue of abortion in contemporary Spain, a subject that is far from being solved, as shown by the controversy caused by Mariano Rajoy's conservative government and its attempts (so far unsuccessful, and seemingly discarded at the time of writing this afterword) to pass one of Europe's most restrictive abortion laws. Meanwhile, references to a new sort of travel with strong bioethical implications have emerged in addition to the "abortion tourism" mentioned by de Lora. According to Saskia Gauthier and others, Switzerland, where assisted suicide is not penalized, is becoming a growing destination for people who desire to die with dignity and are not allowed to by their local governments. In Spain, the Basque Parliament decided to open a debate on euthanasia in October 2014, even though the regulation of assisted suicide does not fall under its competencies. Euthanasia is considered a crime under Spanish law, and there seems to be little political interest in changing that, even after lively public debate on the topic. These debates arose in recent years mostly due to the impact of the case of Ramón Sampedro and the subsequent film *Mar adentro* (The Sea Inside), which Paul Begin deals with in his essay in this volume.

The current Spanish government's indifference has its counterweight in official interest on biotechnology and transgenics, as Sainath Suryanarayanan and Katarzyna Beilin highlight in their essay in this volume. It may be worth considering the issue they bring forth in parallel with the competition to harbor the first "Almacén temporal centralizado" (centralized temporal repository), which will collect the waste produced by Spanish nuclear power plants. Thirteen villages offered their land to build the nuclear depository. In December 2011, José Luis Rodríguez Zapatero's government chose Villar de Cañas (Cuenca). In November 2014, the Consejo de Seguridad Nuclear (Council for Nuclear Safety) questioned this choice of location, due to the geological composition of the land and seismic issues. The local officials who had offered to host this nuclear silo were all interested in receiving it for economic reasons: regional growth, employment, and so on. The Spanish government has become increasingly interested in biotechnology and transgenics for similar motives, according to Suryanarayanan and Beilin. This fact raises important concerns about the type of "progress" or "modernization" that this interest may yield to the nation: is Spain becoming the laboratory, or just the guinea pig? With the recent ban on transgenic crops in Poland and France, Spain stands alone as the only country in Europe with a significant quantity of land devoted to transgenic crops.

Issues such as the ecological crisis or the escalation of social inequality and its effects on life conditions require thinking on a global scale. However, the national cannot be quite yet dismissed as a valid dimension for analyses aimed at illuminating those developments. An area-specific approach continues to hold weight especially if one wishes to interrogate the ways in which cultural manifestations are used to support or undermine the value systems and imaginary constructions that go hand in hand with these issues. In the case of this volume, we are presented a collective, interdisciplinary project under the aegis of Iberian Studies, a rather new umbrella that, in this specific manifestation, effectively overcomes some of the limitations derived from a long-standing fixation on the nation-state as the shaping receptacle of culture. At the same time, this orientation facilitates connections between local production and global problems by superseding the strictures of national systems.

While this sort of epistemological move offers valuable responses to a set of ingrained problems in our discipline (such as its inveterate Castilian-centrism), it also requires a careful consideration of the complexities addressed for it to substantiate a real change in our scholarly practices. One of the risks of this change in perspective is that it may reproduce patterns of inclusion/exclusion and a biased hierarchy of attention analogous to those that have been prevalent within more traditional variants of Hispanism, even when we limit ourselves to the case of Spain. It is worth considering how much thought-provoking work that is being done in, say, Andalusia, Asturias, or the Canary Islands (an area whose study within this new approach would seem problematic from the start, at least nominally) may be left aside in efforts to reconfigure our field vis-à-vis developments that give preeminence to the activity carried out in distinct areas like Catalonia, to mention just one outstanding example.

It should go without saying that attempts to achieve a more balanced, diverse view of the cultural and intellectual spheres at play in Spain (and Portugal) should not generate new blind spots and much less replicate old ones. Future approaches to ecocriticism and related topics from an Iberian Studies perspective must engage works not only in the inescapable Spanish or in Catalan, which fortunately are getting some of the attention they deserve, as is the case in this volume. Our scholarly practice, and not just our explicit intentions, should address work done in Galician (the poetry of Olga Novo can be clearly related to Timothy Morton's notion of the mesh, while Manuel Rivas has acquired a major standing among advocates of environmentalist causes), Portuguese (the works of Miguel Torga or Vergílio Ferreira offer great examples of the relationship between humans and nature), Asturian (it is worth considering Xuan Bello's sense of place, for example), or Basque (one may think of Bernardo Atxaga's problematization of contact zones between the human and the nonhuman.) The fundamental contribution of other arts cannot be ignored either, as Carmen Flys-Junquera and Tonia Raquejo Grado are right to remind us in their essay in this volume, to which more references could be added. Juan Mayorga's theater has explored the implications of giving voice to animals in plays such as *Últimas palabras de Copito de Nieve* (Last Words of Copito de Nieve), *La paz perpetua* (Perpetual Peace) or *La tortuga de Darwin* (Darwin's Tortoise) with demanding intellectualism that has nevertheless been met with great accolades and international success. *Historias Naturales* (Natural Histories/Stories), Miguel Ángel Blanco's intervention at the Prado Museum (2014), confronted taxidermies and other items from the National Museum of Natural Sciences with key paintings of the collection with astonishing results of political as well as aesthetic significance. Blanco's ongoing work *La biblioteca del bosque* (Library of the Forest) also offers new and fascinating approaches to many of these issues. The artist contributed the image for the cover of this volume.

A few fundamental issues can be particularly well assessed from this approach. The cultural and political density of the debates about the treatment of animals in Spain, which are dealt with by several of the authors of this volume, show their complexity when seen through a lens that reveals the richness of positions across regional and intellectual lines. John Beusterien's essay in this volume on the representation of animals in Spanish media casts light on the issue from a perspective that is rarely considered in our disciplines. The controversies around bulls and bullfighting, to mention the most topical example, need to be explored within a context that takes into consideration more than just the recent ban in Catalonia versus the so-called *fiesta nacional*'s subsequent official protection by the Spanish legislature as a heritage of cultural interest for the nation. Older strata of the issue, such as those poeticized by Salvador Espriu in *La pell de brau* (shrewdly read by William Viestenz in this volume) or its very different reception in Galicia or the Canary Islands throughout history, generate an even more complex view of the issue when these different angles are taken into account. It is unquestionable that the disputes on a topic like bullfighting engage arguments—and feelings—that are not limited to ethics. Its political dimension exceeds considerations specific to animal rights advocates and defenders of tradition. This polemic is a clear case of how conflicting interests and projects clash at junctures that would at first seem removed from other contentious issues such as the political nature and future of the nation.

Other consequential experiences, like the disastrous sinking of the oil tanker *Prestige*, emerge as symptomatic of failing structural conditions in Spain. In his essay in this volume, John H. Trevathan pays close attention to movements that join together social and environmental concerns, such as *Nunca Máis*, which emerged in response to the Prestige catastrophe. The disaster was not limited to the shipwreck and the oil spill; the calamitous, irresponsible official response showed that not only developing countries are under what Rob Nixon labels as "slow violence." Slow violence is, in his view, "a violence that occurs gradually and out of sight, a violence of delayed destruction that is dispersed across time and space, an attritional violence that is typically not viewed as violence at all" (2). This violence concerns the poor at first, understanding the poor as "a compendious category subject to almost infinite local variation as well as to fracture along fault lines of ethnicity, gender, race, class, region, religion, and generation" (4). The main goal of this slow violence is amnesia, as "attritional catastrophes that overspill clear boundaries in time and space are marked above all by displacements—temporal, geographical, rhetorical, and technological displacements that simplify violence and underestimate, in advance and in retrospect, the human and environmental costs. Such displacements smooth the way for amnesia, as places are rendered irretrievable to those who once inhabited them, places that ordinarily pass unmourned in the corporate media" (7). Whereas those concepts are usually applied to the state of affairs in countries such as

India, they may be very useful to face certain realities in Spain. The *Nunca Máis* movement was, among other things, an open call for a societal reaction against amnesia and an organized denunciation of slow violence.

Unfortunately, other manifestations of this sort of violence abound in Spain, some of which can be related to the lingering legacy of Francisco Franco's regime—a heritage whose environmental consequences still need to be addressed. Some of them require an exercise of public memory, as in the case of the evacuation of whole villages to build dams under the developmental model of Francoism. Others, such as the massive urban disorder established on the Mediterranean coast (but not only there), were replicated during the greedy years of the so-called "housing bubble." To a great extent, cement (especially when placed at the seaside) has been for many Spaniards what petroleum has been for an equivalent number of Americans: the stuff utopia is made of. That period also saw the radicalization of problems posed by the indiscriminate construction of golf courses in areas with critical and chronic hydrological problems, as is the case in the southern half of Iberia (including Portugal, where many of the issues indicated throughout the volume find parallelisms or even replicas.) In her essay in this volume, Eugenia Afinoguénova analyzes the shift from "quality of life" to "level of life" in Spanish political rhetoric. Indisputably, understanding these problems will entail the decodification of the discourses that are used to make them palatable to the citizenry in their historical depth and symbolic complexity.

The Great Recession whose pre-2008 origins had much to do with the end of the cement dream has radically shaken Spaniards' expectations regarding "level of life." Some of the alternatives that have since emerged are the focus of Matthew I. Feinberg and Susan Larson's essay in this volume. They turn to philosopher José Luis Pardo and architects Iñaki Ábalos and Juan Herreros for responses to the crisis that relate to and re-imagine possibilities for culture and urban space in circumstances where they seem to be discarded by financial capital. This is certainly an old, ongoing tension. As the authors notice, capital regained Madrid's Puerta del Sol a few months after that square became the center of the *indignado* movement: due to a sponsorship agreement between the city council and a communication company the place became, on the subway map, "Vodafone Sol." But socially oriented architects suggest that our cities can also be occupied, or, at least, reshaped for the greater good. As an urban agriculture strategy for Barcelona, Nicola Placella and Magnus Svensson propose using the regular city blocks defined by Idelfons Cerdà as a basis for modular three-story greenhouses or "food shelves" inside which crops are grown. This is but one example of the many alternatives being conceptualized for the new and future needs of our cities. Mitchell Joachim's idea of the Peristaltic City involves "a chimera that combines the technology of an elevator with the biology of a digestive system" (quoted in Carruth 96). Joachim is a member of the Terreform group, one of whose proposals involves using "three-dimensional printed extruded pig cells" (Carruth 98). The cells are grown in-vitro rather than extracted from extant organisms and ecosystems. In any case, those are speculative ideas that seem rather unfeasible—it is still hard to imagine a city made of pig's skin. Nevertheless, Ábalos and Herreros's thoughts on spaces such as latent gardens suggest that more practical alternatives could arise as possible beginnings for an alternative understanding of common spaces and, perhaps, ways of preventing the tragedy of the commons.

Three of the essays contained in this volume deal with issues related to the human perception of animals and the relation between animals and humans. There are noteworthy precedents within Spanish ecocriticism: in 2008 Marta I. González, Jorge Riechmann, Jimena Rodríguez Carreño, and Marta Tafalla put together an influential anthology entitled *Razonar y actuar en defensa de los animales*, while philosopher Jesús Mosterín has recently published two fundamental books about these matters: *El reino de los animales* (2013) and *El triunfo de la compasión* (2014). They all engage in a broader dialogue that has imploded in the last two decades and whose most prominent voices include those of thinkers Jacques Derrida, Giorgio Agamben, and Donna Haraway.

Along these lines, one of the most productive discussions has been that of the reconceptualization of the boundaries, limits, or contact zones between species. Agamben argues that "in our culture, man has always been thought of as the articulation and conjunction of a body and a soul, of a living thing and a *logos*, of a natural (or animal) element and a supernatural or social or divine element. We must learn instead to think of man as what results from the incongruity of these two elements, and investigate not the metaphysical mystery of conjunction, but rather the practical and political mystery of separation" (16). Agamben's ontological inquiry reveals that "the relation between man and animal marks the boundary of an essential domain" (21).

The exploration of that relationship centers Donna Haraway's *When Species Meet*. Haraway observes that Derrida, in his influential *The Animal That Therefore I Am*, questions "whether it is possible to know what *respond* means and how to distinguish a response from a reaction, for human beings as well as for anyone else. He did not fall into the trap of making the subaltern speak: 'It would not be a matter of giving speech back' to animals but perhaps acceding to a thinking . . . that thinks the absence of the name as something other than a privation" (20). However, Haraway blames Derrida's lack of curiosity for his eventual failure at meeting the animal other in a way that could transcend the limits of the division between species; in her view, the philosopher "missed a possible invitation, a possible introduction, to other-worlding" (20). As an alternative, she draws from Mary Louise Pratt's work on the contact zone: a place to meet the other with more interest than prejudices, to descend from the pedestal of the imposed language to the ground where the subaltern can speak her own tongue, and to try to communicate using less talking and more listening—in other words, attempting to understand the other without trying to translate, incorporate, or subalternize her into our own thinking. This position underlines the fact that other species have also been subalternized. One could affirm that subalterns, or the poor, have been animalized, but also that animals have been animalized as well. When Haraway writes that she has "stressed dogs as workers and commodities but rejected the analogies of wage labor, slavery, dependent ward, and nonliving property" (67), she is also suggesting that we have to incorporate other species to class and subaltern discourses but that it must be done carefully, trying to understand the role other species and other individual in those species play, but without carelessly applying to them the same categories we have been using for humans.

The testimonies from those who entered those contact zones (from the jungle to the backyard) to establish exchanges with the other—dog or ape—and our own experiences, show us how other species can be regarded not just with compassion, but, most importantly, with respect. This understanding of animals as interlocutors in the contact zone could create not only a new conception of our relation with other species but also new perspectives about subaltern or persecuted human collectives. Daniel Ares's essay in this volume on the texts (filmic and scholarly) that deal with the life of Marcos Rodríguez Pantoja is a good example, and some others could be mentioned. Those who remained secretly hidden at home after the Spanish civil war because they feared fascist retributions were called "topos" (moles). The hunting missions organized by the Guardia Civil against the anti-Francoist guerrilla in the mountains during the 1940s and 1950s were very similar in armament and strategy to the hunting expeditions that aimed to decimate wolves' packs over the same period. Thinking of humans, moles, and wolves as part of the same mesh could contribute to a better understanding of the implications and consequences of those situations and actions, which can often be related to issues of intra- or inter-species violence, as noted by Sebastiaan Faber in his essay in this volume.

The Spring 2014 issue of the journal *Interdisciplinary Studies in Literature and Environment* opened with a provocative article by Kent C. Ryden entitled "317: An Ecocritical Reading." The purpose of the article is to demonstrate that virtually anything can be illuminated from an ecocritical approach, including numbers chosen at random. Some of the proponents of ecocriticism may envision it as a perspective that could eventually include all the other critical possibilities; one may also wonder about the intellectual sustainability or interest of such a reductive horizon. What is obvious at this point is that ecocriticism is not about looking for funny rabbits and pleasant fountains in poems; it is about understanding our place in the world not as owners and rulers of nature, but as responsible tenants of the commons. It has already contributed to a renewed humanism (for some, a posthumanism) that now involves notions of life that engage existence beyond that of humans. Along these lines, we need to consider how our efforts as scholars of culture continue to have relevance in a cultural and physical landscape in which the main preoccupations are often radically different from those that employed our predecessors. As the French sociologist Razmig Keucheyan puts it, "La nature est un champe de bataille. Elle est déjà à l'heure actuelle, et sera de plus en plus à l'avenir, à mesure que la crise écologique s'approfondira, le théâtre d'affrontements entre des acteurs aux intérêts divergents: mouvements sociaux, États, armées, marchés financiers, assureurs, organisations internationales . . . " (11) (Nature is a battlefield. It is already so, and will increasingly be so in the future. As the ecological crisis deepens, so will the fight between actors with diverging interests: social movements, states, armies, financial markets, insurers, international organizations . . .). Within this context, the term "crisis," which has been ubiquitous in Spain for close to a decade now, may acquire new meanings there as well as in the global scene. Work done in the arts and humanities is at the core of the change in awareness that points to the need for change in a society whose foundations require a radical makeover if it is to safeguard dignified life conditions for generations to come.

Works Cited

Agamben, Giorgio. *The Open: Man and Animal.* Trans. Kevin Attell. Stanford: Stanford University Press, 2004.

Carruth, Allison. "The City Refigured." *Environmental Criticism for the Twenty-First Century.* Ed. Stephanie LeMenager, Teresa Shewry, and Ken Hiltner. New York: Routledge, 2011. 85–102.

Delibes, Miguel. "Mi credo." *Education Ambiental.* Educacion Ambiental. 2014. Web. 22 Dec. 2014.

Derrida, Jacques. *The Animal that Therefore I Am*. Trans. David Wills. New York: Fordham University Press, 2008.

Gauthier, Saskia, Julian Mausbach, Thomas Reisch, and Christine Bartsch. "Suicide tourism: a pilot study on the Swiss phenomenon." *Journal of Medical Ethics* 40.8 (Aug. 2014). Web. 22 Dec. 2014. Published online August 20, 2014.

Guha, Ramachandra. "Radical American Environmentalism and Wilderness Preservation: A Third World Critique." *Social Ethics: Morality and Social Policy*. Eds. Thomas Mappes, Jane Zembaty, and David DeGrazia. New York: McGraw Hill, 1989. 566–74.

Haraway, Donna J. *When Species Meeet*. Minneapolis: University of Minnesota Press, 2008.

Keller, Lynn. "Beyond Imagining, Imagining Beyond." *PMLA* 127.3 (May 2012): 579–85.

Keucheyan, Razmig. *La Nature est un champ de bataille. Essai d'écologie politique*. Paris: Éditions La Découverte, 2014.

Morton, Timothy. "The Mesh." *Environmental Criticism for the Twenty-First Century*. Eds. Stephanie LeMenager, Teresa Shewry, and Ken Hiltner. New York: Routledge, 2011. 19–30.

Nixon, Rob. *Slow Violence and the Environmentalism of the Poor*. Cambridge: Harvard University Press, 2011.

_____. "Neoliberalism, Genre, and 'The Tragedy of the Commons'". *PMLA* 127.3 (2012): 593–99.

Ryden, Kent C. "317: An Ecocritical Reading". *ISLE* 21.2 (Spring 2014): 239–52.

Contributors

Eugenia Afinoguénova is an associate professor of Spanish at Marquette University. She is the author of *El idiota superviviente. Artes y letras españolas frente a la "muerte del hombre," 1969–1990* (2003) and a coeditor (with Jaume Martí-Olivella) of *Spain is (Still) Different: Tourism and Discourse in Spanish Identity* (2008). Her articles and chapters on Spanish tourism and culture have appeared in *The Journal of Tourism History*, *The Journal of Modern History*, *The Hispanic Review*, *Journal of Spanish Cultural Studies*, *Archivos de la Filmoteca*, and many others. Afinoguénova is currently finishing a monograph on the culture and leisure history of the Prado museum.

Daniel Ares López is a doctoral candidate in Hispanic literatures and cultural studies at the University of Wisconsin-Madison. He is presently working on a doctoral project titled *Culturing the Wildness: Nature, Culture and Iberian Wildlife in Twentieth-Century Spain (1940–1980)* that focuses on the relations between humans and wild animals in twentieth-century Spanish cultural history. His main research interest is the study of late-modern and contemporary Iberian literatures and cultures from non-anthropocentric critical approaches, including posthumanist, ecocritical and socioenvironmental ones. His work on Latin American modernist novel and on Spanish animation film in the context of the War on Terror has been published in *Latin American Literary Review* and *Studies in Spanish and Latin American Cinemas.*

Paul Begin is an associate professor of Hispanic studies and chair of International Studies and Languages at Pepperdine University. He has authored various articles on Spanish cinema, especially social issue cinema.

Katarzyna Olga Beilin is a professor in the Department of Spanish and Portuguese at the University of Wisconsin, Madison. She is the author of *Conversaciones literarias con novelistas contemporáneos* (2004), the novel *Meteory* (2005), *Del infierno al cuerpo: otredad en la narrativa y cine peninsular contemporáneo* (2007), and *In Search for Alternative Biopolitics in Contemporary Spain; Anti-Bullfighting, Animality and the Environment in Contemporary Spain* (forthcoming). She is currently working on two new book projects on alternative economies in Spain, as well as on a book on debates on GMOs in the Hispanic world.

John Beusterien is an associate professor at Texas Tech University. His publications include *Canines in Cervantes and Velázquez: An Animal Studies Reading of Early Modern Spain* and *An Eye on Race: Perspectives from Theater in Imperial Spain*. He has also coedited *Touching the Ground: Female Footwear in the Early Modern Hispanic World*; *Sustaining Ecocriticism: Comparative Perspectives*; and *Death and Afterlife: The Case of the Early Modern Hispanic World.*

Sebastiaan Faber is a professor of Hispanic studies at Oberlin College. He is the author of *Exile and Cultural Hegemony: Spanish Intellectuals in Mexico, 1939–1975* (2002), *Anglo-American Hispanists and the Spanish Civil War: Hispanophilia, Commitment, and Discipline* (2008), and coeditor of a collection of essays about Spanish Republican exiles in the United States. He has published widely on topics ranging from the theory of ideology to Latin American literature and the historical memory of the Spanish Civil War, and regularly contributes to the Spanish media. From 2010 until 2015 he served as chair of the Board of Governors of the Abraham Lincoln Brigade Archives (ALBA), and continues to coedit ALBA's quarterly, *The Volunteer.*

Matthew Feinberg is a lecturer in the English department at Case Western Reserve University in Cleveland, Ohio. Interested in the intersection between cultural geography and literary studies, he studies dramatic texts, performance spaces, urban plans, and squatted social centers to explore the relationship between the production of theater and urban space in contemporary Madrid. His work has been published in the *Journal of Spanish Cultural Studies*, *International Journal of Iberian Studies*, the *Journal of Urban Cultural Studies*, *Romance Quarterly*, and *Ometeca*. His current book project examines the theatrical legacy of Madrid's iconic Lavapiés neighborhood within the context of broader struggles over urban space in the years leading up to *la crisis* and the emergence of the 15-M movement.

Carmen Flys-Junquera is an associate professor of American literature in the Department of Modern Languages and at the Franklin Research Institute, both of the University of Alcalá. She founded and coordinates the ecocritical research group GIECO, directs an ecocritical book series, CLYMA, at the Franklin Institute, and was president of EASLSE from 2010–2012. She is the General Editor of *Ecozon@: European Journal of Literature, Culture and Environment*. Her research focuses on ecocriticism, ecofeminism, environmental justice and sense of place, mostly in American contemporary ethnic literature as well as contemporary Spanish literature.

Susan Larson is an associate professor in the Department of Hispanic Studies at the University of Kentucky. Her research deals with discourses of modernity and the cultural implications of urban planning, particularly in Madrid. In addition to her book *Constructing and Resisting Modernity: Madrid 1900–1936* (2011) she has coedited two volumes, *Visualizing Spanish Modernity* (2006, with Eva Woods Peiró) and *Kiosk Literature in Silver Age Spain: Modernity and Mass Culture* (2015, with Jeff Zamostny) and published articles on Spanish film, photography, narrative and popular culture. She is senior editor of the *Arizona Journal of Hispanic Cultural Studies* and coeditor (with Benjamin Fraser) of Palgrave Macmillan's *Hispanic Urban Studies* book series.

Martín López-Vega is the author of *Obreros de la lluz. Los poetes de la duración y la elexía posmoderna* (2012), among other essays on literature. He has edited several books, including *La tierra escrita. Narrativa contemporánea en asturiano* (2002), *Fruta del tiempo. Antología de poesía contemporánea en asturiano* (2002), and *El viajero modernista* (2002). He has translated into Spanish Almeida Garrett, Eça de Queirós, Jorge de Sena, Eugénio Andrade and Lêdo Ivo, among others, and is the author of several books of poems, among them *La eterna cualquiercosa* (2014). He is currently pursuing a doctorate in Spanish literature at the University of Iowa.

Pablo de Lora teaches legal philosophy and bioethics at the Law School of the Universidad Autónoma de Madrid. He has written extensively on topics that lie at the intersection of political and moral philosophy and biomedical ethics. His work has been published in major journals nationally and abroad. He has been a visiting professor at Harvard University and Rutgers University

Luis Martín-Estudillo is an associate professor of Spanish literature at the University of Iowa. His publications include *La mirada elíptica: El trasfondo barroco de la poesía española contemporánea* (2007) and several coedited and edited volumes, including *Hispanic Baroques: Reading Cultures in Context* (2005), *Post-Authoritarian Cultures: Spain and Latin America's Southern Cone* (2008), *Filosofía y tiempo final* (2010), *New Spain, New Literatures* (2010), *Hispanic Literatures and the Question of a Liberal Education* (2011), and *Memory and its Discontents* (2012). He is managing editor of *Hispanic Issues* and *Hispanic Issues Online* (*HIOL*).

Luis I. Prádanos (Iñaki) is an assistant professor of Hispanic contemporary studies at Miami University. His research focuses on ecocritical theory and environmental humanities in relation to contemporary Spanish culture. He is currently working on a book project titled *Postgrowth Imaginaries. Environmental Humanities and Spanish Culture.*

Tonia Raquejo is teaching at the Fine Arts Faculty (UCM, Madrid). She was granted with a Fleming Award to do her doctoral thesis at the Warburg Institute (London University), and published "The Arab Cathedrals: Moorish architecture as seen by British Travellers" (*The Burlington Magazine*, 1986) and *El palacio encantado* (1989). Tracing the Romantic sublime she has published *Los placeres de la imaginación de Joseph Addison* (1991); she has also analysed the connections between nature, art, science and anthropology in several publications, among them *Land Art* (1998). Recently she has coedited a book on art and ecology (2014).

Sainath Suryanarayanan is a postdoctoral scholar in the Department of Community and Environmental Sociology at the University of Wisconsin-Madison. Suryanarayanan is in the midst of developing an interdisciplinary research program at the intersections of science and technology studies (STS), biology and ecology. Initially trained in social insect biology, for the past several years Suryanarayanan has been publishing historically grounded STS research on the phenomenon of dying honey bees. His recent research, supported by the U.S. National Science Foundation, has appeared in refereed journals such as *Social Studies of Science*, *Science, Technology & Human Values*, *Political Power & Social Theory*, and *Insects*. In addition, Suryanarayanan's book on the science and politics of vanishing honey bees, coauthored with Daniel Lee Kleinman, is forthcoming from Rutgers University Press.

John H. Trevathan is a visiting lecturer of Spanish at Indiana University—Bloomington. His book project, *Beyond the Niche: Ecological Cultural Production in the Iberian Peninsula*, is an ecocritical approach to Iberian studies, examining the roles of nature in the construction of cultural identity in a contemporary context.

William Viestenz is an assistant professor of Spanish in the Department of Spanish and Portuguese Studies at the University of Minnesota, Twin Cities, where he also holds an appointment in the Institute for Global Studies. He specializes in modern Iberian literature and culture. He is the author of *By the Grace of God: Franco's Spain and the Sacred Roots of Political Imagination* (2014) and has coedited *The New Ruralism: An Epistemology of Transformed Space* (2013). The volume *Ethics of Life: Contemporary Iberian Debates* is related to his current book-length project that investigates the interrelation of political power in twentieth and twenty-first-century Spain with the processes of regulating life in its human and other-than-human forms. Drawing on analyses of topics including bullfighting, the uses (and abuses) of the environment, and even the popularity of the monstrous in recent Spanish film, the book will show how inscribing death into the concept of life is an inherently political act aimed at asserting, or defusing, sovereign power.

Index

abortion, xx, xxvii, 164–65, 167–70, 171, 173–76, 178n11, 189, 319, 321; abortion tourism, 163, 174, 178n10; "eugenic" abortion, 170, 178n19
actor-network theory (ANT), xiii, 137n5, 156
ADDA (Asociación en Defensa de los Derechos de los Animales), xix–xx
Aguado, Txetxu, 50–51
Almodóvar, Pedro, xi, xvi
Amenábar, Alejandro, xvi, xxviii, 183–85, 187, 193, 196, 200n5
animal rights, x, xvii–xviii, xix–xx, xxiv, 218, 280–83, 286–88, 290, 319, 324
animal studies, xi, xiii, xxiv, xxviii–xxix, 53n10, 233, 280, 282
Anthropocene, 91–92, 94, 97–98, 107, 108, 147
anthropocentrism (also anthropocentric), xi, xiii, xxiv, xxix, 47, 90, 144, 146, 150, 153, 157, 233–35, 238, 239–44, 249–52, 272; non-anthropocentric, xiii, 233–35, 238, 239–44, 251, 252
anthropomorphism, xxix, 47, 244, 248–49
Araújo, Joaquín, 8, 29n8
assisted reproductive technologies (ART), 171
assisted suicide, xxviii, 183–88, 190–93, 196–200, 321
Aznar, José María, 20, 37, 173
Aznar, Manuel, 76–77

Barcelona, xxiv, 52n2, 91–94, 96, 98, 103, 137n3, 195, 197, 208, 218, 325
Barella, Julia, 8–9, 25
Basurama, 21–22, 131, 134–35
beast, the, 260, 266–67, 269–72
bestia negra, 257, 261, 270, 272
bioeconomy, ix, xvii, xxiv, xxvi, 205–7, 209, 216, 219–24
biopolitics, xxvii, 60, 63, 65, 108, 214
bio-power, xxvii, xxix
biotechnology, ix, xiv–xvi, xxviii, 203–16, 222–24, 320, 322
Bull, The, 257, 259–61, 263–72, 274, 276n11, 279–80, 324; fighting, x, xi, xviii–xx, 199, 254, 279–80, 283, 291n1, 300, 324; subsidization of bullfighting, xxi

Casal, Paula, xviii, xx, 8, 217, 218, 220
Castro, Américo, x, 206, 220
chapapote, 41, 46
coexistence, xii, 104–6, 219, 258, 304

dark ecology, 89, 91–92. *See also* ecology
Delibes, Miguel, 5, 320
Der Spiegel (newspaper), 71, 73–74, 285
desarrollismo, xxi, 60, 61, 65, 71, 77–78
disquiet, xxvi, 90–93, 98–102, 104

ecocriticism (also ecocritical), xxiv, xxv, 4, 8–10, 25, 145, 154–57, 320, 323, 326, 328

ecofatigue, xxiii
ecofeminism, 7–9, 144, 151
ecology, 47–48, 52n1, 61, 67, 80, 119, 143, 153, 157; dark, 89, 91–92; political, 47–49, 143, 157; urban, 114–15, 117, 128–29, 134–36, 137n4, 139n13
economics: degrowth, xvii, xxii, xxiii–xxiv, xxvii, xxxn3, 147, 151–57; ecological, 143, 145, 147, 151, 153–57; environmental, 143, 147; environmentalism, xxi, 36, 46, 67, 70, 71, 75, 81, 83, 126, 143–44; financial crisis of 2007–2008 (also Great Recession), 143, 152, 154, 300, 319, 325; neoliberal, ix, xii, xv, xxi, xxiii–xxiv, xxvii, 62, 121–22, 144–46, 148, 151, 205–6, 209, 220, 222; reform, 143, 155
euthanasia, xvi, xxviii, 183–84, 188, 200n1, 321
Exxon Valdez, 37, 40

Falange (also Falangista/Falangist) 60–62, 67, 71–73, 75–79, 219–20
Fernández, Joaquín, xxi, 3
15-M movement, 113–14, 127, 137n2, 154
Foucault, Michel, xxvii, 81, 236, 251
Fraga Iribarne, Manuel, 37, 38, 42, 45, 48, 53n6, 67, 83n1, 220; *fraguismo*, 38, 42
Fuentes, Carlos, xi

Galicia, xxii, xxv, 61, 99, 188, 194–95, 207, 250, 280–81, 323–24; *Costa da Morte*, 36, 38, 53n12; *Galeguidade*, 35, 38, 41–42; identity in, 35, 37, 45–46, 48–50, 52
Garcia Lorca, Federico, x, xvii, 206
genetically modified organisms (GMOs), xv–xvi, 203, 205, 209–16

Haraway, Donna, xiii–xvi, 94, 234–35, 238–40, 242, 247, 251, 253n3, 326–27
hartismo, xxv, 44–45, 48–49, 52
Hispanism, x–xi, xxix, 4, 9, 10, 219, 274, 323
historical memory, xxix, 11, 43, 53n5, 234–35, 300–301, 304, 308, 310
human/nonhuman relations, xii–xiii, xvii, xix, xxv–xxvi, xxviii, 6, 19, 40, 42–43, 50, 52, 91, 93, 97, 100, 116, 144–45, 147, 150–51, 156–57, 233–35, 238–40, 241–44, 247–49, 252, 272, 323

Iberian studies, x–xi, xxiv, xxviii, xxx, 155, 274, 322–23
indignados movement, xxiii–xxiv, 113, 128, 154, 225

Latour, Bruno, xiii–xiv, xxix, 40, 46, 156, 238, 319
leisure, 63, 69–70, 77–78, 80, 83, 151
LeMenager, Stephanie, 45–46
life, ix, xii, xxi 184, 204; quality of, xvi–xvii, xxvi–xxvi, 60–61, 62, 69, 71, 73, 75, 150, 151
Lootz, Eva, 12, 23–25
López de Uralde, Juan, xxii, 3, 36
Luzinterruptus, 22–23

Madrid, xxvii, 25, 49, 54n15, 113–17, 122, 128–29, 133, 137n1, 137n3, 139n11, 207, 218, 301, 325
Molinos, Asunción, 15–17
Montero, Rosa, xx, 8, 25, 183, 200n1
Morton, Timothy, xiii–xiv, 89–91, 97, 319, 323

nature, xiv, 39, 74, 116, 122, 238; and art, 10–11, 27, 28n7, 28n8
Natureculture, xiii, 94, 100, 115, 235, 238, 242, 249, 251

necropastoral, 38–39, 42, 51–52
Nixon, Robert, xxxn4, 44, 53n12, 94, 145, 156–57, 324–25
nonhuman turn, xiii, xxvii, xxviii
Nunca Máis, xxii, xxv, 37, 41–43, 45–46, 49–52, 53n8, 54n15, 324–25

Opus Dei, 62, 67

Partido Popular (PP), 41, 53n6, 153, 168–69
Partido Socialista Obrero Español (PSOE), xxvi, 83, 168–70, 215
pastoral (also pastoralism), 4, 38–39, 132, 247, 252
petro-melancholia, 42, 45–46, 51
posthumanism (also posthumanist), xxiv, xxvii, 144, 145, 150, 155, 156, 157, 319, 328
Prestige (oil tanker), xxii, xxv, 36–37, 39–43, 45–49, 51–52, 54n15, 99, 324

quality of life. *See* life

Rajoy, Mariano, 37, 43, 53n4, 169, 321
reproductive rights, xxvii, 163–65, 170, 172, 175–77; surrogate motherhood, xxvii, 163, 173–77
Riechmann, Jorge, xv–xvi, xx, xxvii–xxviii, xxxn3, 6, 8, 53n10, 143, 150, 203–4, 206, 212, 214–20, 224, 319, 326
Rivas, Manuel, xxv, 35–36, 37–38, 41–48, 52, 53n3, 53n11, 323

Sagrada Familia, 94–95
Sampedro, Ramón, 183–86, 188–89, 193, 195–96, 198, 321
Savater, Fernando, xx–xxi, 183, 200n1
seed, xv, 17, 22, 212, 214
Sepharad, 258–74, 276n11
slow violence, xxxn4, 37, 44–45, 53n12, 94, 156–57, 324–25
Spain: discontent in, 71, 74–75, 218; financial crisis of 2008, 117, 130, 122, 143, 209, 273; Francoism, xxi, xxvi, xxx, 61, 80, 82, 222, 246–47, 257, 262, 271, 298, 302, 325; is different, xxviii, 205–6, 220–21; reproductive rights, 163–65, 168, 170, 172, 175–77; Spanish Civil War, xxx, 3, 53n3, 245–46, 258, 298–301, 305, 308–309, 312, 327; tourism in, 60 62, 67, 69, 71–74 167, 170, 173–74

Tafalla, Marta, 7, 326
trans-corporeality, 144, 156

Unamuno, Miguel de, 206, 219–20
urban (also urbanism), xxvi, 82, 122, 125, 136, 137n3, 138n10; landscape, 120, 122, 128–29; studies, xxiv, 91, 115, 121, 137n5, 137n6, 138n10

Virilio, Paul, 47–49

wellbeing, 65–68, 80,144, 208, 240

zoon politikon, 257, 266, 269, 276n12

VOLUMES IN THE HISPANIC ISSUES SERIES

42 *Ethics of Life: Contemporary Iberian Debates,* edited by Katarzyna Beilin and William Viestenz
41 *In and Of the Mediterranean: Medieval and Early Modern Iberian Studies,* edited by Michelle M. Hamilton and Núria Silleras-Fernández
40 *Coloniality, Religion, and the Law in the Early Iberian World,* edited by Santa Arias and Raúl Marrero-Fente
39 *Poiesis and Modernity in the Old and New Worlds,* edited by Anthony J. Cascardi and Leah Middlebrook
38 *Spectacle and Topophilia: Reading Early (and Post-) Modern Hispanic Cultures,* edited by David R. Castillo and Bradley J. Nelson
37 *New Spain, New Literatures,* edited by Luis Martín-Estudillo and Nicholas Spadaccini
36 *Latin American Jewish Cultural Production,* edited by David William Foster
35 *Post-Authoritarian Cultures: Spain and Latin America's Southern Cone,* edited by Luis Martín-Estudillo and Roberto Ampuero
34 *Spanish and Empire,* edited by Nelsy Echávez-Solano and Kenya C. Dworkin y Méndez
33 *Generation X Rocks: Contemporary Peninsular Fiction, Film, and Rock Culture,* edited by Christine Henseler and Randolph D. Pope
32 *Reason and Its Others: Italy, Spain, and the New World,* edited by David Castillo and Massimo Lollini
31 *Hispanic Baroques: Reading Cultures in Context,* edited by Nicholas Spadaccini and Luis Martín-Estudillo
30 *Ideologies of Hispanism,* edited by Mabel Moraña
29 *The State of Latino Theater in the United States: Hybridity, Transculturation, and Identity,* edited by Luis A. Ramos-García
28 *Latin America Writes Back: Postmodernity in the Periphery (An Interdisciplinary Perspective),* edited by Emil Volek
27 *Women's Narrative and Film in Twentieth-Century Spain: A World of Difference(s),* edited by Ofelia Ferrán and Kathleen M. Glenn
26 *Marriage and Sexuality in Medieval and Early Modern Iberia,* edited by Eukene Lacarra Lanz
25 *Pablo Neruda and the U.S. Culture Industry,* edited by Teresa Longo
24 *Iberian Cities,* edited by Joan Ramon Resina
23 *National Identities and Sociopolitical Changes in Latin America,* edited by Mercedes F. Durán-Cogan and Antonio Gómez-Moriana

22 *Latin American Literature and Mass Media*, edited by Edmundo Paz-Soldán and Debra A. Castillo
21 *Charting Memory: Recalling Medieval Spain*, edited by Stacy N. Beckwith
20 *Culture and the State in Spain: 1550–1850*, edited by Tom Lewis and Francisco J. Sánchez
19 *Modernism and its Margins: Reinscribing Cultural Modernity from Spain and Latin America*, edited by Anthony L. Geist and José B. Monleón
18 *A Revisionary History of Portuguese Literature*, edited by Miguel Tamen and Helena C. Buescu
17 *Cervantes and his Postmodern Constituencies*, edited by Anne Cruz and Carroll B. Johnson
16 *Modes of Representation in Spanish Cinema*, edited by Jenaro Talens and Santos Zunzunegui
15 *Framing Latin American Cinema: Contemporary Critical Perspectives*, edited by Ann Marie Stock
14 *Rhetoric and Politics: Baltasar Gracián and the New World Order*, edited by Nicholas Spadaccini and Jenaro Talens
13 *Bodies and Biases: Sexualities in Hispanic Cultures and Literatures*, edited by David W. Foster and Roberto Reis
12 *The Picaresque: Tradition and Displacement*, edited by Giancarlo Maiorino
11 *Critical Practices in Post-Franco Spain*, edited by Silvia L. López, Jenaro Talens, and Dario Villanueva
10 *Latin American Identity and Constructions of Difference*, edited by Amaryll Chanady
9 *Amerindian Images and the Legacy of Columbus*, edited by René Jara and Nicholas Spadaccini
8 *The Politics of Editing*, edited by Nicholas Spadaccini and Jenaro Talens
7 *Culture and Control in Counter-Reformation Spain*, edited by Anne J. Cruz and Mary Elizabeth Perry
6 *Cervantes's Exemplary Novels and the Adventure of Writing*, edited by Michael Nerlich and Nicholas Spadaccini
5 *Ortega y Gasset and the Question of Modernity*, edited by Patrick H. Dust
4 *1492–1992: Re/Discovering Colonial Writing*, edited by René Jara and Nicholas Spadaccini
3 *The Crisis of Institutionalized Literature in Spain*, edited by Wlad Godzich and Nicholas Spadaccini
2 *Autobiography in Early Modern Spain*, edited by Nicholas Spadaccini and Jenaro Talens
1 *The Institutionalization of Literature in Spain*, edited by Wlad Godzich and Nicholas Spadaccini

www.ingramcontent.com/pod-product-compliance
Lightning Source LLC
LaVergne TN
LVHW101456200926
848483LV00001B/145